FIELD-BOOK

FOR

RAILROAD ENGINEERS

FIELD-BOOK

FOR

RAILROAD ENGINEERS.

CONTAINING

FORMULÆ

FOR LAYING OUT CURVES, DETERMINING FROG ANGLES, LEVELLING, CALCULATING EARTH-WORK, ETC., ETC.,

TOGETHER WITH

TABLES

OF RADII, ORDINATES, DEFLECTIONS, LONG CHORDS, MAGNETIC VARIATION, LOGARITHMS, LOGARITHMIC AND NATURAL SINES, TANGENTS, ETC., ETC.

BY

JOHN B. HENCK, A.M.,

CIVIL ENGINEER.

NEW YORK:

D. APPLETON AND COMPANY,

443 AND 445 BROADWAY.

LONDON, 16 LITTLE BRITAIN.

1866.

PREFACE.

The object of the present work is to supply a want very generally felt by Assistant Engineers on Railroads. Books of convenient form for use in the field, containing the ordinary logarithmic tables, are common enough; but a book combining with these tables others peculiar to railroad work, and especially the necessary formulæ for laying out curves, turnouts, crossings, &c., is yet a desideratum. These formulæ, after long disuse perhaps, the engineer is often called upon to apply at a moment's notice in the field, and he is, therefore, obliged to carry with him in manuscript such methods as he has been able to invent or collect, or resort to what has received the very appropriate name of "fudging." This the intelligent engineer always considers a reproach; and he will, therefore, it is hoped, receive with favor any attempt to make a resort to it inexcusable.

Besides supplying the want just alluded to, it was thought that some improvements upon former methods might be made, and some entirely new methods introduced. Among the processes believed to be original may be specified those in §§ 41 – 48, on Compound Curves, in Chapter II., on Parabolic Curves, in §§ **106 – 109**, on Vertical Curves, and in the article on Excavation and Embankment. It is but just to add, that a great part of what is said on Reversed Curves, Turnouts, and Crossings, and most of the Miscellaneous Problems, are the result of original investigations. In the remaining portions, also, many simplifications have been made. In all parts the object has been to reduce the operation necessary in the field to a single process, indicated by a formula standing on a line by itself, and distinguished by a ☞. This could not be done in all cases, as will be readily seen on examination. Certain preliminary steps were sometimes necessary, and these, whenever it was practicable, have been indicated by *words in italics.*

Of the methods given for Compound Curves, that in § 46 will be found particularly useful, from the great variety of applications of which it is susceptible.

Methods of laying out Parabolic Curves are here given, that those so disposed may test their reputed advantages. Two things are certainly in their favor; they are adapted to unequal as well as equal tangents, and their curvature generally decreases towards both extremities, thus making the transition to and from a straight line easier. Some labor has been given to devising convenient ways of laying out these curves. The method of determining the radius of curvature at certain points is believed to be entirely new. Better processes, however, may already exist, particularly in France, where these curves are said to be in general use.

The mode of calculating Excavation and Embankment here presented, will, it is thought, be found at least as simple and expeditious as those commonly used, with the advantage over most of them in point of accuracy. The usual Tables of Excavation and Embankment have been omitted. To include all the varieties of slope, width of road-bed, and depth of cutting, they must be of great extent, and unfitted for a field-book. Even then they apply only to ground whose cross-section is level, though often used in a manner shown to be erroneous in § 128. When the cross-section of the ground is level, the place of the tables is supplied by the formula of § 119, and when several sections are calculated together, as is usually the case, and the work is arranged in tabular form, as in § 120, the calculation is believed to be at least as short as by the most extended tables. The correction in excavation on curves (§ 129) is not known to have been introduced elsewhere.

In a work of this kind, brevity is an essential feature. The form of "Problem" and "Solution" has, therefore, been adopted, as presenting most concisely the thing to be done and the manner of doing it. Every solution, however, carries with it a demonstration, which is deemed an equally essential feature. These demonstrations, with a few unavoidable exceptions, principally in Chapter II., presuppose a knowledge of nothing beyond Algebra, Geometry, and Trigonometry. The result is in general expressed by an algebraic formula, and not in words. Those familiar with algebraic symbols need not be told how much more intelligible and quickly apprehended a process becomes when thus expressed. Those not familiar with these symbols should lose no time in acquiring the ready use of a language so direct and expressive. It may be remarked that it was no part of the author's design to furnish a collection of mere "rules," professing to require only an ability to read for their successful application. Rules can seldom be safely applied without a thorough understanding of the principles on which they rest, and such an understanding, in the present case, implies a knowledge of algebraic formulæ.

The tables here presented will, it is hoped, prove relia

ble. Those specially prepared for this work have been computed with great care. The values have in some cases been carried out farther than ordinary practice requires, in order that interpolated values may be obtained from them more accurately. For the greater part of the material composing the Table of Magnetic Variation the author is indebted to Professor Bache, whose distinguished ability in conducting the operations of the Coast Survey is equalled only by his desire to diffuse its results. The remaining tables have been carefully examined by comparing them with others of approved reputation for accuracy. Many errors have in this way been detected in some of the tables of corresponding extent in general use, particularly in the Table of Squares, Cubes, &c., and the Tables of Logarithmic and Natural Sines, Cosines, &c. The number of tables might have been greatly increased, but for an unwillingness to insert any thing not falling strictly within the plan of the work or not resting on sufficient authority.

J. B. H.

BOSTON, *February*, 1854.

TABLE OF CONTENTS.

CHAPTER I.

CIRCULAR CURVES.

Article I. — Simple Curves.

C. *Ordinates.*

D. *Curving Rails.*

ARTICLE II.—REVERSED AND COMPOUND CURVES.

A. *Reversed Curves.*

B. *Compound Curves.*

Article III. — Turnouts and Crossings.

Article IV. — Miscellaneous Problems.

CHAPTER II.

PARABOLIC CURVES.

Article I.—Locating Parabolic Curves.

Article II.—Radius of Curvature.

CHAPTER III.

LEVELLING.

Article I. — Heights and Slope Stakes.

Article II. — Correction for the Earth's Curvature and for Refraction.

Article III. — Vertical Curves.

Article IV. — Elevation of the Outer Rail on Curves.

CHAPTER IV.

EARTH-WORK.

Article I. — Prismoidal Formula.

Article II. — Borrow-Pits.

TABLES.

EXPLANATION OF SIGNS.

The sign $+$ indicates that the quantities between which it is placed are to be *added* together.

The sign $-$ indicates that the quantity before which it is placed is to be *subtracted*.

The sign $\times$ indicates that the quantities between which it is placed are to be *multiplied* together.

The sign $\div$ or : indicates that the first of two quantities between which it is placed is to be *divided* by the second.

The sign $=$ indicates that the quantities between which it is placed are *equal*.

The sign $\backsim$ indicates that the *difference* of the two quantities between which it is placed is to be taken.

The sign $\therefore$ stands for the word "hence" or "therefore."

The *ratio* of one quantity to another may be regarded as the quotient of the first *divided* by the second. Hence, the ratio of a to b is expressed by $a : b$, and the ratio of c to d by $c : d$. A *proportion* expresses the *equality* of two ratios. Hence, a proportion is represented by placing the sign $=$ between two ratios; as, $a : b = c : d$.

In the text and in the tables the foot has been taken as the unit of measure when no other unit is specified.

FIELD-BOOK.

CHAPTER I.

CIRCULAR CURVES.

ARTICLE I. — SIMPLE CURVES.

1. THE railroad curves here considered are either Circular or Parabolic. Circular curves are divided into Simple, Reversed, and Compound Curves. We begin with Simple Curves.

2. Let the arc $ADEFB$ (fig. 1) represent a railroad curve, unit-

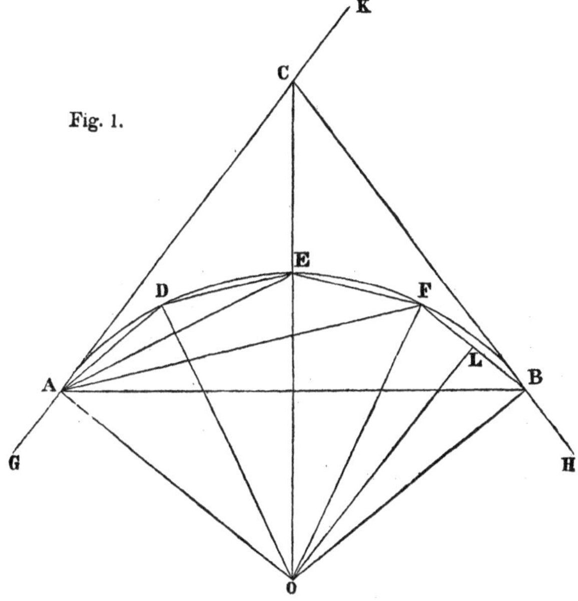

Fig. 1.

ing the straight lines GA and BH. The length of such a curve is measured by chords, each 100 feet long.* Thus, if the chords AD, DE, EF, and FB are each 100 feet in length, the whole curve is said to be 400 feet long. The straight lines GA and BH are always tangent to the curve at its extremities, which are called *tangent points*. If GA and BH are produced, until they meet in C, AC and BC are called the *tangents* of the curve. If AC is produced a little beyond C to K, the angle KCB, formed by one tangent with the other produced, is called the *angle of intersection*, and shows the *change of direction* in passing from one tangent to the other.

The following propositions relating to the circle are derived from Geometry.

I. A tangent to a circle is perpendicular to the radius drawn through the tangent point. Thus, AC is perpendicular to AO, and BC to BO.

II. Two tangents drawn to a circle from any point are equal, and if a chord be drawn between the two tangent points, the angles between this chord and the tangents are equal. Thus $AC = BC$, and the angle $BAC = ABC$.

III. An acute angle between a tangent and a chord is equal to half the central angle subtended by the same chord. Thus, $CAB = \frac{1}{2} AOB$.

IV. An acute angle subtended by a chord, and having its vertex in the circumference of a circle, is equal to half the central angle subtended by the same chord. Thus, $DAE = \frac{1}{2} DOE$.

V. Equal chords subtend equal angles at the centre of a circle, and also at the circumference, if the angles are inscribed in similar segments. Thus, $AOD = DOE$, and $DAE = EAF$.

VI. The angle of intersection of two tangents is equal to the central angle subtended by the chord which unites the tangent points. Thus, $KCB = AOB$.

3. In order to unite two straight lines, as GA and BH, by a curve, the angle of intersection is measured, and then a radius for the curve may be assumed, and the tangents calculated, or the tangents may be assumed of a certain length, and the radius calculated.

---

* Some engineers prefer a chain 50 feet in length, and measure the length of a curve by chords of 50 instead of 100 feet. The chord of 100 feet has been adopted throughout this article; but the formulæ deduced may be very readily modified to suit chords of any length. See also § 13.

4 **Problem.** *Given the angle of intersection $K\,C\,B = I$ (fig. 1), and the radius $A\,O = R$, to find the tangent $A\,C = T$.*

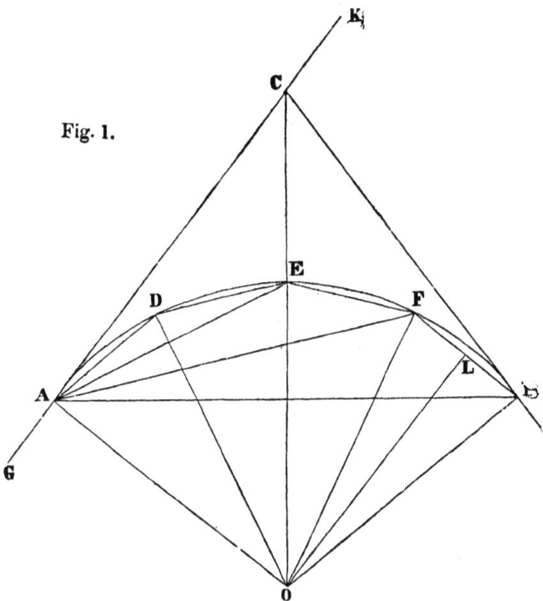

Fig. 1.

Solution. Draw $C\,O$. Then in the right triangle $A\,O\,C$ we have (Tab. X. 3) $\frac{A\,C}{A\,O} = \tan.\ A\,O\,C$, or, since $A\,O\,C = \frac{1}{2} I$ (§ 2, VI.), $\frac{T}{R} = \tan.\ \frac{1}{2} I$;

☞ $\therefore\ T = R \tan.\ \frac{1}{2} I.$

Example. Given $I = 22° 52'$, and $R = 3000$, to find T. Here

$R = 3000$	3.477121
$\frac{1}{2} I = 11° 26'$	tan. 9.305869
$T = 606.72$	2.782990

5. **Problem.** *Given the angle of intersection $K\,C\,B = I$ (fig. 1), and the tangent $A\,C = T$, to find the radius $A\,O = R$.*

Solution. In the right triangle $A\,O\,C$ we have (Tab. X. 6) $\frac{A\,O}{A\,C} = \text{cot. } A\,O\,C$, or $\frac{R}{T} = \text{cot. } \frac{1}{2} I$;

☞ $\therefore R = T \text{ cot. } \frac{1}{2} I.$

Example. Given $I = 31° 16'$ and $T = 950$, to find R. Here

$T = 950$	2.977724
$\frac{1}{2} I = 15° 38'$	cot. 0.553102
$R = 3394.89$	3.530826

6. The *degree* of a curve is determined by the angle subtended at its centre by a chord of 100 feet. Thus, if $A\,O\,D = 6°$ (fig. 1), $A\,D\,E\,F\,B$ is a 6° curve.

7. The *deflection angle* of a curve is the acute angle formed at any point between a tangent and a chord of 100 feet. The deflection angle is, therefore (§ 2, III.), half the degree of the curve. Thus, $C\,A\,D$ or $C\,B\,F$ is the deflection angle of the curve $A\,D\,E\,F\,B$, and is half $A\,O\,D$ or half $F\,O\,B$.

A. *Method by Deflection Angles.*

8. The usual method of laying out a curve on the ground is by means of deflection angles.

9. **Problem.** *Given the radius $A\,O = R$ (fig. 1), to find the deflection angle $C\,B\,F = D$.*

Solution. Draw $O\,L$ perpendicular to $B\,F$. Then the angle $B\,O\,L = \frac{1}{2} B\,O\,F = D$, and $B\,L = \frac{1}{2} B\,F = 50$. But in the right triangle $O\,B\,L$ we have (Tab. X. 1) $\text{sin. } B\,O\,L = \frac{B\,L}{B\,O}$;

☞ $\therefore \text{sin. } D = \frac{50}{R}.$

Example. Given $R = 5729.65$, to find D. Here

50	1.698970
$R = 5729.65$	3.758128
$D = 30'$	sin. 7.940842

Hence a curve of this radius is a 1° curve, and its deflection angle is 30′.

10. **Problem.** *Given the deflection angle $C\,B\,F = D$ (fig. 1), to find the radius $A\,O = R$.*

Solution. By the preceding section we have $\sin. D = \frac{50}{R}$, whence $R \sin. D = 50$;

☞ $$\therefore R = \frac{50}{\sin. D}.$$

By this formula the radii in Table I. are calculated.

Example. Given $D = 1°$, to find R. Here

50	1.698970
$D = 1°$	sin. 8 241855
$R = 2864.93$	3.457115

11. **Problem.** *Given the angle of intersection $KCB = I$ (fig. 1), and the tangent $AC = T$, to find the deflection angle $CAD = D$.*

Solution. From § 9 we have $\sin. D = \frac{50}{R}$, and from § 5, $R = T \cot. \frac{1}{2} I$. Substituting this value of R in the first equation, we get $\sin. D = \frac{50}{T \cot. \frac{1}{2} I}$;

☞ $$\therefore \sin. D = \frac{50 \tan. \frac{1}{2} I}{T}.$$

Example. Given $I = 21°$ and $T = 424.8$, to find D. Here

50	1.698970
$\frac{1}{2} I = 10° 30'$	tan. 9.267967
	0.966937
$T = 424 8$	2.628185
$D = 1° 15'$	sin. 8.338752

12. **Problem.** *Given the angle of intersection $KCB = I$ (fig. 1) and the deflection angle $CAD = D$, to find the tangent $AC = T$.*

Solution. From the preceding section we have $\sin. D = \frac{50 \tan. \frac{1}{2} I}{T}$. Hence, $T \sin. D = 50 \tan. \frac{1}{2} I$;

☞ $$\therefore T = \frac{50 \tan. \frac{1}{2} I}{\sin. D}$$

Example. Given $I = 28°$ and $D = 1°$, to find T. Here

$$T = \frac{50 \tan. 14°}{\sin. 1°} = 714 31.$$

13. **Problem.** *Given the angle of intersection $KCB = I$ (fig. 1), and the deflection angle $CAD = D$, to find the length of the curve.*

Solution. By § 2 the length of a curve is measured by chords of 100 feet applied around the curve. Now the first chord AD makes with the tangent AC an angle $CAD = D$, and each succeeding chord DE, EF, &c. subtends at A an additional angle DAE, EAF, &c., each equal to D; since each of these angles (§ 2, IV.) is half of a central angle subtended by a chord of 100 feet. The angle $CAB = \frac{1}{2} AOB = \frac{1}{2} I$ is, therefore, made up of as many times D, as there are chords around the curve. Then if n represents the number of chords, we have $nD = \frac{1}{2} I$;

☞ $$\therefore n = \frac{\frac{1}{2} I}{D}.$$

If D is not contained an even number of times in $\frac{1}{2} I$, the quotient above will still give the length of the curve. Thus, in fig. 2, suppose D is contained $4\frac{3}{8}$ times in $\frac{1}{2} I$. This shows that there will be four whole chords and $\frac{3}{8}$ of a chord around the curve from A to B. The angle GAB, the fraction of D, is called a *sub-deflection angle*, and GB, the fraction of a chord, is called a *sub-chord* *

The length of the curve thus found is not the actual length of the arc, but the length required in locating a curve. If the actual length of the arc is required, it may be found by means of Table VI.

Example. Given $I = 16° 52'$ and $D = 1° 20'$, to find the length of the curve. Here $n = \frac{\frac{1}{2} I}{D} = \frac{8° 26'}{1° 20'} = \frac{506'}{80'} = 6.325$, that is, the curve is 632.5 feet long.

To find the arc itself in this example, we take from Table VI. the length of an arc of 16° 52′, since the central angle of the whole curve is equal to I (§ 2, VI.), and multiply this length by the radius of the curve.

Arc	10°	= .1745329
"	6°	= .1047198
"	50′	= .0145444
"	2′	= .0005818
"	16° 52′	= .2943789

* This method of finding the length of a sub-chord is not mathematically accurate; for, by geometry, angles inscribed in a circle are proportional to the *arcs* on which they stand; whereas this method supposes them to be proportional to the *chords* of these arcs. In railroad curves, the error arising from this supposition is too small to be regarded.

The radius of the curve is found from Table I. to be 2148.79, and this multiplied by .2943789 gives 632.558 feet for the length of the arc.

14. **Problem.** *Given the deflection angle D, to lay out a curve from a given tangent point.*

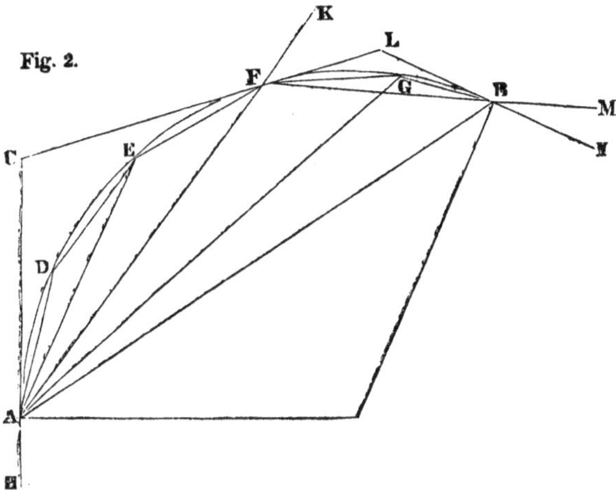

Fig. 2.

Solution. Let A (fig. 2) be the given tangent point in the tangent $H\,C$. Set the instrument at A, and lay off the given deflection angle D from $A\,C$. This will give the direction $A\,D$, and 100 feet being measured from A in this direction, the point D will be determined Lay off in succession the additional angles $D\,A\,E$, $E\,A\,F$, &c., each equal to D, and make $D\,E$, $E\,F$, &c. each 100 feet, and the points E, F, &c. will be determined. The points D, E, F, &c., thus determined, are points on the required curve (§ 7, and § 2, III., IV.), and are called *stations.*

If there is a sub-chord at the end, as $G\,B$, the sub-deflection angle $G\,A\,B$ must be the same part of D that $G\,B$ is of a whole chord (§ 13).

15. It is often impossible to lay out the whole of a curve, without removing the instrument from its first position, either on account of the great length of the curve, or because some obstruction to the sight may be met with. In this case, after determining as many stations as possible, and removing the instrument to the last of these stations, we ought to be able to find the tangent to the curve at this station; for

then the curve could be continued by deflections from the new tangent, in precisely the same way as it was begun from the first tangent.

16. **Problem.** *After running a curve a certain number of stations, to find a tangent to the curve at the last station.*

Solution. Suppose that the curve (fig. 2) has been run three stations to F, and that $F L$ is the tangent required. Produce $A F$ to K, and we have the angle $K F L = A F C$. But (§ 2, II.) $A F C = F A C$. Therefore $K F L = F A C$. Now $F A C$ is the sum of all the deflection angles laid off from the tangent at A, that is, in this case, $F A C = 3 D$, and the tangent $F L$ is, therefore, obtained by laying off from $A F$ produced an angle $K F L$ equal to the total deflection from the preceding tangent.

If the curve is afterwards continued beyond F, as, for instance, to B, a tangent $B N$ at B is obtained by laying off from $F B$ produced an angle $M B N = L B F = L F B$, the total deflection from the preceding tangent $F L$.

B. *Method by Tangent and Chord Deflections.*

17. Let $A B C D$ (fig. 3) be a curve between the two tangents $E A$ and $D L$, having the chords $A B$, $B C$, and $C D$ of the same length

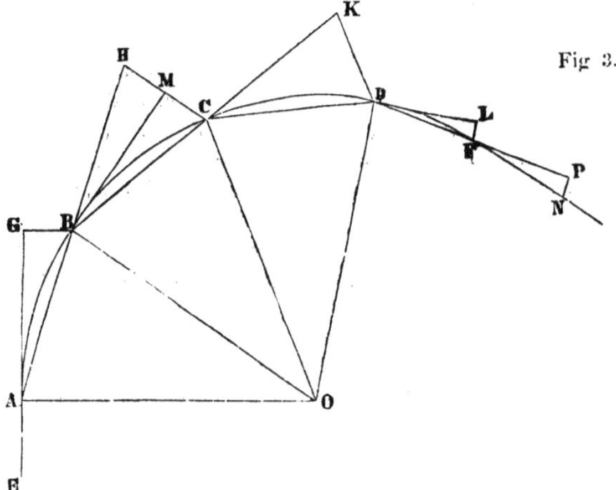

Fig. 3.

Produce the tangent $E A$, and from B draw $B G$ perpendicular to $A G$. Produce also the chords $A B$ and $B C$, and make the produced parts $B\,H$ and $C\,K$ of the same length as the chords. Draw $C\,H$ and $D\,K$. $B\,G$ is called the *tangent deflection*, and $C\,H$ or $D\,K$ the *chord deflection*.

18. **Problem.** *Given the radius $A\,O = R$ (fig. 3), to find the tangent deflection $B\,G$, and the chord deflection $C\,H$.*

Solution. The triangle $C\,B\,H$ is similar to $B\,O\,C$; for the angle $B\,O\,C = 180° - (O\,B\,C + B\,C\,O)$, or, since $B\,C\,O = A\,B\,O$, $B\,O\,C = 180° - (O\,B\,C + A\,B\,O) = C\,B\,H$, and, as both the triangles are isosceles, the remaining angles are equal. The homologous sides are, therefore, proportional, that is, $B\,O : B\,C = B\,C : C\,H$, or, representing the chord by c and the chord deflection by d, $R : c = c : d$;

☞ $$\therefore\ d = \frac{c^2}{R}.$$

To find the tangent deflection, draw $B\,M$ to the middle of $C\,H$, bisecting the angle $C\,B\,H$, and making $B\,M\,C$ a right angle. Then the right triangles $B\,M\,C$ and $A\,G\,B$ are equal; for $B\,C = A\,B$, and the angle $C\,B\,M = \frac{1}{2}\,C\,B\,H = \frac{1}{2}\,B\,O\,C = \frac{1}{2}\,A\,O\,B = B\,A\,G$ (§ 2, III.). Therefore $B\,G = C\,M = \frac{1}{2}\,C\,H = \frac{1}{2}\,d$, that is, *the tangent deflection is half the chord deflection.*

19. **Problem.** *Given the deflection angle D of a curve, to find the chord deflection d.*

Solution. By the preceding section we have $d = \frac{c^2}{R}$, and by § 10, $R = \frac{50}{\sin. D}$ Substituting this value of R in the first equation, we find

☞ $$d = \frac{c^2 \sin. D}{50}.$$

This formula gives the chord deflection for a chord c of any length, though D is the deflection angle for a chord of 100 feet (§ 7). When $c = 100$, the formula becomes $d = 200 \sin D$, or for the tangent deflection $\frac{1}{2}\,d = 100 \sin. D$. By these formulæ the tangent and chord deflections in Table I. may be easily obtained from the table of natural sines

20. The length of the curve may be found by first finding D (§ 9 or § 11), and then proceeding as in § 13.

21. **Problem.** *To draw a tangent to the curve at any station, as B (fig. 3).*

Solution. Bisect the chord deflection $H\,C$ of the next station in M.

2

A line drawn through B and M will be the tangent required; for it has been proved (§ 18) that the angle $C B M$ is in this case equal to $\frac{1}{2} B O C$, and $B M$ is consequently (§ 2, III.) a tangent at B.

If B is at the end of the curve, the tangent at B may be found without first laying off $H C$. Thus, if a chain equal to the chord is extended to H on $A B$ produced, the point H marked, and the chain then swung round, keeping the end at B fixed, until $H M = \frac{1}{2} d$, $B M$ will be the direction of the required tangent.*

22. **Problem.** *Given the chord deflection d, to lay out a curve from a given tangent point.*

Solution. Let A (fig. 3) be the given tangent point, and suppose d has been calculated for a chord of 100 feet. Stretch a chain of 100 feet from A to G on the tangent $E A$ produced, and mark the point G. Swing the chain round towards $A B$, keeping the end at A fixed, until $B G$ is equal to the tangent deflection $\frac{1}{2} d$, and B will be the first station on the curve. Stretch the chain from B to H on $A B$ produced, and having marked this point, swing the chain round, until $H C$ is equal to the chord deflection d. C is the second station on the curve. Continue to lay off the chord deflection from the preceding chord produced, until the curve is finished.

Should a sub-chord $D F$ occur at the end of the curve, find the tangent $D L$ at D (§ 21), lay off from it the proper tangent deflection $L F$ for the given sub-chord, making $D F$ of the given length, and F will be a point on the curve. The proper tangent deflection for the sub-chord may be found thus. Represent the sub-chord by c', and the corresponding chord deflection by d', and we have (§ 18) $\frac{1}{2} d' = \frac{c'^2}{2 R}$; but since $\frac{1}{2} d = \frac{c^2}{2 R}$, we have $\frac{1}{2} d' : \frac{1}{2} d = c'^2 : c^2$. Therefore $\frac{1}{2} d' = \frac{1}{2} d \left(\frac{c'}{c}\right)^2$.

Example. Given the intersection angle I between two tangents equal to $16° 30'$, and $R = 1250$, to find T, d, and the length of the curve in stations. Here

(§ 4) $T = R \tan. \frac{1}{2} I = 1250 \tan. 8° 15' = 181.24$;

(§ 18) $d = \frac{c^2}{R} = \frac{100^2}{1250} = 8$;

---

* The distance $B M$ is not exactly equal to the chord, but the error arising from taking it equal is too small to be regarded in any curves but those of very small radius. If necessary, the true length of $B M$ may be calculated; for $B M = \sqrt{B H^2 - H M^2}$

(§ 9) sin. $D = \frac{50}{R} = \frac{50}{1250} = .04 =$ nat. sin. $2° 17\frac{1}{2}'$;

(§ 13) $n = \frac{\frac{1}{2} I}{D} = \frac{8° 15'}{2° 17\frac{1}{2}'} = \frac{495'}{137.5'} = 3.60.$

These results show, that the tangent point A (fig. 3) on the first tangent is 181 24 feet from the point of intersection, — that the tangent deflection $GB = \frac{1}{2} d = 4$ feet, — that the chord deflection HC or KD = 8 feet, — and that the curve is 360 feet long. The three whole stations B, C, and D having been found, and the tangent DL drawn, the tangent deflection for the sub-chord of 60 feet will be, as shown above, $\frac{1}{2} d' = 4\left(\frac{60}{100}\right)^2 = 4 \times .6^2 = 4 \times 36 = 1\ 44.$ $LF = 1.44$ feet being laid off from DL, the point F will, if the work is correct, fall upon the second tangent point. A tangent at F may be found (§ 21) by producing DF to P, making $FP = DF = 60$ feet, and laying off $PN = 1.44$ feet. FN will be the direction of the required tangent, which should, of course, coincide with the given tangent.

23. Curves may be laid out with accuracy by tangent and chord deflections, if an instrument is used in producing the lines. But if an instrument is not at hand, and accuracy is not important, the lines may be produced by the eye alone. The radius of a curve to unite two given straight lines may also be found without an instrument by § 73, or, having assumed a radius, the tangent points may be found by § 74.

C. *Ordinates.*

24. The preceding methods of laying out curves determine points 100 feet distant from each other. These points are usually sufficient for grading a road; but when the track is laid, it is desirable to have intermediate points on the curve accurately determined. For this purpose the chord of 100 feet is divided into a certain number of equal parts, and the perpendicular distances from the points of division to the curve are calculated. These distances are called *ordinates.* If the chord is divided into eight equal parts, we shall have points on the curve at every 12.5 feet, and this will be often enough, if the rails, which are seldom shorter than 15 feet, have been properly curved (§ 28).

25. **Problem.** *Given the deflection angle D or the radius R of a curve, to find the ordinates for any chord.*

Solution. I. To find the middle ordinate. Let AEB (fig 4) be a portion of a curve, subtended by a chord AB, which may be denoted by c. Draw the middle ordinate ED, and denote it by m. Produce ED to the centre F, and join AF and AE. Then (Tab. X 3)

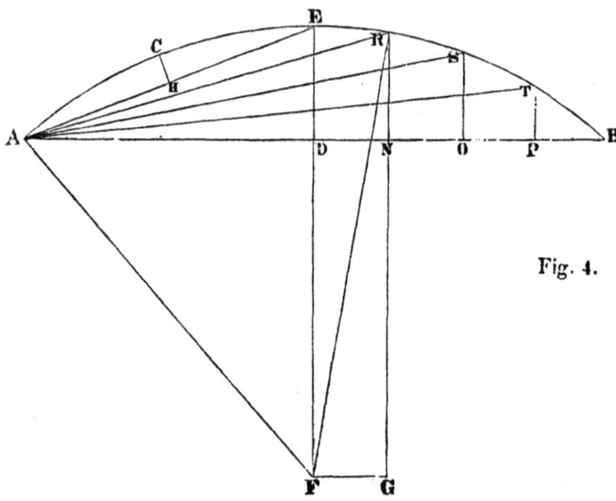

Fig. 4.

$\frac{ED}{AD} = \tan. EAD$, or $ED = AD \tan. EAD$. But, since the angle EAD is measured by half the arc BE, or by half the equal arc AE, we have $EAD = \frac{1}{2} AFE$. Therefore $ED = AD \tan. \frac{1}{2} AFE$, or

☞ $$m = \tfrac{1}{2} c \tan. \tfrac{1}{2} AFE.$$

When $c = 100$, $AFE = D$ (§ 7), and $m = 50 \tan. \frac{1}{2} D$, whence m may be obtained from the table of natural tangents, by dividing tan $\frac{1}{2} D$ by 2, and removing the decimal point two places to the right.

The value of m may be obtained in another form thus. In the triangle ADF we have $DF = \sqrt{AF^2 - AD^2} = \sqrt{R^2 - \frac{1}{4}c^2}$. Then $m = EF - DF = R - DF$, or

☞ $$m = R - \sqrt{R^2 - \tfrac{1}{4}c^2}.$$

II. To find any other ordinate, as RN, at a distance $DN = b$ from the centre of the chord. Produce RN until it meets the diameter parallel to AB in G, and join RF. Then $RG = \sqrt{RF^2 - FG^2} = \sqrt{R^2 - b^2}$, and $RN = RG - NG = RG - DF$. Substituting the value of RG and that of DF found above, we have

☞ $$RN = \sqrt{R^2 - b^2} - \sqrt{R^2 - \tfrac{1}{4}c^2}.$$

By these formulæ the ordinates in Table I are calculated.

The other ordinates may also be found from the middle ordinate by the following shorter, but not strictly exact method. It is founded on the supposition, that, if the half-chord BD be divided into any number of equal parts, the ordinates at these points will divide the arc EB into the same number of equal parts, and upon the further supposition, that the tangents of small angles are proportional to the angles themselves. These suppositions give rise to no material error in finding the ordinates of railroad curves for chords not exceeding 100 feet. Making, for example, four divisions of the chord on each side of the centre, and joining AR, AS, and AT, we have the angle $RAN = \frac{3}{4} EAD$, since RB is considered equal to $\frac{3}{4} EB$. But $EAD = \frac{1}{2} AFE$. Therefore, $RAN = \frac{3}{8} AFE$. In the same way we should find $SAO = \frac{1}{4} AFE$, and $TAP = \frac{1}{8} AFE$. We have then for the ordinates, $RN = AN \tan. RAN = \frac{5}{8} c \tan. \frac{3}{8} AFE$, $SO = AO \tan. SAO = \frac{3}{4} c \tan. \frac{1}{4} AFE$, and $TP = AP \tan. TAP = \frac{7}{8} c \tan. \frac{1}{8} AFE$. But, by the second supposition, $\tan. \frac{3}{8} AFE = \frac{3}{4} \tan. \frac{1}{2} AFE$, $\tan. \frac{1}{4} AFE = \frac{1}{2} \tan. \frac{1}{2} AFE$, and $\tan. \frac{1}{8} AFE = \frac{1}{4} \tan. \frac{1}{2} AFE$. Substituting these values, and recollecting that $\frac{1}{2} c \tan. \frac{1}{2} AFE = m$, we have

☞
$$\begin{cases} RN = \frac{15}{16} \times \frac{1}{2} c \tan. \frac{1}{2} AFE = \frac{15}{16} m, \\ SO = \frac{3}{4} \times \frac{1}{2} c \tan. \frac{1}{2} AFE = \frac{3}{4} m, \\ TP = \frac{7}{16} \times \frac{1}{2} c \tan. \frac{1}{2} AFE = \frac{7}{16} m. \end{cases}$$

In general, if the number of divisions of the chord on each side of the centre is represented by n, we should find for the respective ordinates, beginning nearest the centre, $\frac{(n+1)(n-1)m}{n^2}$, $\frac{(n+2)(n-2)m}{n^2}$, $\frac{(n+3)(n-3)m}{n^2}$, &c.

Example Find the ordinates of an 8° curve to a chord of 100 feet. Here $m = 50 \tan. 2° = 1.746$, $RN = \frac{15}{16} m = 1.637$, $SO = \frac{3}{4} m = 1.310$, and $TP = \frac{7}{16} m = 0.764$.

26. An approximate value of m also may be obtained from the formula $m = R - \sqrt{R^2 - \frac{1}{4} c^2}$. This is done by adding to the quantity under the radical the very small fraction $\frac{c^4}{64 R^2}$, making it a perfect

square, the root of which will be $R - \frac{c^2}{8R}$. We have, then, $m = R - \left(R - \frac{c^2}{8R}\right)$;

☞ $$\therefore m = \frac{c^2}{8R}.$$

27. From this value of m we see that the middle ordinates of any two chords in the same curve are to each other nearly as the squares of the chords. If, then, AE (fig. 4) be considered equal to $\frac{1}{2}AB$, its middle ordinate $CH = \frac{1}{4}ED$. Intermediate points on a curve may, therefore, be very readily obtained, and generally with sufficient accuracy, in the following manner. Stretch a cord from A to B, and by means of the middle ordinate determine the point E. Then stretch the cord from A to E, and lay off the middle ordinate $CH = \frac{1}{4}ED$, thus determining the point C, and so continue to lay off from the successive half-chords one fourth the preceding ordinate, until a sufficient number of points is obtained.

D. *Curving Rails.*

28. The rails of a curve are usually curved before they are laid. To do this properly, it is necessary to know the middle ordinate of the curve for a chord of the length of a rail.

29. **Problem.** *Given the radius or deflection angle of a curve, to find the middle ordinate for curving a rail of given length.*

Solution. Denote the length of the rail by l, and we have (§ 25) the exact formula $m = R - \sqrt{R^2 - \frac{1}{4}l^2}$, and (§ 26) the approximate formula

☞ $$m = \frac{\frac{1}{4}l^2}{2R}.$$

This formula is always near enough for chords of the length of a rail If we substitute for R its value (§ 10) $R = \frac{50}{\sin D}$, we have,

☞ $$m = \frac{1}{4}l^2 \times \frac{\sin. D}{100}.$$

Example. In a 1° curve find the ordinate for a rail of 18 feet in length. Here R is found by Table I. to be 5729 65, and therefore, by the first formula, $m = \frac{9^2}{11459.3} = .00707$. By the second formula, $m = .81 \sin. 30' = .00707$. The exact formula would give the same result even to the fifth decimal.

By keeping in mind, that the ordinate for a rail of 18 feet in a 1° curve is .007, the corresponding ordinate in a curve of any other degree may be found with sufficient accuracy, by multiplying this decimal by the number expressing the degree of the curve. Thus, for a curve of 5° 36′ or 5.6°, the ordinate would be .007 × 5.6 = .039 ft. = .468 in.

For a rail of 20 feet we have $\frac{1}{4} l^2 = 100$, and, consequently, $m = \sin. D$. This gives for a 1° curve, $m = .0087$. The corresponding ordinate in a curve of any other degree may be found with sufficient accuracy, by multiplying this decimal by the number expressing the degree of the curve.

By the above formula for m, the ordinates for curving rails in Table I. are calculated.

Article II. — Reversed and Compound Curves.

30. Two curves often succeed each other having a common tangent at the point of junction. If the curves lie on *opposite* sides of the common tangent, they form a *reversed* curve, and their radii may be the same or different. If they lie on the *same* side of the common tangent,

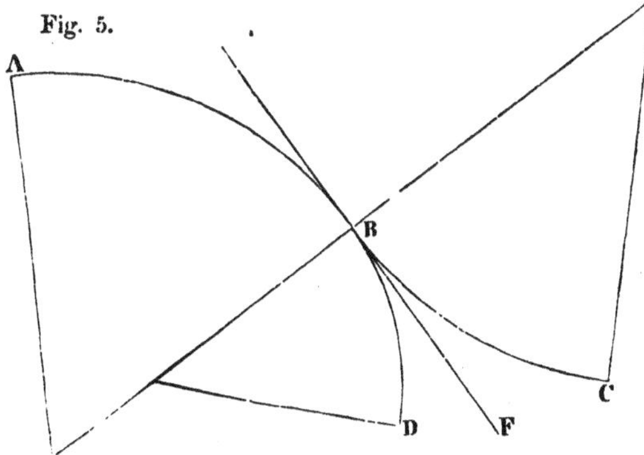

they have different radii, and form a *compound* curve. Thus $A\,B\,C$ (fig. 5) is a reversed curve, and $A\,B\,D$ a compound curve.

31. **Problem.** *To lay out a reversed or a compound curve, when the radii or deflection angles and the tangent points are known.*

Solution. Lay out the first portion of the curve from A to B (fig. 5), by one of the usual methods. Find $B\,F$, the tangent to $A\,B$, at the point B (§ 16 or § 21). Then $B\,F$ will be the tangent also of the second portion $B\,C$ of a reversed, or $B\,D$ of a compound curve, and from this tangent either of these portions may be laid off in the usual manner.

A. *Reversed Curves.*

32 **Theorem.** *The reversing point of a reversed curve between parallel tangents is in the line joining the tangent points.*

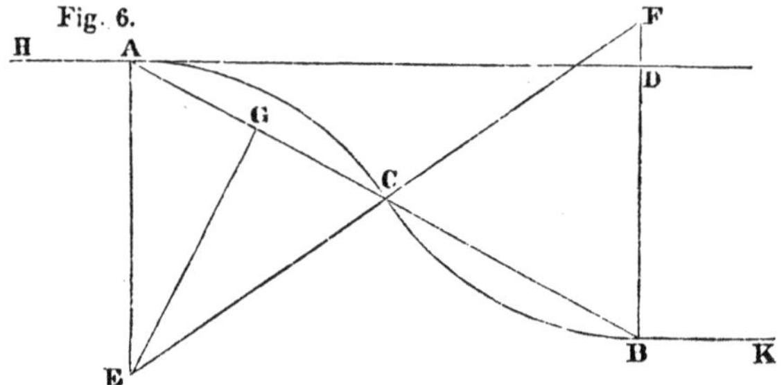

Demonstration. Let $A\,C\,B$ (fig. 6) be a reversed curve, uniting the parallel tangents $H\,A$ and $B\,K$, having its radii equal or unequal, and reversing at C. If now the chords $A\,C$ and $C\,B$ are drawn, we have to prove that these chords are in the same straight line. The radii $E\,C$ and $C\,F$, being perpendicular to the common tangent at C (§ 2, I.), are in the same straight line, and the radii $A\,E$ and $B\,F$, being perpendicular to the parallel tangents $H\,A$ and $B\,K$, are parallel. Therefore, the angle $A\,E\,C = C\,F\,B$, and, consequently, $E\,C\,A$, the half supplement of $A\,E\,C$, is equal to $F\,C\,B$, the half supplement of $C\,F\,B$; but these angles cannot be equal, unless $A\,C$ and $C\,B$ are in the same straight line.

33. **Problem.** *Given the perpendicular distance between two parallel tangents $B\,D = b$ (fig 6), and the distance between the two tangent points $A\,B = a$, to determine the reversing point C and the common radius $E\,C = C\,F = R$ of a reversed curve uniting the tangents $H\,A$ and $B\,K$.*

Solution. Let $A\,C\,B$ be the required curve. Since the radii are equal, and the angle $AEC = BFC$, the triangles AEC and BFC are equal, and $AC = CB = \frac{1}{2}a$. *The reversing point C is, therefore, the middle point of A B.*

To find R, draw EG perpendicular to AC. Then the right triangles AEG and BAD are similar, since (§ 2, III.) the angle $BAD = \frac{1}{2}AEC = AEG$. Therefore $AE : AG = AB : BD$, or $R : \frac{1}{4}a = a : b$;

☞ $$\therefore R = \frac{a^2}{4b}.$$

Corollary. If R and b are given, to find a, the equation $R = \frac{a^2}{4b}$ gives $a^2 = 4Rb$;

☞ $$\therefore a = 2\sqrt{Rb}.$$

Examples. Given $b = 12$, and $a = 200$, to determine R. Here $R = \frac{200^2}{4 \times 12} = \frac{10000}{12} = 833\frac{1}{3}$.

Given $R = 675$, and $b = 12$, to find a. Here $a = 2\sqrt{675 \times 12} = 2\sqrt{8100} = 2 \times 90 = 180$.

34. **Problem.** *Given the perpendicular distance between two parallel tangents* $BD = b$ *(fig. 7), the distance between the two tangent points* $AB = a$, *and the first radius* $EC = R$ *of a reversed curve uniting the tangents* HA *and* BK, *to find the chords* $AC = a'$ *and* $CB = a''$, *and the second radius* $CF = R'$.

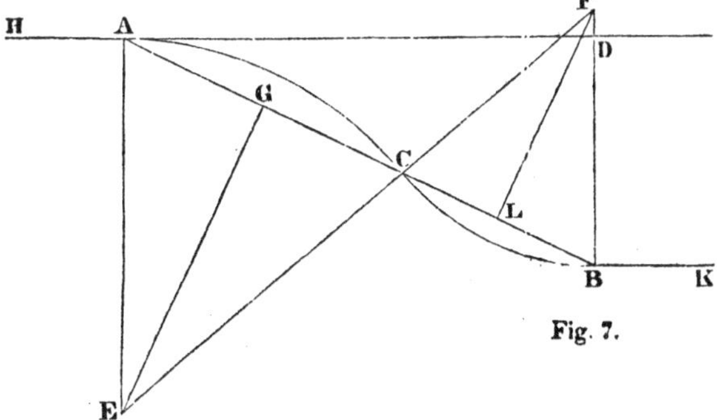

Fig. 7.

Solution. Draw the perpendiculars EG and FL. Then the right triangles ABD and EAG are similar, since the angle $BAD =$

$\frac{1}{2} AEC = AEG$. Therefore $AB : BD = EA : AG$, or $a : b = R : \frac{1}{2} a'$;

☞ $$\therefore a' = \frac{2Rb}{a}.$$

Since a' and a'' are (§ 32) parts of a, we have

☞ $$a'' = a - a'.$$

To find R' the similar triangles ABD and FBL give $AB : BD = FB : BL$, or $a : b = R' : \frac{1}{2} a''$;

☞ $$\therefore R' = \frac{a a''}{2b}.$$

Example. Given $b = 8$, $a = 160$, and $R = 900$, to find a', a'', and R'. Here $a' = \frac{2 \times 900 \times 8}{160} = 90$, $a'' = 160 - 90 = 70$, and $R' = \frac{160 \times 70}{2 \times 8} = 700$.

35. **Corollary 1.** If b, a', and a'' are given, to find a, R, and R', we have (§ 34)

☞ $$a = a' + a''; \qquad R = \frac{a a'}{2b}; \qquad R' = \frac{a a''}{2b}.$$

Example. Given $b = 8$, $a' = 90$, and $a'' = 70$, to find a, R, and R Here $a = 90 + 70 = 160$, $R = \frac{160 \times 90}{2 \times 8} = 900$, and $R' = \frac{160 \times 70}{2 \times 8} = 700$.

36. **Corollary 2.** If R, R', and b are given, to find a, a', and a'', we have (§ 35), $R + R' = \frac{a a' + a a''}{2b} = \frac{a(a' + a'')}{2b} = \frac{a^2}{2b}$. Therefore $a^2 = 2b(R + R')$;

☞ $$\therefore a = \sqrt{2b(R + R')}.$$

Having found a, we have (§ 34)

☞ $$a' = \frac{2Rb}{a}; \qquad a'' = \frac{2R'b}{a}.$$

Example. Given $R = 900$, $R' = 700$, and $b = 8$, to find a, a', and a''. Here $a = \sqrt{2 \times 8(900 + 700)} = \sqrt{16 \times 1600} = 160$, $a' = \frac{2 \times 900 \times 8}{160} = 90$, and $a'' = \frac{2 \times 700 \times 8}{160} = 70$.

37. **Problem.** *Given the angle $A\ K\ B = K$, which shows the change of direction of two tangents $H\ A$ and $B\ K$ (fig. 8), to unite these tangents by a reversed curve of given common radius R, starting from a given tangent point A.*

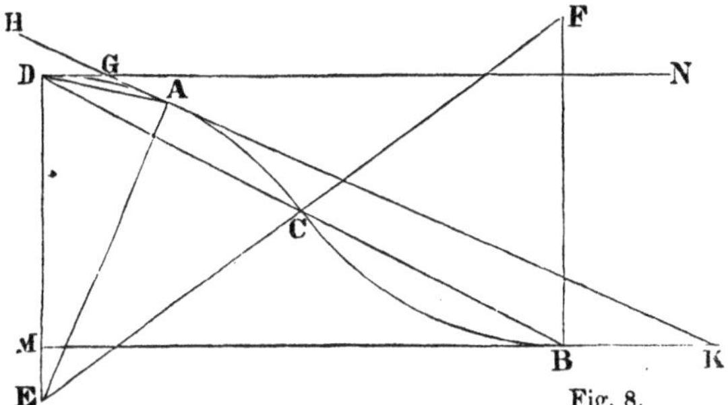

Fig. 8.

Solution. With the given radius run the curve to the point D, where the tangent $D\ N$ becomes parallel to $B\ K$. The point D is found thus. Since the angle $N\ G\ K$, which is double the angle $H\ A\ D$ (§ 2, II.), is to be made equal to $A\ K\ B = K$, lay off from $H\ A$ the angle $H\ A\ D = \frac{1}{2}\ K$. Measure in the direction thus found the chord $A\ D = 2\ R$ sin. $\frac{1}{2}\ K$. This will be shown (§ 69) to be the length of the chord for a deflection angle $\frac{1}{2}\ K$. Having found the point D, *measure the perpendicular distance $D\ M = b$ between the parallel tangents.*

The distance $D\ B = 2\ D\ C = a$ may then be obtained from the formula (§ 33, *Cor.*)

☞ $$a = 2\sqrt{R\ b}\,.$$

The second tangent point B and the reversing point C are now determined. The direction of $D\ B$ or the angle $B\ D\ N$ may also be obtained; for sin. $B\ D\ N =$ sin. $D\ B\ M = \frac{D\ M}{D\ B}$, or

☞ $$\text{sin. } B\ D\ N = \frac{b}{a}\,.$$

38. **Problem.** *Given the line $A\ B = a$ (fig. 9) which joins the fixed tangent points A and B, the angles $H\ A\ B = A$ and $A\ B\ L = B$, and the first radius $A\ E = R$, to find the second radius $B\ F = R'$ of a reversed curve to unite the tangents $H'\ A$ and $B\ K$.*

First Solution. With the given radius run the curve to the point D, where the tangent $D\ N$ becomes parallel to $B\ K$. The point D is found thus. Since the angle $H G N$, which is double $H A D$ (§ 2, II.), is equal to $A \backsim B$, lay off from $H A$ the angle $H A D = \frac{1}{2} (A \backsim B)$, and measure in this direction the chord $A D = 2 R$ sin. $\frac{1}{2}$ $(A \backsim B)$ (§ 69).

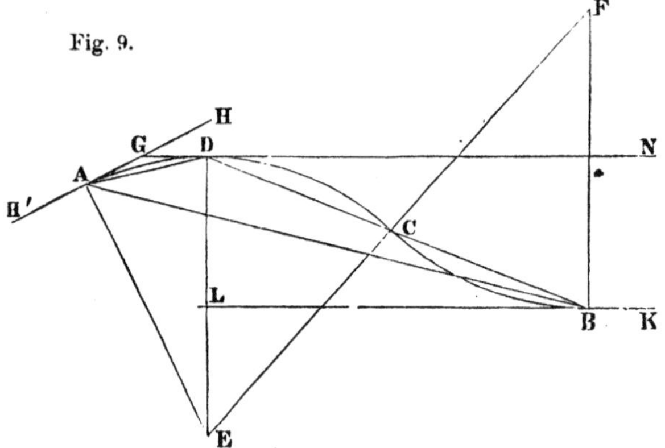

Fig. 9.

Setting the instrument at D, run the curve to the reversing point C in the line from D to B (§ 32), and measure D C and C B. Then the similar triangles $D E C$ and $B F C$ give $D C : D E = C B : B F$, or $D C : R = C B : R'$;

☞ $$\therefore R' = \frac{C B}{D C} \times R.$$

Second Solution. By this method the second radius may be found by calculation alone. The figure being drawn as above, we have, in the triangle $A B D$, $A B = a$, $A D = 2 R$ sin. $\frac{1}{2}$ $(A - B)$, and the included angle $D A B = H A B - H A D = A - \frac{1}{2} (A - B) = \frac{1}{2} (A + B)$. *Find in this triangle* (Tab. X. 14 and 12) *$B D$ and the angle $A B D$. Find also the angle $D B L = B + A B D$.*

Then the chord $C B = 2 R'$ sin. $\frac{1}{2}$ $B F C = 2 R'$ sin. $D B L$, and the chord $D C = 2 R$ sin. $\frac{1}{2}$ $D E C = 2 R$ sin. $D B L$ (§ 69). But $C B = B D - D C$; whence $2 R'$ sin. $D B L = B D - 2 R$ sin $D B L$;

☞ $$\therefore R' = \frac{B D}{2 \text{ sin. } D B L} - R.$$

When the point D falls on the other side of A, that is, when the angle B is greater than A, the solution is the same, except that the angle $D A B$ is then $180° - \frac{1}{2} (A + B)$, and the angle $D B L = B - A B D$.

39. **Problem.** *Given the length of the common tangent $D\,G = a$, and the angles of intersection I and I' (fig. 10), to determine the common radius $C\,E = C\,F = R$ of a reversed curve to unite the tangents $H\,A$ and $B\,L$.*

Fig. 10.

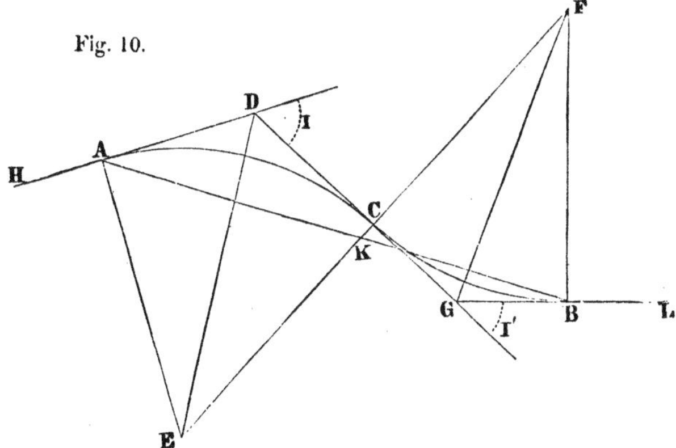

Solution. By § 4 we have $D\,C = R \tan. \frac{1}{2} I$, and $C\,G = R \tan \frac{1}{2} I'$; whence $R\,(\tan. \frac{1}{2} I + \tan. \frac{1}{2} I') = D\,C + C\,G = a$, or

☞ $$R = \frac{a}{\tan. \frac{1}{2} I + \tan. \frac{1}{2} I'}.$$

This formula may be adapted to calculation by logarithms; for we have (Tab. X. 35) $\tan. \frac{1}{2} I + \tan. \frac{1}{2} I' = \frac{\sin. \frac{1}{2}(I + I')}{\cos. \frac{1}{2} I \cos. \frac{1}{2} I'}$. Substituting this value, we get

☞ $$R = \frac{a \cos. \frac{1}{2} I \cos. \frac{1}{2} I'}{\sin. \frac{1}{2}(I + I')}.$$

The tangent points A and B are obtained by measuring from D a distance $A\,D = R \tan. \frac{1}{2} I$, and from G a distance $B\,G = R \tan. \frac{1}{2} I'$.

Example. Given $a = 600$, $I = 12°$, and $I' = 8°$, to find R. Here

$a = 600$	2.778151
$\frac{1}{2} I = 6°$	cos. 9.997614
$\frac{1}{2} I' = 4°$	cos. 9.998941
	2.774706
$\frac{1}{2}(I + I') = 10°$	sin. 9.239670
$R = 3427.96$	3.535036

10 **Problem.** *Given the line $A B = a$ (fig. 10), which joins the fixed tangent points A and B, the angle $D A B = A$, and the angle $A B G = B$, to find the common radius $E C = C F = R$ of a reversed curve to unite the tangents $H A$ and $B L$.*

Fig. 10.

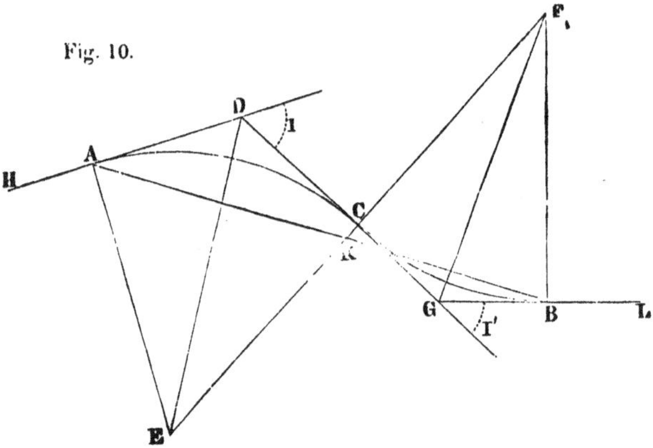

Solution. Find first the auxiliary angle $A K E = B K F$, which may be denoted by K. For this purpose the triangle $A E K$ gives $A E : E K = \sin. K : \sin. E A K$. Therefore $E K \sin. K = A E \sin. E A K = R \cos. A$, since $E A K = 90° - A$. In like manner, the triangle $B F K$ gives $F K \sin K = B F \sin. F B K = R \cos. B$. Adding these equations, we have $(E K + F K) \sin. K = R (\cos. A + \cos. B)$, or, since $E K + F K = 2 R$, $2 R \sin. K = R (\cos. A + \cos. B)$ Therefore, $\sin. K = \frac{1}{2} (\cos. A + \cos. B)$. For calculation by logarithms, this becomes (Tab. X. 28)

☞ $$\sin K = \cos. \tfrac{1}{2} (A + B) \cos. \tfrac{1}{2} (A - B).$$

Having found K, we have the angle $A E K = E = 180° - K - E A K = 180° - K - (90° - A) = 90° + A - K$, and the angle $B F K = F = 180° - K - F B K = 180° - K - (90° - B) = 90° + B - K$. Moreover, the triangle $A E K$ gives $A E : A K = \sin. K : \sin. E$, or $R \sin. E = A K \sin K$, and the triangle $B F K$ gives $B F : B K = \sin. K : \sin F$, or $R \sin. F = B K \sin. K$. Adding these equations, we have $R (\sin. E + \sin. F) = (A K + B K) \sin. K = a \sin. K$ Substituting for $\sin. E + \sin. F$ its value $2 \sin. \frac{1}{2} (E + F)$

cos. $\frac{1}{2}(E - F)$ (Tab. X. 26), we have $2R \sin. \frac{1}{2}(E+F) \cos. \frac{1}{2}(E-F) = a \sin. K$. Therefore $R = \frac{\frac{1}{2} a \sin. K}{\sin. \frac{1}{2}(E+F) \cos. \frac{1}{2}(E-F)}$. Finally, substituting for E its value $90° + A - K$, and for F its value $90° + B - K$, we get $\frac{1}{2}(E+F) = 90° - [K - \frac{1}{2}(A+B)]$, and $\frac{1}{2}(E-F) = \frac{1}{2}(A-B)$; whence

☞ $$R = \frac{\frac{1}{2} a \sin. K}{\cos. [K - \frac{1}{2}(A+B)] \cos. \frac{1}{2}(A-B)}$$

Example. Given $a = 1500$, $A = 18°$, and $B = 6°$, to find R. Here

$\frac{1}{2}(A+B) = 12°$		cos. 9.990404
$\frac{1}{2}(A-B) = 6°$		cos 9.997614
$K = 76° 36' 10''$		sin. 9.988018
$\frac{1}{2} a = 750$		2.875061
		2.863079
$K - \frac{1}{2}(A+B) = 64° 36' 10'$	cos. 9.632347	
$\frac{1}{2}(A-B) = 6°$	cos. 9.997614	
		9.629961
$R = 1710.48$		3.233118

B. *Compound Curves.*

41. **Theorem.** *If one branch of a compound curve be produced, until the tangent at its extremity is parallel to the tangent at the extremity of the second branch, the common tangent point of the two arcs is in the straight line produced, which passes through the tangent points of these parallel tangents.*

Demonstration. Let ACB (fig. 11) be a compound curve, uniting the tangents HA and BK. The radii CE and CF, being perpendicular to the common tangent at C (§ 2, I.), are in the same straight line. Continue the curve AC to D, where its tangent OD becomes parallel to BK, and consequently the radius DE parallel to BF. Then if the chords CD and CB be drawn, we have the angle $CED = CFB$; whence ECD, the half-supplement of CED, is equal to FCB, the half-supplement of CFB. But ECD cannot be equal to FCB, unless CD coincides with CB. Therefore the line BD produced passes through the common tangent point C

42. **Problem.** *To find a limit in one direction of each radius of a compound curve.*

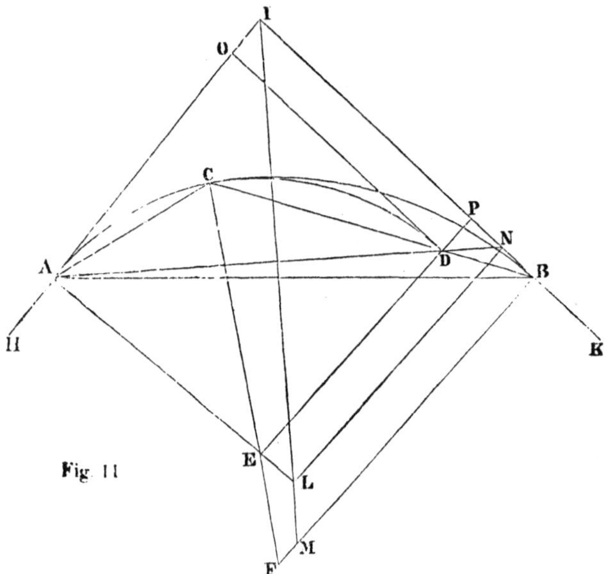

Fig. 11

Solution. Let $A\,I$ and $B\,I$ (fig. 11) be the tangents of the curve. Through the intersection point I, draw $I\,M$ bisecting the angle $A\,I\,B$. Draw $A\,L$ and $B\,M$ perpendicular respectively to $A\,I$ and $B\,I$, meeting $I\,M$ in L and M. Then the radius of the branch commencing on the shorter tangent $A\,I$ must be *less* than $A\,L$, and the radius of the branch commencing on the longer tangent $B\,I$ must be *greater* than $B\,M$. For suppose the shorter radius to be made equal to $A\,L$, and make $I\,N = A\,I$, and join $L\,N$. Then the equal triangles $A\,I\,L$ and $N\,I\,L$ give $A\,L = L\,N$; so that the curve, if continued, will pass through N, where its tangent will coincide with $I\,N$. Then (§ 41) the common tangent point would be the intersection of the straight line through B and N with the first curve; but in this case there can be no intersection, and therefore no common tangent point. Suppose next, that this radius is *greater* than $A\,L$, and continue the curve, until its tangent becomes parallel to $B\,I$. In this case the extremity of the curve will fall outside the tangent $B I$ in the line $A N$ produced, and a straight line through B and this extremity will again fail to intersect the curve already drawn. As no common tangent point can be found when this radius is taken equal to $A L$ or greater than $A L$, no compound curve is possible. This radius must, therefore, be *less* than $A L$. In a similar manner it might be shown, that the radius of the other branch of the curve must be *greater* than $B M$. If we suppose the tangents $A I$ and $B I$ and the intersection angle I to be known, we have (§ 5) $A L = A I \cot. \frac{1}{2} I$, and $B M = B I \cot. \frac{1}{2} I$. These values are therefore, the limits of the radii in one direction.

43. If nothing were given but the position of the tangents and the tangent points, it is evident that an indefinite number of different compound curves might connect the tangent points; for the shorter radius might be taken of any length less than the limit found above, and a corresponding value for the greater could be found. Some other condition must, therefore, be introduced, as is done in the following problems.

44. **Problem.** *Given the line $A B = a$ (fig. 11), which joins the fixed tangent points A and B, the angle $B A I = A$, the angle $A B I = B$, and the first radius $A E = R$, to find the second radius $B F = R'$ of a compound curve to unite the tangents $H A$ and $B K$.*

Solution. Suppose the first curve to be run with the given radius from A to D, where its tangent $D O$ becomes parallel to $B I$, and the angle $I A D = \frac{1}{2}(A + B)$. Then (§ 41) the common tangent point C is in the line $B D$ produced, and the chord $C B = C D + B D$. Now in the triangle $A B D$ we have $A B = a$, $A D = 2 R \sin. \frac{1}{2}(A + B)$ (§ 69), and the included angle $D A B = I A B - I A D = A - \frac{1}{2}(A + B) = \frac{1}{2}(A - B)$. *Find in this triangle* (Tab. X. 14 and 12) *the angle $A B D$ and the side $B D$. Find also the angle $C B I = B - A B D$.*

Then (§ 69) the chord $CB = 2 R' \sin. C B I$, and the chord $CD = 2 R \sin\ C D O = 2 R \sin. C B I$. Substituting these values of $C B$ and $C D$ in the equation found above, $C B = C D + B D$, we have $2 R' \sin. C B I = 2 R \sin. C B I + B D$;

☞ $$\therefore R' = R + \frac{B D}{2 \sin. C B I}.$$

When the angle B is greater than A, that is, when the greater radius is given, the solution is the same, except that the angle $D A B =$

$\frac{1}{2}(B - A)$, and $C B I$ is found by subtracting the *supplement* of $A B D$ from B. We shall also find $C B = C D - B D$, and consequently $R' = R - \frac{B D}{2 \sin. C B I}$.

If more convenient, the point D may be determined in the field, by laying off the angle $I A D = \frac{1}{2}(A + B)$, and measuring the distance $A D = 2 R \sin. \frac{1}{2}(A + B)$. $B D$ and $C B I$ may then be measured, instead of being calculated as above.

Example. Given $a = 950$, $A = 8°$, $B = 7°$, and $R = 3000$, to find R'. Here $A D = 2 \times 3000 \sin. \frac{1}{2}(8° + 7°) = 783.16$, and $D A B = \frac{1}{2}(8° - 7°) = 30'$. Then to find $A B D$ we have

$A B - A D =$	166 84	2.222300
$\frac{1}{2}(A D B + A B D) =$	89° 45′	tan. 2.360180
		4.582480
$A B + A D =$	1733.16	3.238839
$\frac{1}{2}(A D B - A B D) =$	87° 24′ 17″	tan. 1.343641
$\therefore A B D =$	2° 20′ 43″	

Next, to find $B D$,

$A D =$	783.16	2.893849
$D A B =$	30′	sin. 7.940842
		0.834691
$A B D =$	2° 20′ 43″	sin. 8.611948
$B D =$	167.01	2.222743
$B - A B D = C B I =$	4° 39′ 17″	sin. 8.909292
$2(R' - R) =$	2058.03	3.313451
$\therefore R' - R =$	1029.01	
$\therefore R' = 3000 + 1029.01 =$	4029.01	

To find the central angle of each branch, we have $C F B = 2 C B I = 9° 18' 34''$, which is the central angle of the second branch; and $A E C = A E D - C E D = A + B - 2 C B I = 5° 41' 26''$, which is the central angle of the first branch

45. **Problem.** *Given (fig. 11) the tangents $A I = T$, $B I = T'$, the angle of intersection $= I$, and the first radius $A E = R$, to find the second radius $B F = R'$.*

Solution. Suppose the first curve to be run with the given radius from A to D, where its tangent $D O$ becomes parallel to $B I$. Through

D draw $D P$ parallel to $A I$, and we have $I P = D O = A O = R \tan. \frac{1}{2} I$ (§ 4). Then in the triangle DPB we have $D P = I O = A I - A O = T - R \tan. \frac{1}{2} I$, $BP = BI - IP = T' - R \tan. \frac{1}{2} I$, and the included angle $DPB = AIB = 180° - I$. *Find in this triangle the angle CBI, and the side BD. The remainder of the solution is the same as in* § 44. The determination of the point D in the field is also the same, the angle IAD being here $= \frac{1}{2} I$. When B is greater than A, that is, when the greater radius is given, the solution is the same, except that $DP = R \tan. \frac{1}{2} I - T$, and $BP = R \tan. \frac{1}{2} I - T'$.

Example. Given $T = 447.32$, $T' = 510.84$, $I = 15°$, and $R = 3000$, to find R'. Here $R \tan. \frac{1}{2} I = 3000 \tan. 7\frac{1}{2}° = 394.96$, $DP = 447.32 - 394.96 = 52.36$, $BP = 510.84 - 394.96 = 115.88$, and $DPB = 180° - 15° = 165°$. Then (Tab. X. 14 and 12)

$BP - DP = 63.52$		1.802910
$\frac{1}{2}(BDP + PBD) = 7° 30'$	tan.	9.119429
		0.922339
$BP + DP = 168.24$		2.225929
$\frac{1}{2}(BDP - PBD) = 2° 50' 44''$	tan.	8.696410
$\therefore PBD = CBI = 4° 39' 16''$		

Next, to find BD,

$DP = 52.36$		1.719000
$DPB = 15°$	sin	9.412996
		1.131996
$PBD = 4° 39' 16''$	sin.	8.909266
$BD = 167.005$		2.222730

The tangents in this example were calculated from the example in § 44. The values of CBI and BD here found differ slightly from those obtained before. In general, the triangle DBP is of better form for accurate calculation than the triangle ADB.

46. If no circumstance determines either of the radii, the condition may be introduced, that the common tangent shall be parallel to the line joining the tangent points.

Problem. *Given the line $AB = a$ (fig. 12), which unites the fixed tangent points A and B, the angle $IAB = A$, and the angle $ABI = B$, to find the radii $AE = R$ and $BF = R'$ of a compound curve, having the common tangent DG parallel to AB.*

Solution Let $A\,C$ and $B\,C$ be the two branches of the required curve, and draw the chords $A\,C$ and $B\,C$. These chords bisect the

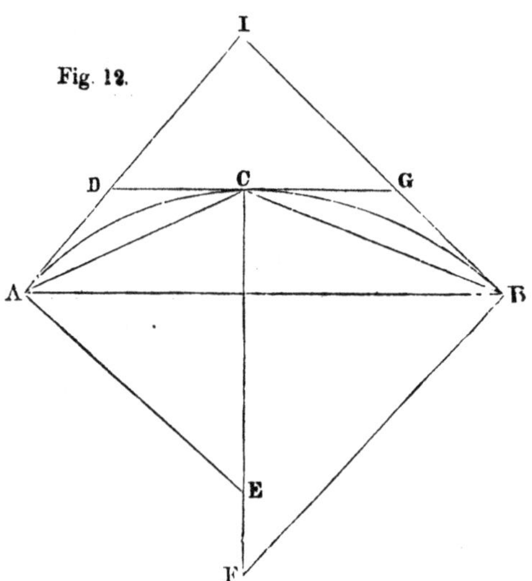

Fig. 12.

angles A and B; for the angle $D\,A\,C = \frac{1}{2}\,I\,D\,G = \frac{1}{2}\,I\,A\,B$, and the angle $G\,B\,C = \frac{1}{2}\,D\,G\,I = \frac{1}{2}\,A\,B\,I$. Then in the triangle $A\,C\,B$ we have $A\,C : A\,B = \sin.\ A\,B\,C : \sin.\ A\,C\,B$. But $A\,C\,B = 180° - (C\,A\,B + C\,B\,A) = 180° - \frac{1}{2}\,(A + B)$, and as the sine of the supplement of an angle is the same as the sine of the angle itself, $\sin.\ A\,C\,B = \sin.\ \frac{1}{2}\,(A + B)$. Therefore $A\,C : a = \sin.\ \frac{1}{2}\,B : \sin.\ \frac{1}{2}\,(A + B)$, or $A\,C = \frac{a \sin\ \frac{1}{2}\,B}{\sin.\ \frac{1}{2}\,(A + B)}$. In a similar manner we should find $B\,C = \frac{a \sin\ \frac{1}{2}\,A}{\sin.\ \frac{1}{2}\,(A + B)}$. Now we have (§ 68) $R = \frac{\frac{1}{2}\,A\,C}{\sin\ \frac{1}{2}\,A}$, and $R' = \frac{\frac{1}{2}\,B\,C}{\sin.\ \frac{1}{2}\,B}$, or, substituting the values of $A\,C$ and $B\,C$ just found.

$$R = \frac{\frac{1}{2}\,a \sin.\ \frac{1}{2}\,B}{\sin.\ \frac{1}{2}\,A \sin.\ \frac{1}{2}\,(A + B)}; \quad R' = \frac{\frac{1}{2}\,a \sin.\ \frac{1}{2}\,A}{\sin.\ \frac{1}{2}\,B \sin.\ \frac{1}{2}\,(A + B)}.$$

Example Given $a = 950$, $A = 8°$, and $B = 7°$, to find R and R' Here

$\frac{1}{2}a = 475$		2.676694
$\frac{1}{2}B = 3° 30'$		sin 8 785675
		1.462369
$\frac{1}{2}A = 4°$	sin. 8.843585	
$\frac{1}{2}(A + B) = 7° 30'$	sin. 9.115698	
		7.959283
$R = 3184.83$		3.503086

Transposing these same logarithms according to the formula for R' we have

$\frac{1}{2}a = 475$		2.676694
$\frac{1}{2}A = 4°$		sin. 8.843585
		1.520279
$\frac{1}{2}B = 3° 30'$	sin. 8.785675	
$\frac{1}{2}(A + B) = 7° 30'$	sin. 9 115698	
		7.901373
$R' = 4158.21$		3.618906

47. **Problem.** *Given the line $A B = a$ (fig. 12), which unites the fixed tangent points A and B, and the tangents $A I = T$ and $B I = T'$, to find the tangents $A D = x$ and $B G = y$ of the two branches of a compound curve, having its common tangent $D G$ parallel to $A B$.*

Solution. Since $D C = A D = x$, and $C G = B G = y$, we have $D G = x + y$. Then the similar triangles $I D G$ and $I A B$ give $ID : IA = DG : AB$, or $T - x : T = x + y : a$. Therefore $a T - a x = T x + T y$ (1). Also $A D : A I = B G : B I$, or $x : T = y \,.\, T'$. Therefore $T y = T' x$ (2). Substituting in (1) the value of $T y$ in (2), we have $a T - a x = T x + T' x$, or $a x + T x + T' x = a T$;

$$\therefore x = \frac{a T}{a + T + T'};$$

and, since from (2), $y = \frac{T' x}{T}$,

$$y = \frac{a T'}{a + T + T'}.$$

The intersection points D and G and the common tangent point C are now easily obtained on the ground, and the radii may be found by the usual methods. Or, if the angles $I A B = A$ and $A B I = B$

have been measured or calculated, we have (§ 5) $R = x \cot. \frac{1}{2} A$, and $R' = y \cot. \frac{1}{2} B$. Substituting the values of x and y found above, we have $R = \frac{a\ T \cot\ \frac{1}{2} A}{a + T + T'}$, and $R' = \frac{a\ T' \cot. \frac{1}{2} B}{a + T + T'}$.*

Example. Given $a = 500$, $T = 250$, and $T' = 290$, to find x and y. Here $a + T + T' = 500 + 250 + 290 = 1040$, whence $x = 500 \times 250 \div 1040 = 120.19$, and $y = 500 \times 290 \div 1040 = 139.42$.

48. **Problem.** *Given the tangents $A I = T$, $B I = T'$, and the angle of intersection I, to unite the tangent points A and B (fig. 13) by a compound curve, on condition that the two branches shall have their angles of intersection $I D G$ and $I G D$ equal.*

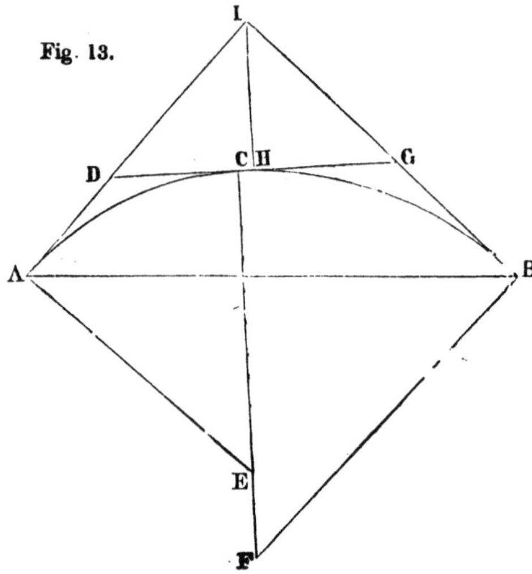

Fig. 13.

Solution. Since $I D G = I G D = \frac{1}{2} I$, we have $I D = I G$. *Represent the line $I D = I G$ by x.* Then if the perpendicular $I H$ be let

* The radii of an oval of given length and breadth, or of a three-centre arch of given span and rise, may also be found from these formulæ In these cases $A + B = 90°$, and the values of R and R' may be reduced to $R = \frac{a\ T}{a + T' - T}$ and $R' = \frac{a\ T'}{a + T - T'}$. These values admit of an easy construction, or they may be readily calculated

fall from I, we have (Tab. X. 11) $D H = I D \cos. I D G = x \cos \frac{1}{2} I$, and $D G = 2 x \cos. \frac{1}{2} I$. But $D G = D C + C G = A D + B G = T - x + T' - x = T + T' - 2 x$ Therefore $2 x \cos. \frac{1}{2} I = T + T' - 2 x$, or $2 x + 2 x \cos. \frac{1}{2} I = T + T'$; whence $x = \frac{\frac{1}{2}(T + T')}{1 + \cos. \frac{1}{2} I}$, or (Tab. X. 25)

☞ $$x = \frac{\frac{1}{4}(T + T')}{\cos.^2 \frac{1}{4} I}$$

The tangents $A D = T - x$ and $B G = T' - x$ are now readily found. With these and the known angles of intersection, the radii or deflection angles may be found (§ 5 or § 11). This method answers very well, when the given tangents are nearly equal; but in general the preceding method is preferable.

Example. Given $T = 480$, $T' = 500$, and $I = 18°$, to find x. Here

$\frac{1}{4}(T + T') = 245$		2.389166
$\frac{1}{4} I = 4° 30'$	2 cos.	9.997318
$x = 246.52$		2.391848

Then $A D = 480 - 246.52 = 233.48$, and $B G = 500 - 246.52 = 253.48$. The angle of intersection for both branches of the curve being 9°, we find the radii $A E = 233.48 \cot. 4° 30' = 2966.65$, and $B F = 253.48 \cot. 4° 30' = 3220.77$.

Article III. — Turnouts and Crossings.

49. The usual mode of turning off from a main track is by switching a pair of rails in the main track, and putting in a turnout curve tangent to the switched rails, with a frog placed where the outer rail of the turnout crosses the rail of the main track. $A B$ (fig. 14) represents one of the rails of the main track switched, $B F$ represents the outer rail of the turnout curve, tangent to $A B$, and F shows the position of the frog The switch angle, denoted by S, is the angle $D A B$, formed by the switched rail $A B$ with $A D$, its former position in the main track. The frog angle, denoted by F, is the angle $G F M$ made by the crossing rails, the direction of the turnout rail at F being the tangent $F M$ at that point. In the problems of this article the gauge of the track $D C$, denoted by g, and the distance $D B$, denoted by d, are supposed to be known. The switch angle S is also supposed to be known, since its sine (Tab. X. 1) is equal to d divided by the length

of the switched rail. If, for example, the rail is 18 feet in length and $d = .42$, we have $S = 1° 20'$.

A. *Turnout from Straight Lines.*

50. **Problem.** *Given the radius R of the centre line of a turnout (fig. 14), to find the frog angle $G F M = F$ and the chord $B F$.*

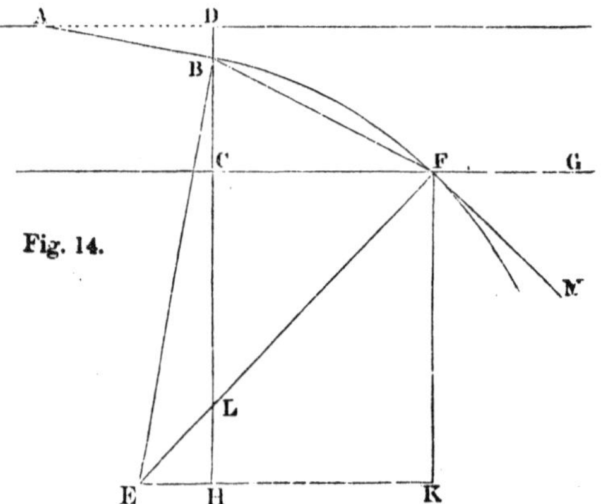

Solution. Through the centre E draw $E K$ parallel to the main track. Draw $B H$ and $F K$ perpendicular to $E K$, and join $E F$. Then, since $E F$ is perpendicular to $F M$ and $F K$ is perpendicular to $F G$, the angle $E F K = G F M = F$; and since $E B$ and $B H$ are respectively perpendicular to $A B$ and $A D$, the angle $E B H = D A B = S$. Now the triangle $E F K$ gives (Tab. X. 2) $\cos. E F K = \frac{F K}{E F}$. But $E F$, the radius of the outer rail, is equal to $R + \frac{1}{2} g$, and $F K = C H = B H - B C = B E \cos. E B H - B C = (R + \frac{1}{2} g) \cos. S - (g - d)$. Substituting these values, we have $\cos. E F K = \frac{(R + \frac{1}{2} g) \cos. S - (g - d)}{R + \frac{1}{2} g}$, or

$$\cos. F = \cos. S - \frac{g - d}{R + \frac{1}{2} g}.$$

From this formula F may be found by the table of natural cosines. To adapt it to calculation by logarithms, we may consider $g - d$ to be equal to $(g - d) \cos. S$, which will lead to no material error, since

$g - d$ is very small, and cos. S almost equal to unity The value of cos. F then becomes

☞ $$\cos. F = \frac{(R - \frac{1}{2} g + d) \cos. S}{R + \frac{1}{2} g}.$$

To find $B F$, the right triangle $B C F$ gives (Tab. X. 9) $B F = \frac{B C}{\sin. B F C}$. But $B C = g - d$ and the angle $B F C = B F E - C F E = (90° - \frac{1}{2} B E F) - (90° - F) = F - \frac{1}{2} B E F$. But $B E F = B L F - E B L = F - S$. Therefore $B F C = F - \frac{1}{2}(F - S) = \frac{1}{2}(F + S)$. Substituting these values in the formula for $B F$, we have

☞ $$B F = \frac{g - d}{\sin. \frac{1}{2}(F + S)}.$$

By the above formulæ the columns headed F and $B F$ in Table V are calculated.

Example. Given $g = 4.7$, $d = .42$, $S = 1° 20'$, and $R = 500$, to find F and $B F$. Here nat. cos. $S = .999729$, $g - d = 4.28$, $R + \frac{1}{2} g = 502.35$, and $4.28 \div 502.35 = .008520$. Therefore nat. cos. $F = .999729 - .008520 = .991209$, which gives $F = 7° 36' 10''$. Next, to find $B F$,

$g - d$	$= 4.28$	0.631444
$\frac{1}{2}(F + S)$	$= 4° 28' 5''$	sin. 8.891555
$B F$	$= 54.94$	1.739889

51. **Problem.** *Given the frog angle $G F M = F$ (fig. 14), to find the radius R of the centre line of a turnout, and the chord $B F$.*

Solution. From the preceding solution we have $\cos. F = \frac{(R + \frac{1}{2} g) \cos. S - (g - d)}{R + \frac{1}{2} g}$. Therefore $(R + \frac{1}{2} g) \cos. F = (R + \frac{1}{2} g) \cos. S - (g - d)$, or

☞ $$R + \frac{1}{2} g = \frac{g - d}{\cos. S - \cos. F}.$$

For calculation by logarithms this becomes (Tab. X. 29)

☞ $$R + \frac{1}{2} g = \frac{\frac{1}{2}(g - d)}{\sin. \frac{1}{2}(F + S) \sin. \frac{1}{2}(F - S)}.$$

Having thus found $R + \frac{1}{2} g$, we find R by subtracting $\frac{1}{2} g$. $B F$ is found, as in the preceding problem, by the formula

☞ $$B F = \frac{g - d}{\sin. \frac{1}{2}(F + S)}.$$

Example. Given $g = 4.7$, $d = .42$, $S = 1^\circ\ 20'$, and $F = 7^\circ$, to find R. Here

$\frac{1}{2}(g - d) = 2.14$		0.330414
$\frac{1}{2}(F + S) = 4^\circ\ 10'$	sin. 8.861283	
$\frac{1}{2}(F - S) = 2^\circ\ 50'$	sin. 8.693998	
		7 555281
$R + \frac{1}{2}g = 595\ 85$		2.775133
$\therefore R = 593.5$		

52. **Problem.** *To find mechanically the proper position of a given frog.*

Solution. Denote the length of the switch rail by l, the length of the frog by f, and its width by w. From B as a centre with a radius $BH = 2l$, describe on the ground an arc GHK (fig. 15), and from the inside of the rail at G measure $GH = 2d$, and from H measure HK such that $HK : BH = \frac{1}{2}w : f$, or $HK : 2l = \frac{1}{2}w : f$; that is, $HK = \frac{wl}{f}$. Then a straight line through B and the point K will strike the inside of the other rail at F, the place for the point of the

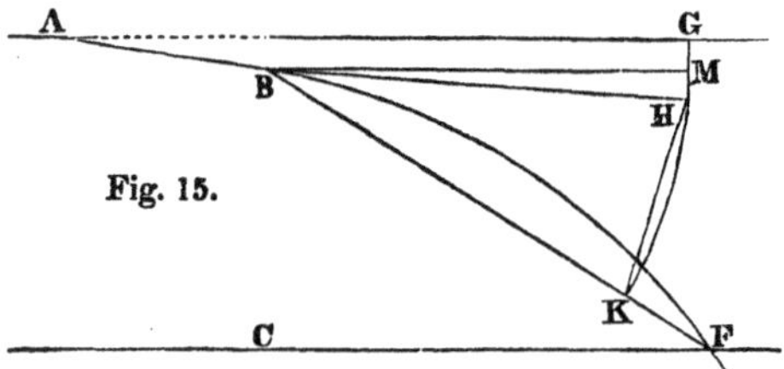

Fig. 15.

frog. For the angle HBK has been made equal to $\frac{1}{2}F$, and if BM be drawn parallel to the main track, the angle MBH is seen to be equal to $\frac{1}{2}S$. Therefore, $MBK = BFC = \frac{1}{2}(F + S)$, and this was shown (§ 50) to be the true value of BFC.

53. If the turnout is to reverse, and become parallel to the main track, the problems on reversed curves already given will in general be sufficient. Thus, if the tangent points of the required curve are fixed, the common radius may be found by § 40. If the tangent point at the switch is fixed, and the common radius given, the reversing oint and the other tangent point may be found by § 37, the change of direction of the two tangents being here equal to S. But when the

frog angle is given, or determined from a given first radius, and the point of the frog is taken as the reversing point, the radius of the second portion may be found by the following method.

54. **Problem.** *Given the frog angle F and the distance $HB = b$ (fig. 16) between the main track and a turnout, to find the radius R' of the second branch of the turnout, the reversing point being taken opposite F, the point of the frog.*

Fig. 16.

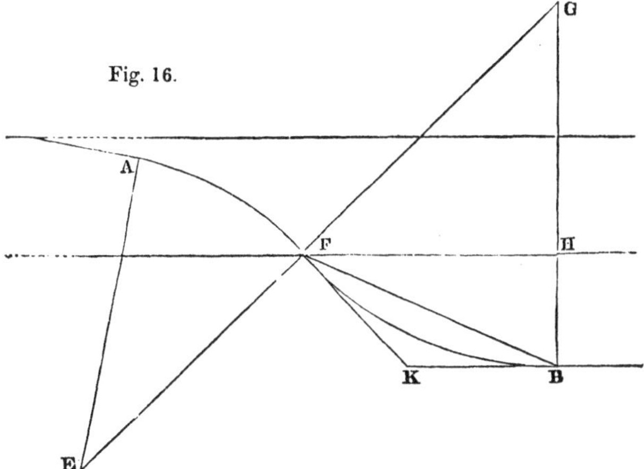

Solution. Let the arc FB be the inner rail of the second branch, $FG = R' - \frac{1}{2}g$ its radius, and B the tangent point where the turnout becomes parallel to the main track. Now since the tangent FK is one side of the frog produced, the angle $HFK = F$, and since the angle of intersection at K is also equal to F, $BFK = \frac{1}{2}F$ (§ 2, II.); whence $BFH = \frac{1}{2}F$. Then (§ 68) $FG = \frac{\frac{1}{2}BF}{\sin. BFK}$, or $R' - \frac{1}{2}g = \frac{\frac{1}{2}BF}{\sin. \frac{1}{2}F}$. But $BF = \frac{HB}{\sin. BFH}$ (Tab. X. 9), or $\frac{1}{2}BF = \frac{\frac{1}{2}b}{\sin. \frac{1}{2}F}$. Substituting this value of $\frac{1}{2}BF$, we have

☞ $$R' - \frac{1}{2}g = \frac{\frac{1}{2}b}{\sin.^2 \frac{1}{2}F}.$$

In measuring the distance $HB = b$, it is to be observed, that the widths of both rails must be included.

Example. Given $b = 6.2$ and $F = 8°$, to find R'. Here

$\frac{1}{2} b = 3.1$	0.491362
$\frac{1}{2} F = 4°$	sin. 8.843585
$\frac{1}{2} B F = 44.44$	1.647777
$\frac{1}{2} F = 4°$	sin. 8.843585
$R' - \frac{1}{2} g = 637.08$	2.804192
$\therefore R' = 639.43$	

B. *Crossings on Straight Lines.*

55. When a turnout enters a parallel main track by a second switch it becomes a *crossing*. As the switch angle is the same on both tracks a crossing on a straight line is a reversed curve between parallel tangents. Let $H D$ and $N K$ (fig. 17) be the centre lines of two parallel tracks, and $H A$ and $B K$ the direction of the switched rails. If now the tangent points A and B are fixed, the distance $A B = a$ may be measured, and also the perpendicular distance $B P = b$ between the tangents $H P$ and $B K$. Then the common radius of the crossing $A C B$ may be found by § 33; or if the radius of one part of the crossing is fixed, the second radius may be found by § 34. But if both frog angles are given, we have the two radii or the common radius of a crossing given, and it will then be necessary to determine the distance $A B$ between the two tangent points.

56. **Problem.** *Given the perpendicular distance $G N = b$ (fig. 17) between the centre lines of two parallel tracks, and the radii $E C = R$ and $C F = R'$ of a crossing, to find the chords $A C$ and $B C$.*

Solution. Draw $E G$ perpendicular to the main track, and $A L$, $C M$, and $B L'$ parallel to it. *Denote the angle $A E C$ by E.* Then, since the angle $A E L = A H G = S$, we have $C E L = E + S$, and in the right triangle $C E M$ (Tab. X. 2), $C E \cos. C E M = R \cos. (E + S) = E M = E L - L M$. But $E L = A E \cos. A E L = R \cos. S$, and $L M : L' M = A C : B C$. Now $A C : B C = E C : C F = R : R'$. Therefore, $L M : L' M = R : R'$, or $L M : L M + L' M = R : R + R'$; that is, $L M : b - 2 d = R : R + R'$, whence $L M = \frac{R (b - 2 d)}{R + R'}$. Substituting these values of $E L$ and $L M$ in the equation for $R \cos. (E + S)$, we have $R \cos. (E + S) = R \cos. S - \frac{R (b - 2 d)}{R + R'}$,

☞ $$\therefore \cos.(E + S) = \cos. S - \frac{b - 2d}{R + R'}.$$

Having thus found $E + S$, we have the angle E and also its equal CFB. Then (§ 69)

☞ $$A\,C = 2\,R \sin. \tfrac{1}{2}\,E; \qquad B\,C = 2\,R' \sin. \tfrac{1}{2}\,E.$$

We have also $A\,B = A\,C + B\,C$, since $A\,C$ and $B\,C$ are in the same straight line (§ 32), or $A\,B = 2\,(R + R') \sin\, \tfrac{1}{2}\,E$.

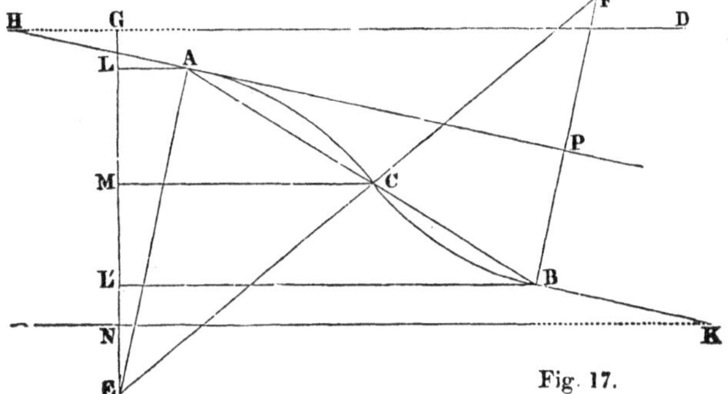

Fig. 17.

When the two radii are equal, the same formulæ apply by making $R' = R$. In this case, we have

☞ $$\cos.(E + S) = \cos. S - \frac{b - 2d}{2\,R};$$

☞ $$A\,C = B\,C = 2\,R \sin. \tfrac{1}{2}\,E.$$

Example. Given $d = .42$, $g = 4.7$, $S = 1^\circ\, 20'$, $b = 11$, and the angles of the two frogs each 7°, to find $A\,C = B\,C = \tfrac{1}{2}\,A\,B$. The common radius R, corresponding to $F = 7^\circ$, is found (§ 51) to be 593.5. Then $2\,R = 1187$. $b - 2\,d = 10.16$, and $10.16 \div 1187 = .00856$. Therefore, nat. cos. $(E + S) = .99973 - .00856 = .99117$; whence $E + S = 7^\circ\, 37'\, 15''$. Subtracting S, we have $E = 6^\circ\, 17'\, 15''$. Next

$2\,R = 1187$	3.074451
$\tfrac{1}{2}\,E = 3^\circ\, 8'\, 37\tfrac{1}{2}''$	sin. 8.739106
$A\,C = 65.1$	1.813557

C. *Turnout from Curves.*

57. **Problem.** *Given the radius R of the centre line of the main track and the frog angle F, to determine the position of the frog by means of the chord B F (figs. 18 and 19), and to find the radius R' of the centre line of the turnout.*

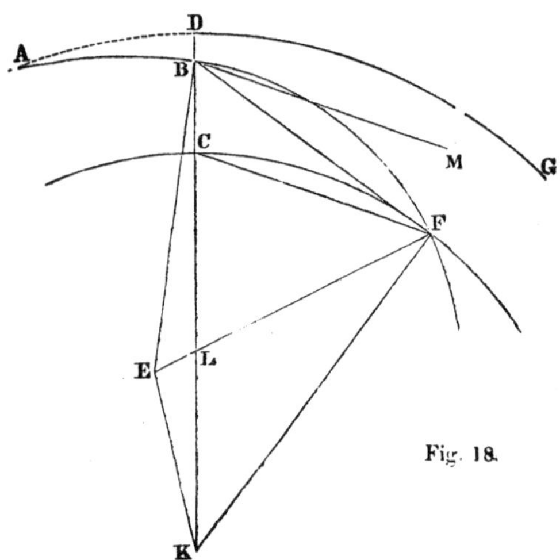

Fig. 18.

Solution. I. When the turnout is from the *inside* of the curve (fig. 18). Let $A\,G$ and $C\,F$ be the rails of the main track, $A\,B$ the switch rail, and the arc $B\,F$ the outer rail of the turnout, crossing the inside rail of the main track at F. Then, since the angle $E\,F\,K$ has its sides perpendicular to the tangents of the two curves at F, it is equal to the acute angle made by the crossing rails, that is, $E\,F\,K = F$. Also $E\,B\,L = S.$ *The first step is to find the angle $B\,K\,F$ denoted by K.* To find this angle, we have in the triangle $B\,F\,K$ (Tab. X. 14), $B\,K + K\,F : B\,K - K\,F = \tan.\frac{1}{2}(B\,F\,K + F\,B\,K) : \tan.\frac{1}{2}(B\,F\,K - F\,B\,K)$. But $B\,K = R + \frac{1}{2}g - d$, and $K\,F = R - \frac{1}{2}g$. Therefore, $B\,K + K\,F = 2\,R - d$, and $B\,K - K\,F = g - d$. Moreover, $B\,F\,K = B\,F\,E + E\,F\,K = B\,F\,E + F$, and $F\,B\,K = E\,B\,F - E\,B\,K = B\,F\,E - S$. Therefore, $B\,F\,K - F\,B\,K = F + S$. Lastly, $B\,F\,K + F\,B\,K = 180° - K$. Substituting these values in the preceding proportion, we have $2\,R - d : g - d = \tan.(90° - \frac{1}{2}K) : \tan.\frac{1}{2}(F + S)$.

or $\tan.(90^\circ - \frac{1}{2}K) = \frac{(2R - d)\tan.\frac{1}{2}(F+S)}{g-d}$. But $\tan.(90^\circ - \frac{1}{2}K) = \cot.\frac{1}{2}K = \frac{1}{\tan.\frac{1}{2}K}$;

☞ $$\therefore \tan.\tfrac{1}{2}K = \frac{g-d}{(2R-d)\tan.\frac{1}{2}(F+S)}.$$

Next, to find the chord BF, we have, in the triangle BFC (Tab. X. 12), $BF = \frac{BC \sin. BCF}{\sin. BFC}$. But $BC = g - d$, and $BCF = 180^\circ - FCK = 180^\circ - (90^\circ - \frac{1}{2}K) = 90^\circ + \frac{1}{2}K$, or $\sin. BCF = \cos.\frac{1}{2}K$. Moreover, $BFC = \frac{1}{2}(F+S)$; for $BFK = KFC + BFC$, and $FBK = KCF - BFC = KFC - BFC$. Therefore, $BFK - FBK = 2BFC$. But, as shown above, $BFK - FBK = F + S$. Therefore, $2BFC = F+S$, or $BFC = \frac{1}{2}(F+S)$. Substituting these values in the expression for BF, we have

☞ $$BF = \frac{(g-d)\cos.\frac{1}{2}K}{\sin.\frac{1}{2}(F+S)}.$$

Lastly, to find R', we have (§ 68) $R' + \frac{1}{2}g = EF = \frac{\frac{1}{2}BF}{\sin.\frac{1}{2}BEF}$. But $BEF = BLF - EBL$, and $BLF = LFK + LKF = F + K$. Therefore, $BEF = F + K - S$, and

☞ $$R' + \tfrac{1}{2}g = \frac{\frac{1}{2}BF}{\sin.\frac{1}{2}(F+K-S)}.$$

II. When the turnout is from the *outside* of the curve, the preceding solution requires a few modifications. In the present case, the angle $EFK' = F$ (fig. 19) and $EBL = S$. To find K, we have in the triangle BFK, $KF + BK : KF - BK = \tan.\frac{1}{2}(FBK + BFK) : \tan.\frac{1}{2}(FBK - BFK)$. But $KF = R + \frac{1}{2}g$, and $BK = R - \frac{1}{2}g + d$. Therefore, $KF + BK = 2R + d$, and $KF - BK = g - d$. Moreover, $FBK = 180^\circ - FBL = 180^\circ - (EBF - EBL) = 180^\circ - (EBF - S)$, and $BFK = 180^\circ - BFK' = 180^\circ - (BFE + EFK') = 180^\circ - (EBF + F)$. Therefore, $FBK - BFK = F + S$. Lastly, $FBK + BFK = 180^\circ - K$. Substituting these values in the preceding proportion, we have $2R + d : g - d = \tan.(90^\circ - \frac{1}{2}K) : \tan.\frac{1}{2}(F+S)$, or $\tan.(90^\circ - \frac{1}{2}K) = \frac{(2R+d)\tan.\frac{1}{2}(F+S)}{g-d}$. But $\tan.(90^\circ - \frac{1}{2}K) = \cot.\frac{1}{2}K = \frac{1}{\tan\frac{1}{2}K}$;

☞ $$\therefore \tan.\tfrac{1}{2}K = \frac{g-d}{(2R+d)\tan.\frac{1}{2}(F+S)}.$$

Next to find BF, we have, in the triangle B[illegible] BF = $\frac{BC \sin.\ BCF}{\sin.\ BFC}$. But $BC = g - d$, and $BCF = 90°$ [illegible]

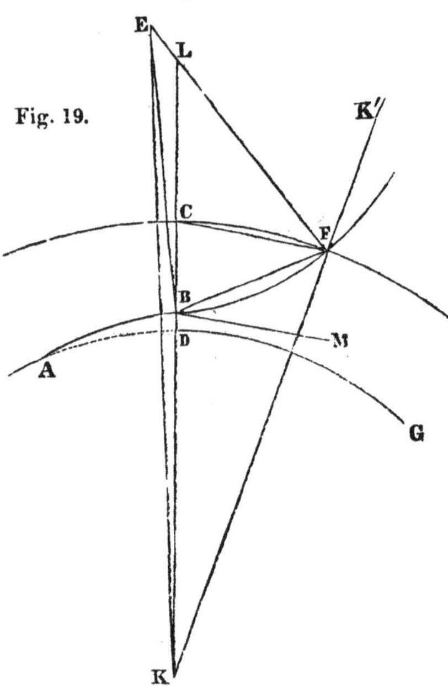

Fig. 19.

sin. $BCF = \cos.\ \frac{1}{2} K$. Moreover, $BFC = \frac{1}{2}(F + S)$; for $BFK = KFC - BFC$, and $FBK = KCF + BFC = KFC + BFC$. Therefore, $FBK - BFK = 2\,BFC$. But, as shown above, $FBK - BFK = F + S$. Therefore, $2\,BFC = F + S$, or $BFC = \frac{1}{2}(F + S)$. Substituting these values in the expression for BF, we have, as before.

☞ $$BF = \frac{(g - d) \cos.\ \frac{1}{2} K}{\sin\ \frac{1}{2}(F + S)}.^{*}$$

Lastly, to find R', we have (§ 68) $R' + \frac{1}{2} g = EF = \frac{\frac{1}{2} BF}{\sin.\ \frac{1}{2} BEF}$.

---

* Since $\frac{1}{2} K$ is generally very small, an approximate value of BF may be obtained by making $\cos.\ \frac{1}{2} K = 1$. This gives $BF = \frac{g - d}{\sin.\ \frac{1}{2}(F + S)}$, which is identical with the formula for BF in § 50. Table V. will, therefore, give a close approximation to the value of BF on curves also, for any value of F contained in the table.

But $BEF = BLF - EBL$, and $BLF = LFK - LKF = F - K$. Therefore, $BEF = F - K - S$, and

$$☞ \qquad R' + \tfrac{1}{2} g = \frac{\tfrac{1}{2} B F}{\sin. \tfrac{1}{2} (F - K - S)}.$$

Example. Given $g = 4.7$, $d = .42$, $S = 1° 20'$, $R = 4583.75$, and $F = 7°$, to find the chord BF and the radius R' of a turnout from the *outside* of the curve. Here

$g - d = 4.28$		0.631444	0.631444
$2R + d = 9167.92$	3.962271		
$\tfrac{1}{2}(F + S) = 4° 10'$	tan. 8.862433		sin. 8.861283
		2.824704	1.770161
$\tfrac{1}{2} K = 22' 1.8''$		tan. 7.806740	cos. 9.999991
$BF = 58.905$			1.770152
2		0.301030	
$\tfrac{1}{2}(F - K - S) = 2° 27' 58.2''$		sin. 8.633766	
			8.934796
$R' + \tfrac{1}{2} g = 684.47$			2.835356
$\therefore R' = 682.12$			

58. **Problem.** *To find mechanically the proper position of a given frog.*

Solution. The method here is similar to that already given, when the turnout is from a straight line (§ 52). Draw BM (figs. 18 and 19) parallel to FC, and we have $FBM = BFC = \tfrac{1}{2}(F + S)$, as just shown (§ 57). This angle is to be laid off from BM; but as F is the point to be found, the chord FC can be only estimated at first, and BM taken parallel to it, from which the angle $\tfrac{1}{2}(F + S)$ may be laid off by the method of § 52. In this case, however, the first measure on the arc is d, and not $2d$, since we have here to start from BM, and not from the rail Having thus determined the point F approximately, BM may be laid off more accurately, and F found anew.

59. When frogs are cast to be kept on hand, it is desirable to have them of such a pattern that they will fall at the beginning or end of a certain rail; that is, the chord BF is known, and the angle F is required.

Problem. *Given the position of a frog by means of the chord B F (figs. 14, 18, and 19), to determine the frog angle F.*

Solution. The formula $BF = \frac{g-d}{\sin. \frac{1}{2}(F+S)}$, which is exact on straight lines (§ 50), and near enough on ordinary curves (§ 57, note), gives

☞ $$\sin. \tfrac{1}{2}(F+S) = \frac{g-d}{BF}.$$

By this formula $\frac{1}{2}(F+S)$ may be found, and consequently F.

60 **Problem.** *Given the radius R of the centre line of the main track, and the radius R' of the centre line of a turnout, to find the frog angle F, and the chord B F (figs. 18 and 19).*

Solution. I. When the turnout is from the *inside* of the curve (fig. 18). *In the triangle B E K find the angle B E K and the side E K.* For this purpose we have $BE = R' + \frac{1}{2}g$, $BK = R + \frac{1}{2}g - d$, and the included angle $EBK = S$. Then in the triangle EFK we have EK, as just found, $EF = R' + \frac{1}{2}g$, and $FK = R - \frac{1}{2}g$. The frog angle $EFK = F$ may, therefore, be found by formula 15, Tab X., which gives

☞ $$\tan. \tfrac{1}{2}F = \sqrt{\frac{(s-b)(s-c)}{s(s-a)}},$$

where s is the half sum of the three sides, a the side EK, and b and c the remaining sides.

Find also in the triangle EFK the angle FEK, and we have the angle BEF = BEK − FEK. Then in the triangle BEF we have (§ 69)

☞ $$BF = 2(R' + \tfrac{1}{2}g)\sin. \tfrac{1}{2}BEF.^*$$

II. When the turnout is from the *outside* of the curve (fig. 19). *In the triangle B E K find the angle B E K and the side E K.* For this purpose we have $BE = R' + \frac{1}{2}g$, $BK = R - \frac{1}{2}g + d$, and the included angle $EBK = 180° - S$. Then in the triangle EFK we have EK, as just found, $EF = R' + \frac{1}{2}g$, and $FK = R + \frac{1}{2}g$. The angle EFK may, therefore, be found by formula 15, Tab. X, which gives $\tan. \frac{1}{2}EFK = \sqrt{\frac{(s-b)(s-c)}{s(s-a)}}$. But the angle $EFK' = F$

* The value of BF may be more easily found by the approximate formula $BF = \frac{g-d}{\sin. \frac{1}{2}(F+S)}$, and generally with sufficient accuracy. See note to § 57. This remark applies also to BF in the second part of this solution.

$= 180° - EFK$. Therefore $\frac{1}{2} F = 90° - \frac{1}{2} EFK$, and $\cot \frac{1}{2} F =$ $\tan. \frac{1}{2} EFK$;

☞ $$\therefore \cot. \tfrac{1}{2} F = \sqrt{\frac{(s-b)(s-c)}{s(s-a)}},$$

where s is the half sum of the three sides, a the side EK, and b and c the remaining sides.

Find also in the triangle EFK the angle FEK, and we have the angle $BEF = FEK - BEK$. Then in the triangle BEF we have (§ 69)

☞ $$BF = 2 (R' + \tfrac{1}{2} g) \sin. \tfrac{1}{2} BEF.$$

Example. Given $g = 4.7$, $d = .42$, $S = 1° 20'$, $R = 4583.75$, and $R' = 682.12$, to find F and the chord BF of a turnout from the *outside* of the curve. Here in the triangle BEK (fig. 19) we have $BE = R' + \frac{1}{2} g = 684.47$, $BK = R - \frac{1}{2} g + d = 4581.82$, and the angles $BEK + BKE = S = 1° 20'$. Then

$BK - BE = 3897.35$		3.590769
$\frac{1}{2}(BEK + BKE) = 40'$	tan.	8.065806
		1.656575
$BK + BE = 5266.29$		3.721505
$\frac{1}{2}(BEK - BKE)$* $= 29.6029'$	tan.	7.935070
$\therefore BEK = 1° 9.6029'$		

EK is now found by the formula $EK = \frac{BK \sin EBK}{\sin BEK}$, or $\log. EK = \log. 4581.82 + \log. \sin. 178° 40' - \log \sin. 1° 9\ 6029' = 3.721491$ whence $EK = 5266.12$.

Then to find F, we have, in the triangle EFK, $s = \frac{1}{2}$ (5266 12 + 684.47 + 4586.10) = 5268.34, $s - a = 2.22$, $s - b = 4583.87$, and $s - c = 682.24$.

$s - b = 4583.87$		3.661233
$s - c = 682.24$		2.833937
		6.495170
$s = 5268.34$	3.721674	
$s - a = 2.22$	0.346353	
		4.068027
		2)2.427143
$\frac{1}{2} F = 3° 30'$		cot. 1.213571
$\therefore F = 7°$		

* This angle and the sine of 1° 9 6029′ below, are found by the method given in connection with Table XIII. If the ordinary interpolations had been used, we should have found $F = 7° 7'$, whereas it should be 7°, since this example is the converse of that in § 57.

To find $F E K$, we have s as before, but as a is here the side $F K$ opposite the angle sought, we have $s - a = 682.24$, $s - b = 4583.87$, and $s - c = 2.22$. Then by means of the logarithms just used, we find $\frac{1}{2} F E K = 3^\circ\ 2'\ 45''$. Subtracting $\frac{1}{2} B E K = 34'\ 48''$, we have $\frac{1}{2} B E F = 2^\circ\ 27'\ 57''$. Lastly, $B F = 1368.94$ sin. $2^\circ\ 27'\ 57'' = 58.897$.

The formula $B F = \frac{g - d}{\text{sin.} \frac{1}{2} (F + S)}$ (§ 57, note) would give $B F = 58.906$, and this value is even nearer the truth than that just found, owing, however, to no error in the formulæ, but to inaccuracies incident to the calculation.

61. If the turnout is to reverse, in order to join a track parallel to the main track, as $A C B$ (fig. 20), it will be necessary to determine the reversing points C and B. These points will be determined, if we find the angles $A E C$ and $B F C$, and the chords $A C$ and $C B$

62. **Problem.** *Given the radius $D K = R$ (fig 20) of the centre line of the main track. the common radius $E C = C F = R'$ of the centre line of a turnout, and the distance $B G = b$ between the centre lines of the parallel tracks, to find the central angles $A E C$ and $B F C$ and the chords $A C$ and $B C$.*

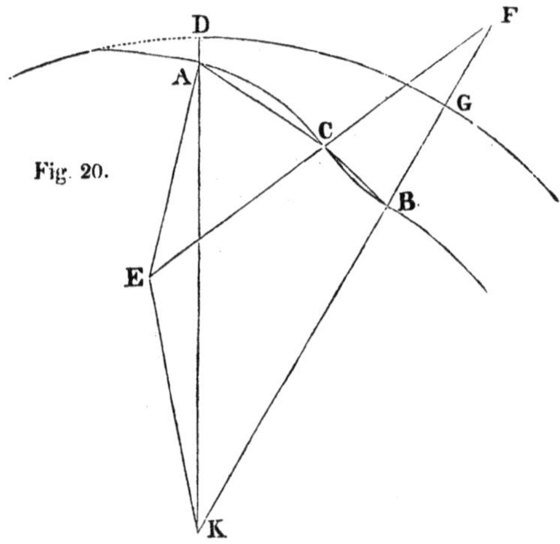

Fig. 20.

Solution. *In the triangle $A E K$ find the angle $A E K$ and the side*

$E K$. For this purpose we have $A E = R'$, $A K = R - d$, and the included angle $E A K = S$. Or, if the frog angle has been previously calculated by § 60, the values of $A E K$ and $E K$ are already known.*

Find in the triangle $E F K$ the angles $E F K$ and $F E K$. For this purpose we have $E K$, as just found, $E F = 2 R'$, and $F K = R + R' - b$. *Then* $A E C = A E K - F E K$, *and* $B F C = E F K$. Lastly, (§ 69)

$$A C = 2 R' \sin \tfrac{1}{2} A E C; \quad C B = 2 R' \sin. \tfrac{1}{2} B F C.$$

This solution, with a few obvious modifications, will apply, when the turnout is from the outside of a curve.

D. *Crossings on Curves.*

63. When a turnout enters a parallel main track by a second switch, it becomes a *crossing*. Then if the tangent points A and B (fig. 21) are fixed, the distance $A B$ must be measured, and also the angles which $A B$ makes with the tangents at A and B. The common radius of the crossing may then be found by § 40; or if one radius of the crossing is given, the other may be found by § 38. But if one tangent point A is fixed, and the common radius of the crossing is given, it will be necessary to determine the reversing point C and the tangent point B. These points will be determined, if we find the angles $A E C$ and $B F C$, and the chords $A C$ and $C B$.

64. **Problem.** *Given the radius $D K = R$ (fig. 21) of the centre line of the main track, the common radius $E C = C F = R'$ of the centre line of a crossing, and the distance $D G = b$ between the centre lines of the parallel tracks, to find the central angles $A E C$ and $B F C$ and the chords $A C$ and $C B$.*

Solution. In the triangle $A E K$ find the angle $A E K$ and the side $E K$. For this purpose we have $A E = R'$, $A K = R - d$, and the included angle $E A K = S$.

Find in the triangle $B F K$ the angle $B F K$ and the side $F K$. For this purpose we have $B F = R'$, $B K = R - b + d$, and the included angle $F B K = 180° - S$.

Find in the triangle $E F K$ the angles $F E K$ and $E F K$. For this

* The triangle $A E K$ does not correspond precisely with $B E K$ in § 60, A being on the centre line and B on the outer rail; but the difference is too slight to affect the calculations.

purpose we have EK and FK as just found, and $EF = 2R'$. Then $AEC = AEK - FEK$, and $BFC = EFK - BFK$. Lastly (§ 69,)

☞ $AC = 2R' \sin. \frac{1}{2} AEC$; $CB = 2R' \sin. \frac{1}{2} BFC$.

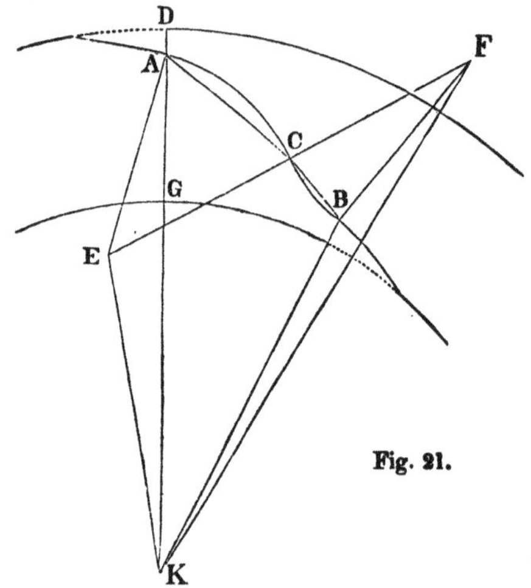

Fig. 21.

Article IV.—Miscellaneous Problems.

65. **Problem.** *Given $AB = a$ (fig. 22) and the perpendicular $BC = b$, to find the radius of a curve that shall pass through C and the tangent point A.*

Solution. Let O be the centre of the curve, and draw the radii AO and CO and the line CD parallel to AB. Then in the right triangle COD we have $OC^2 = CD^2 + OD^2$. But $OC = R$, $CD = a$, and $OD = AO - AD = R - b$. Therefore, $R^2 = a^2 + (R - b)^2 = a^2 + R^2 - 2Rb + b^2$, or $2Rb = a^2 + b^2$;

☞ $$\therefore R = \frac{a^2}{2b} + \frac{1}{2}b.$$

Example. Given $a = 204$ and $b = 24$, to find R. Here $R = \frac{204^2}{2 \times 24} + \frac{24}{2} = 867 + 12 = 879.$

66. **Corollary 1.** If R and b are given to find $A B = a$, that is, to determine the tangent point from which a curve of given radius

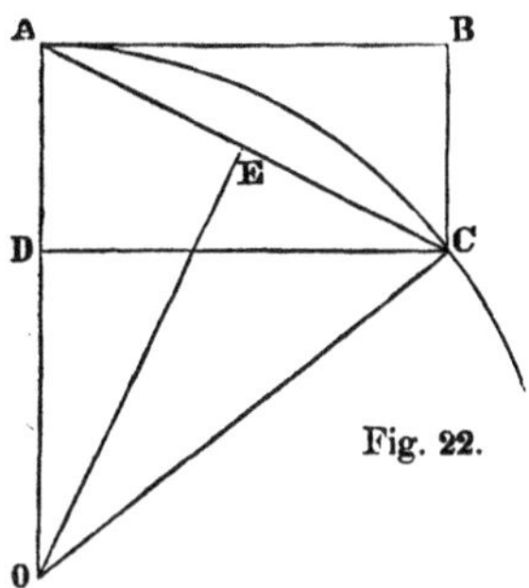

Fig. 22.

must start to pass through a given point, we have (§ 65) $2 R b = a^2 + b^2$, or $a^2 = 2 R b - b^2$;

☞ $$\therefore a = \sqrt{b\ (2 R - b)}.$$

Example. Given $b = 24$ and $R = 879$, to find a. Here $a = \sqrt{24\ (1758 - 24)} = \sqrt{41616} = 204$.

67. **Corollary 2.** If R and a are given, and b is required, we have (§ 65) $2 R b = a^2 + b^2$, or $b^2 - 2 R b = - a^2$. Solving this equation, we find for the value of b here required,

☞ $$b = R - \sqrt{R^2 - a^2}.$$

68. **Problem.** *Given the distance $A\ C = c$ (fig. 22) and the angle $B\ A\ C = A$, to find the radius R or deflection angle D of a curve, that shall pass through C and the tangent point A.*

Solution. Draw $O\ E$ perpendicular to $A\ C$. Then the angle $A\ O\ E = \frac{1}{2} A\ O\ C = B A\ C = A$ (§ 2, III.), and the right triangle $A\ O E$ gives (Tab. X. 9) $A\ O = \frac{A\ E}{\sin.\ A\ O\ E}$;

☞ $$\therefore R = \frac{\frac{1}{2} c}{\sin.\ A}.$$

To find D, we have (§ 9) $\sin.\ D = \frac{50}{R}$. Substituting for R its value just found, we have $\sin.\ D = 50 \div \frac{\frac{1}{2} c}{\sin.\ A}$;

☞ $$\therefore \sin. D = \frac{100 \sin. A}{c}.$$

Example. Given $c = 285.4$ and $A = 5°$, to find R and D. Here $R = \frac{142.7}{\sin. 5°} = 1637.3$; and $\sin. D = \frac{100 \sin. 5°}{285.4} = \frac{\sin. 5°}{2.854} = \sin. 1° 45'$, or $D = 1° 45'$.

69. **Problem.** *Given the radius R or the deflection angle D of a curve, and the angle $B A C = A$ (fig. 22), made by any chord with the tangent at A, to find the length of the chord $A C = c$.*

Solution. If R is given, we have (§ 68) $R = \frac{\frac{1}{2} c}{\sin. A}$;

☞ $$\therefore c = 2 R \sin. A.$$

If D is given, we have (§ 68) $\sin. D = \frac{100 \sin. A}{c}$;

☞ $$\therefore c = \frac{100 \sin. A}{\sin. D}.$$

This formula is useful for finding the length of chords, when a curve is laid out by points two, three, or more stations apart. Thus, suppose that the curve $A C$ is four stations long, and that we wish to find the length of the chord $A C$. In this case the angle $A = 4 D$ and $c = \frac{100 \sin. 4 D}{\sin. D}$. By this method Table II. is calculated.

Example. Given $R = 2455.7$ or $D = 1° 10'$, and $A = 4° 40'$, to find c. Here, by the first formula, $c = 4911.4 \sin. 4° 40' = 399.59$. By the second formula, $c = \frac{100 \sin. 4° 40'}{\sin. 1° 10'} = 399.59$.

70. **Problem.** *Given the angle of intersection $K C B = I$ (fig. 23), and the distance $C D = b$ from the intersection point to the curve in the direction of the centre, to find the tangent $A C = T$, and the radius $A O = R$.*

Solution. In the triangle $A D C$ we have $\sin. C A D : \sin. A D C = C D : A C$. But $C A D = \frac{1}{2} A O D = \frac{1}{4} I$ (§ 2, III. and VI.), and as the sine of an angle is the same as the sine of its supplement, $\sin A D C = \sin A D E = \cos. D A E = \cos. \frac{1}{4} I$. Moreover, $C D = b$ and $A C = T$. Substituting these values in the preceding proportion, we have $\sin. \frac{1}{4} I : \cos. \frac{1}{4} I = b : T$, or $T = \frac{b \cos. \frac{1}{4} I}{\sin. \frac{1}{4} I}$; whence (Tab. X. 33)

☞ $$T = b \cot. \tfrac{1}{4} I.$$

To find R, we have (§ 5) $R = T \cot. \frac{1}{2} I$. Substituting for T its value just found, we have

☞ $$R = b \cot. \tfrac{1}{4} I \cot. \tfrac{1}{2} I$$

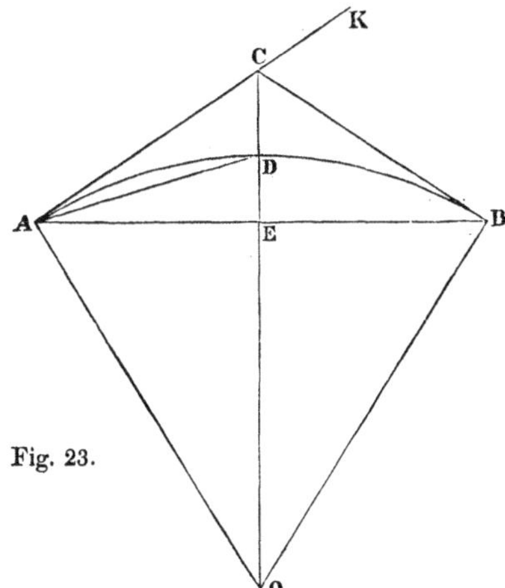

Fig. 23.

Example. Given $I = 30°$, $b = 130$, to find T and R. Here

$b = 130$		2.113943
$\frac{1}{4} I = 7° 30'$	cot.	0.880571
$T = 987.45$		2.994514
$\frac{1}{2} I = 15°$	cot.	0.571948
$R = 3685.21$		3.566462

71. **Problem.** *Given the angle of intersection $KCB = I$ (fig. 23), and the tangent $AC = T$, or the radius $AO = R$, to find $CD = b$.*

Solution. If T is given, we have (§ 70) $T = b \cot. \frac{1}{4} I$, or $b = \frac{T}{\cot. \frac{1}{4} I}$;

☞ $$\therefore b = T \tan. \tfrac{1}{4} I.$$

If R is given, we have (§ 70) $R = b \cot. \frac{1}{4} I \cot. \frac{1}{2} I$, or $b = \frac{R}{\cot. \frac{1}{4} I \cot. \frac{1}{2} I}$;

☞ $$\therefore b = R \tan. \tfrac{1}{4} I \tan. \tfrac{1}{2} I.$$

Example. Given $I = 27°$, $T = 600$ or $R = 2499.18$, to find b. Here $b = 600$ tan. 6° 45′ $= 71.01$, or $b = 2499.18$ tan. 6° 45′ tan. 13° 30′ $= 71.01$.

72. **Problem.** *Given the angle of intersection I of two tangents A C and B C (fig. 24), to find the tangent point A of a curve, that shall pass through a point E, given by C D = a, D E = b, and the angle C D E = ½ I.*

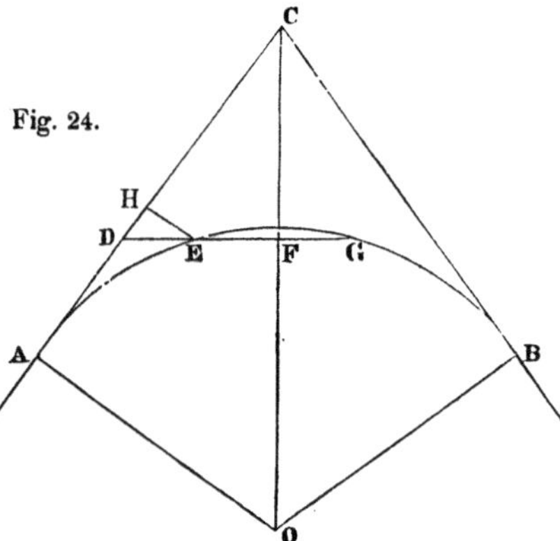

Fig. 24.

Solution. Produce DE to the curve at G, and draw CO to the centre O. *Denote* DF *by* c. Then in the right triangle CDF we have (Tab. X. 11) $DF = CD$ cos. CDF, or

☞ $$c = a \cos. \tfrac{1}{2} I.$$

Denote the distance AD *from* D *to the tangent point by* x. Then, by Geometry, $x^2 = DE \times DG$. But $DG = DF + FG = DF + EF = 2DF - DE = 2c - b$. Therefore, $x^2 = b(2c - b)$, and

☞ $$x = \sqrt{b(2c - b)}.$$

Having thus found AD, we have the tangent $AC = AD + DC = x + a$. Hence, R or D may be found (§ 5 or § 11).

If the point E is given by EH and CH perpendicular to each other, a and b may be found from these lines. For $a = CH + DH = CH + EH \cot. \frac{1}{2} I$ (Tab. X. 9), and $b = DE = \frac{EH}{\sin. \frac{1}{2} I}$.

Example. Given $I = 20° 16'$, $a = 600$, and $b = 80$, to find x and R. Here $c = 600$ cos. $10° 8' = 590.64$, $2c - b = 1101 28$, and $x = \sqrt{80 \times 1101.28} = 296.82$. Then $T = 600 + 296.82 = 896.82$, and $R = 896.82$ cot. $10° 8' = 5017.82$.

73. **Problem.** *Given the tangent $A\,C$ (fig. 25), and the chord $A\,B$, uniting the tangent points A and B, to find the radius $A\,O = R$.*

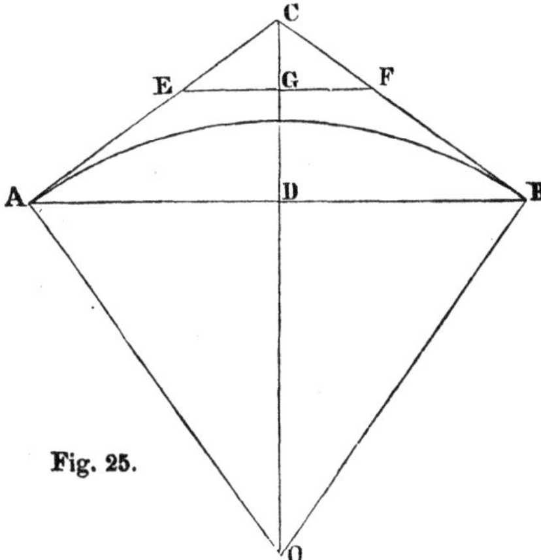

Fig. 25.

Solution. Measure or calculate the perpendicular $C\,D$. Then if $C\,D$ be produced to the centre O, the right triangles $A\,D\,C$ and $C\,A\,O$, having the angle at C common, are similar, and give $C\,D : A\,D = A\,C : A\,O$, or

☞ $$R = \frac{A\,D \times A\,C}{C\,D}.$$

If it is inconvenient to measure the chord $A\,B$, a line $E\,F$, parallel to it, may be obtained by laying off from C equal distances $C\,E$ and $C\,F$. Then measuring $E\,G$ and $G\,C$, we have, from the similar triangles $E\,G\,C$ and $C\,A\,O$, $C\,G : G\,E = A\,C : A\,O$, or $R = \frac{G\,E \times A\,C}{C\,G}$.

Example. Given $A\,C = 246$ and $A\,D = 240$, to find R. Here $C\,D = 54$, and $R = \frac{240 \times 246}{54} = 1093.33$.

74. **Problem.** *Given the radius* $A\,O = R$ (*fig* 25), *to find the tangent* $A\,C = T$ *of a curve to unite two straight lines given on the ground*

Solution. Lay off from the intersection C *of the given straight lines any equal distances* $C\,E$ *and* $C\,F$. *Draw the perpendicular* $C\,G$ *to the middle of* $E\,F$, *and measure* $G\,E$ *and* $C\,G$. Then the right triangles $E\,G\,C$ and $C\,A\,O$, having the angle at C common, are similar, and give $G\,E : C\,G = A\,O : A\,C$, or

☞ $$T = \frac{C\,G \times A\,O}{G\,E}.$$

By this problem and the preceding one, the radius or tangent points of a curve may be found without an instrument for measuring angles.

Example. Given $R = 1093\frac{1}{3}$, $G\,E = 80$, and $C\,G = 18$, to find T. Here $T = \frac{18 \times 1093\frac{1}{3}}{80} = 246$.

75 **Problem.** *To find the angle of intersection* I *of two straight lines, when the point of intersection is inaccessible, and to determine the tangent points, when the length of the tangents is given*

Solution. I. To find the angle of intersection I Let $A\,C$ and $C\,V$ (fig. 26) be the given lines *Sight from some point* A *on one line to a point* B *on the other, and measure the angles* $C\,A\,B$ *and* $T\,B\,V$. These angles make up the change of direction in passing from one tangent to the other. But the angle of intersection (§ 2) shows the change of direction between two tangents, and it must, therefore, be equal to the sum of $C\,A\,B$ and $T\,B\,V$, that is,

☞ $$I = C\,A\,B + T\,B\,V$$

But if obstacles of any kind render it necessary to pass from $A\,C$ to $B\,V$ by a broken line, as $A\,D\,E\,F\,B$, *measure the angles* $C\,A\,D$, $N\,D\,E$, $P\,E\,F$, $R\,F\,B$, *and* $S\,B\,V$, *observing to note those angles as minus which are laid off contrary to the general direction of these angles*. Thus the general direction of the angles in this case is to the *right;* but the angle $P\,E\,F$ lies to the *left* of $D\,E$ produced, and is therefore to be marked *minus*. The angles to be measured show the successive changes of direction in passing from one tangent to the other. Thus $C\,A\,D$ shows the change of direction between the first tangent and $A\,D$, $N\,D\,E$ shows the change between $A\,D$ produced and $D\,E$, $P\,E\,F$ the change between $D\,E$ produced and $E\,F$, $R\,F\,B$ the change between $E\,F$ produced and $F\,B$, and, lastly, $S\,B\,V$ the change between $B\,F$ produced and the second tangent. But the angle of intersection (§ 2) shows the change of direction in passing from one tangent to another, and it must, therefore, be equal to the sum of the partial changes measured, that is,

☞ $I = CAD + NDE - PEF + RFB + SBV.$

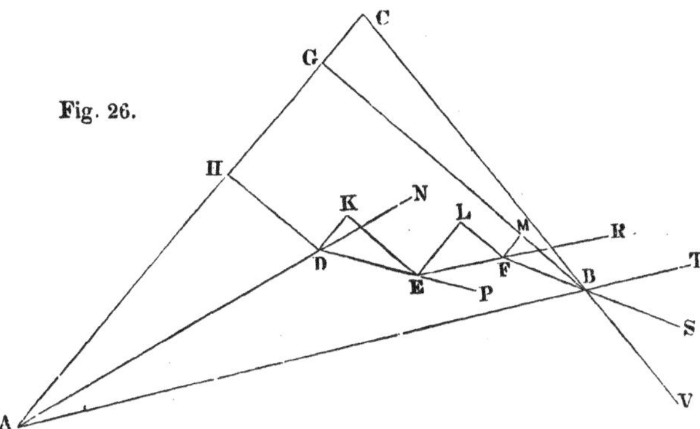

Fig. 26.

II. To determine the tangent points. This will be done if we find the distances AC and BC; for then any other distances from C may be found. It is supposed that the distance AB, or the distances AD, DE, EF, and FB have been measured.

If one line AB connects A and B, find AC and BC in the triangle ABC. For this purpose we have one side AB and all the angles.

If a broken line $ADEFB$ connects A and B, let fall a perpendicular BG from B upon AC, produced if necessary, and find AG and BG by the usual method of working a traverse. Thus, if AC is taken as a meridian line, and DK, EL, and FM are drawn parallel to AC, and DH, EK, and FL are drawn parallel to BG, the difference of latitude AG is equal to the sum of the partial differences of latitude AH, DK, EL, and FM, and the departure BG is equal to the sum of the partial departures DH, EK, FL, and BM. To find these partial differences of latitude and departures, we have the distances AD, DE, EF, and FB, and the bearings may be obtained from the angles already measured. Thus the bearing of AD is CAD, the bearing of DE is $KDE = KDN + NDE = CAD + NDE$, the bearing of EF is $LEF = LEP - PEF = KDE - PEF$, and the bearing of FB is $MFB = MFR + RFB = LEF + RFB$; that is, the bearing of each line is equal to the *algebraic sum* of the preceding bearing and its own change of direction. The differences of latitude and the departures may now be obtained from a traverse table, or more correctly by the formulæ:

Diff. of lat. = dist. $\times$ cos. of bearing ; dep. = dist. $\times$ sin. of bearing

Thus, $A H = A D \cos. C A D$, and $D H = A D \sin. C A D$.

Having found $A G$ and $B G$, we have, in the right triangle $B G C$, (Tab. X. 9) $G C = B G \cot. B C G$, and $B C = \frac{B G}{\sin. B C G}$. But $B C G = 180° - I$. Therefore, $\cot. B C G = -\cot. I$, and $\sin. B C G = \sin. I$. Hence $G C = -B G \cot. I$, and $B C = \frac{B G}{\sin. I}$. Then, since $A C = A G + G C$, we have

☞ $$A C = A G - B G \cot. I; \qquad B C = \frac{B G}{\sin. I}.$$

When I is between 90° and 180°, as in the figure, cot. I is negative, and $-B G \cot. I$ is, therefore, positive. When I is less than 90°, G will fall on the other side of I; but the same formula for $A C$ wil still apply ; for cot. I is now positive, and consequently, $-B G \cot. I$ is negative, as it should be, since, in this case, $A C$ would equal $A G$ *minus* $G C$.

Example. Given $A D = 1200$, $D E = 350$, $E F = 300$, $F B = 310$, $C A D = 20°$, $N D E = 44°$, $P E F = -25°$, $R F B = 31°$, and $S B V = 30°$, to find the angle of intersection I, and the distances $A C$ and $B C$.

Here $I = 20° + 44° - 25° + 31° + 30° = 100°$. To find $A G$ and $B G$, the work may be arranged as in the following table :—

Angles to the Right.	Bearings.	Distances.	N.	E.
° 20	° N. 20 E.	1200	1127.63	410.42
44	64	350	153.43	314.58
—25	39	300	233.14	188.80
31	70	310	106.03	291.30
			1620.23	1205.10

The first column contains the observed angles. The second contains the bearings, which are found from tne angles of the first column, in the manner already explained. $A C$ is considered as running north from A, and the bearings are, therefore, marked N. E. The other columns require no explanation. We find $A G = 1620.23$, and $B G = 1205.10$ Then $G C = - B G$ cot. $I = - 1205.1 \times$ cot. $100^\circ = 212.49$. This value is positive, because it is the product of two negative factors, cot. 100° being the same as $-$cot. 80°, a negative quantity. Then $A C = A G + G C = 1620.23 + 212.49 = 1832.72$, and $B C = \frac{1205.1}{\sin. 100^\circ} = 1223.69$. Having thus found the distances of A and B from the point of intersection, we can easily fix the tangent points for tangents of any given length.

76. **Problem.** *To lay out a curve, when an obstruction of any kind prevents the use of the ordinary methods.*

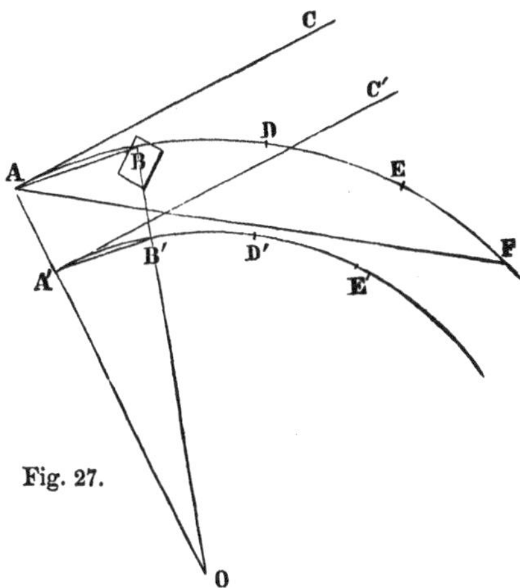

Fig. 27.

Solution. First Method. Suppose the instrument to be placed at A (fig. 27), and that a house, for instance, covers the station at B, and also obstructs the view from A to the stations at D and E. Lay off from $A C$, the tangent at A, such a multiple of the deflection angle D, as will be sufficient to make the sight clear the obstruction. In the figure it is supposed that $4 D$ is the proper angle. The sight will then pass through F, the fourth station from A, and this station will be determined by measuring from A the length of the chord $A F$, found by

§ 69 or by Table II. From the station at F the stations at D and E may afterwards be fixed, by laying off the proper deflections from the tangent at F.

Second Method. This consists in running an auxiliary curve parallel to the true curve, either inside or outside of it. For this purpose lay off perpendicular to $A\ C$, the tangent at A, a line $A\ A'$ of any convenient length, and from A' a line $A'\ C'$ parallel to $A\ C$. Then $A'\ C'$ is the tangent from which the auxiliary curve $A'\ E'$ is to be laid off. The stations on this curve are made to correspond to stations of 100 feet on the true curve, that is, a radius through B' passes through B, a radius through D' passes through D, &c. The chord $A'\ B'$ is, therefore, parallel to $A\ B$, and the angle $C'\ A'\ B' = C\ A\ B$; that is, the deflection angle of the auxiliary curve is equal to that of the true curve It remains to find the length of the auxiliary chords $A'\ B'$, $B'\ D'$, &c. Call the distance $A\ A' = b$. Then the similar triangles $A\ B\ O$ and $A'\ B'\ O$ give $A\ O : A'\ O = A\ B : A'\ B'$, or $R : R - b = 100 : A'\ B'$. Therefore, $A'\ B' = \frac{100\ (R - b)}{R} = 100 - \frac{100\ b}{R}$. If the auxiliary curve were on the outside of the true curve, we should find in the same way $A'\ B' = 100 + \frac{100\ b}{R}$. It is well to make b an aliquot part of R; for the auxiliary chord is then more easily found. Thus, if n is any whole number, and we make $b = \frac{R}{n}$, we have $A'\ B' = 100 \pm \frac{100\ b}{R}$ $= 100 \pm \frac{100}{n}$. If, for example, $b = \frac{R}{100}$, we have $n = 100$, and $A'\ B$ $= 100 \pm 1 = 101$ or 99. When the auxiliary curve has been run, the corresponding stations on the true curve are found, by laying off in the proper direction the distances $B\ B'$, $D\ D'$, &c., each equal to b.

77. **Problem.** *Having run a curve $A\ B$ (fig. 28), to change the tangent point from A to C, in such a way that a curve of the same radius may strike a given point D.*

Solution. Measure the distance $B\ D$ from the curve to D in a direction parallel to the tangent $C\ E$. This direction may be sometimes judged of by the eye, or found by the compass. A still more accurate way is to make the angle $D\ B\ E$ equal to the intersection angle at E, or to twice $B\ A\ E$, the total deflection angle from A to B; or if A can be seen from B, the angle $D\ B\ A$ may be made equal to $B\ A\ E$.

Measure on the tangent (backward or forward, as the case may be) a distance $A\ C = B\ D$, and C will be the new tangent point required. For, if $C\ H$ be drawn equal and parallel to $A\ F$, we have $F\ H$ equal and parallel to $A\,C$, and therefore equal and parallel to $B\,D$. Hence $D\,H = B\,F = A\,F = C\,H$, and $D\,H$ being equal to $C\,H$, a curve of radius $C\,H$ from the tangent point C must pass through D.

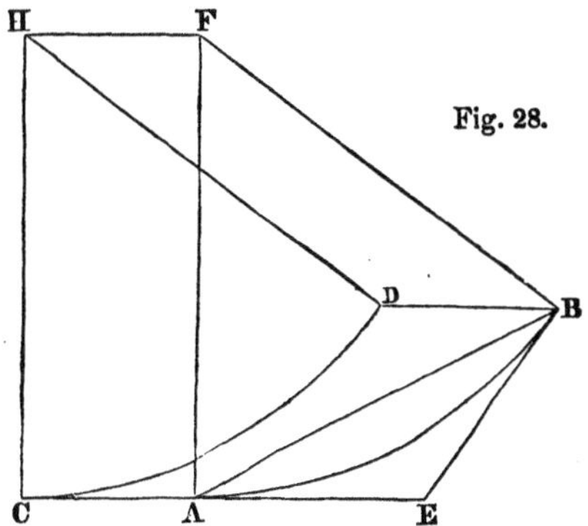

Fig. 28.

78. **Problem.** *Having run a curve $A\,B$ (fig. 29) of radius R or deflection angle D, terminating in a tangent $B\,D$, to find the radius R' or deflection angle D' of a curve $A\,C$, that shall terminate in a given parallel tangent $C\,E$.*

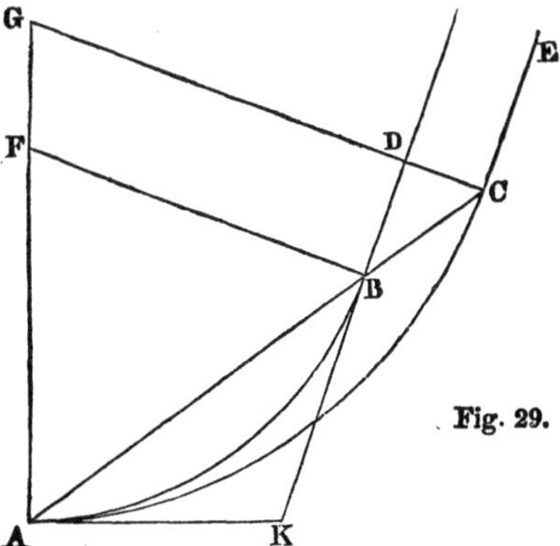

Fig. 29.

Solution. Since the radii $B\,F$ and $C\,G$ are perpendicular to the parallel tangents $C\,E$ and $B\,D$, they are parallel, and the angle $A\;G\;C = A\,F\,B$. Therefore, $A\;C\,G$, the half-supplement of $A\;G\;C$, is equal to

$A\,B\,F$, the half-supplement of $A\,F\,B$. Hence $A\,B$ and $B\,C$ are in the same straight line, and the new tangent point C is the intersection of $A\,B$ produced with $C\,E$.

Represent $A\,B$ *by* c, *and* $A\,C = c + B\,C$ *by* c'. *Measure* $B\,C$, *or, if more convenient, measure* $D\,C$ *and find* $B\,C$ *by calculation.* To calculate $B\,C$ from $D\,C$, we have $B\,C = \frac{D\,C}{\sin.\,D\,B\,C}$ (Tab. X. 9), and the angle $D\,B\,C = A\,B\,K = B\,A\,K$, the total deflection from A to B. Then the triangles $A\,F\,B$ and $A\,G\,C$ give $A\,B : A\,C = B\,F : C\,G$, or $c : c' = R : R'$;

☞ $$\therefore R' = \frac{c'}{c}\,R.$$

To find D', we have (§ 10) $R' = \frac{50}{\sin.\,D'}$, and $R = \frac{50}{\sin\,D}$. Substituting these values in the equation for R', we have $\frac{50}{\sin.\,D'} = \frac{c'}{c} \times \frac{50}{\sin.\,D}$;

☞ $$\therefore \sin.\,D' = \frac{c}{c'}\sin.\,D.$$

79. **Problem.** *Given the length of two equal chords* $A\,C$ *and* $B\,C$ (*fig.* 30); *and the perpendicular* $C\,D$, *to find the radius* R *of the curve.*

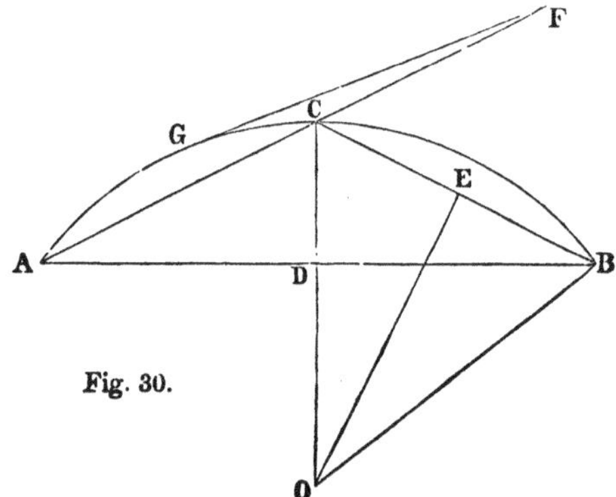

Fig. 30.

Solution. From O, the centre of the curve, draw the perpendicular $O\,E$. Then the similar triangles $O\,B\,E$ and $B\,C\,D$ give $B\,O : B\,E = B\,C : C\,D$, or $R : \frac{1}{2}B\,C = B\,C : C\,D$. Hence

☞ $$R = \frac{B\,C^2}{2\,C\,D}.$$

This problem serves to find the radius of a curve on a track already laid. For if from any point C on the curve we measure two equal chords $A\,C$ and $B\,C$, and also the perpendicular $C\,D$ from C upon the whole chord $A\,B$, we have the data of this problem.

80. **Problem.** *To draw a tangent $F\,G$ (fig. 30) to a given curve from a given point F.*

Solution. On any straight line $F\,A$, which cuts the curve in two points, measure $F\,C$ and $F\,A$, the distances to the curve. Then, by Geometry,

☞ $$F\,G = \sqrt{F\,C \times F\,A}.$$

This length being measured from F, will give the point G. When $F\,G$ exceeds the length of the chain, the direction in which to measure it, so that it will just touch the curve, may be found by one or two trials.

81. **Problem.** *Having found the radius $A\,O = R$ of a curve (fig. 31), to substitute for it two radii $A\,E = R_1$ and $D\,F = R_2$, the longer of which $A\,E$ or $B\,E'$ is to be used for a certain distance only at each end of the curve.*

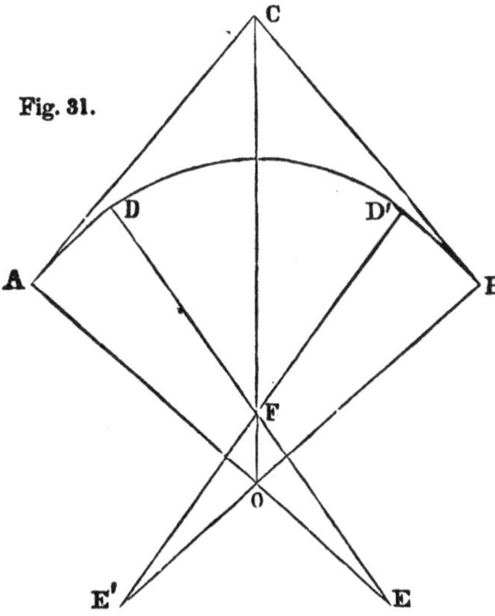

Solution. Assume the longer radius of any length which may be thought

proper, and find (§ 9) *the corresponding deflection angle* D_1. Suppose that each of the curves $A\,D$ and $B\,D'$ is 100 feet long. Then drawing $C\,O$, we have, in the triangle $F\,O\,E$, $O\,E : F\,E = \sin. O\,F\,E : \sin. F\,O\,E$. But the side $O\,E = A\,E - A\,O = R_1 - R$, $F\,E = D\,E - D\,F = R_1 - R_2$, the angle $F\,O\,E = 180° - A\,O\,C = 180° - \frac{1}{2}\,I$, and the angle $O\,F\,E = A\,O\,F - O\,E\,F = \frac{1}{2}\,I - 2\,D_1$, since $O\,E\,F = 2\,D_1$ (§ 7). Substituting these values, and recollecting that $\sin. (180° - \frac{1}{2}\,I) = \sin. \frac{1}{2}\,I$, we have $R_1 - R : R_1 - R_2 = \sin. (\frac{1}{2}\,I - 2\,D_1) : \sin. \frac{1}{2}\,I$ Hence

$$R_1 - R_2 = \frac{(R_1 - R)\sin. \frac{1}{2}\,I}{\sin. (\frac{1}{2}\,I - 2\,D_1)}.$$

R_2 is then easily found, and this will be the radius from D to D', or until the central angle $D\,F\,D' = I - 4\,D_1$.

The object of this problem is to furnish a method of flattening the extremities of a sharp curve. It is not necessary that the first curve should be just 100 feet long; in a long curve it may be longer, and in a short curve shorter. The value of the angle at E will of course change with the length of $A\,D$, and this angle must take the place of $2\,D_1$ in the formula. The longer the first curve is made, the shorter the second radius will be. It must also be borne in mind, in choosing the first radius, that the longer the first radius is taken, the shorter will be the second radius.

Example. Given $R = 1146.28$ and $I = 45°$, to find R_2, if R_1 is assumed $= 1910.08$, and $A\,D$ and $B\,D'$ each 100. Here, by Table I., $D_1 = 1°\,30'$. Then

$R_1 - R = 763.8$	2.882980
$\frac{1}{2}\,I = 22°\,30'$	sin. 9.582840
	2.465820
$\frac{1}{2}\,I - 2\,D_1 = 19°\,30'$	sin. 9.523495
$R_1 - R_2 = 875.64$	2.942325
$\therefore R_2 = R_1 - 875.64 = 1034.44$	

82. **Problem.** *To locate the second branch of a compound or reversed curve from a station on the first branch.*

Solution. Let $A\,B$ (fig 32) be the first branch of a compound curve, and D its deflection angle, and let it be required to locate the second branch $A\,B'$, whose deflection angle is D', from some station B on $A\,B$.

Let n be the number of stations from A to B, and n' the number of stations from A to any station B' on the second branch. Represent by V the angle A B B', which it is necessary to lay off from the chord B A to strike B'. Let the corresponding angle $A\,B'\,B$ on the other curve be repre-

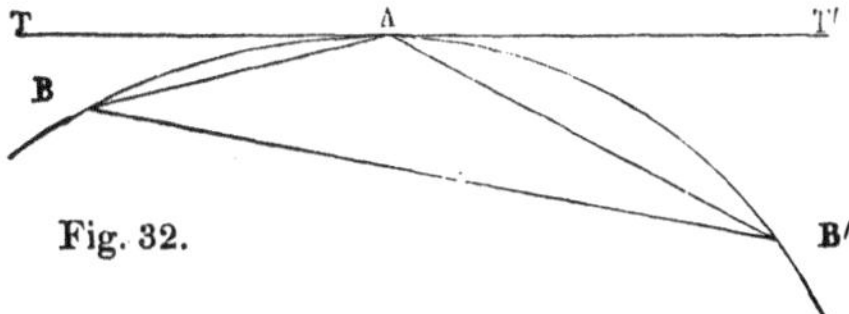

Fig. 32.

sented by V'. Then we have $V + V' = 180° - B\,A\,B'$. But if $T\,T'$ be the common tangent at A, we have $T\,A\,B + T'\,A\,B' = n\,D + n'\,D' = 180° - B\,A\,B'$. Therefore, $V + V' = n\,D + n'\,D'$. Next in the triangle $A\,B\,B'$ we have sin. V' : sin. $V = A\,B : A\,B'$. But $A\,B : A\,B' = n : n'$, *nearly*, and sin. V' : sin. $V = V' : V$, *nearly*. Therefore we have approximately $V' : V = n : n'$, or $V' = \frac{n}{n'}\,V$. Substituting this value of V' in the equation for $V + V'$, we have $V + \frac{n}{n'}\,V = n\,D + n'\,D'$. Therefore, $n'\,V + n\,V = n'\,(n\,D + n'\,D')$, or

☞ $$V = \frac{n'\,(n\,D + n'\,D')}{n + n'}.$$

The same reasoning will apply to *reversed curves*, the only change being that in this case $V + V' = n\,D - n'\,D'$, and consequently

☞ $$V = \frac{n'\,(n\,D - n'\,D')}{n + n'}.$$

When in this formula $n'\,D'$ becomes greater than $n\,D$, V becomes *minus*, which signifies that the angle V is to be laid off above $B\,A$ instead of below.

This problem is particularly useful, when the tangent point of a curve is so situated, that the instrument cannot be set over it. The same method is applicable, *when the curve A B' starts from a straight line;* for then we may consider $A\,B'$ as the second branch of a compound curve, of which the straight line is the first branch, having its radius equal to infinity, and its deflection angle $D = 0$. Making $D = 0$, the formula for V becomes

$$V = \frac{n'^2 D'}{n + n'}.$$

When n and n' are each 1, the formula for V is in all cases exact; for then the supposition that $V' : V = n : n'$ is strictly true, since $A B$ will equal $A B'$ and V and V', being angles at the base of an isosceles triangle, will also be equal. Making n and n' equal to 1, we have

$$V = \tfrac{1}{2} (D + D').$$

When the curve starts from a straight line, this formula becomes, by making $D = 0$,

$$V = \tfrac{1}{2} D'.$$

We have seen that when n or n' is more than 1, the value of V is only approximate. It is, however, so near the truth, that when neither n nor n' exceeds 3, the error in curves up to 5° or 6° varies from a fraction of a second to less than half a minute. The exact value of V might of course be obtained by solving the triangle $A B B'$, in which the sides $A B$ and $A B'$ may be found from Table II., and the included angle at A is known. The extent to which these formulæ may be safely used may be seen by the following table, which gives the approximate values of V for several different values of n, n', D, and D', and also the error in each case.

Compound Curves.						Reversed Curves.					
n.	D.	n'.	D'.	V.	Error.	n.	D.	n'.	D'.	V.	Error.
	°		°	° ′	″		°		°	° ′	″
1	0	5	1	4 10	0.9	1	3	4	3	7 12	27.2
1	0	5	3	12 30	25.3	2	3	4	3	4 0	23.5
2	0	3	3	5 24	22.1	3	3	4	3	1 $42\frac{6}{7}$	8.3
3	0	3	3	4 30	29.7	3	$\frac{1}{2}$	3	3	3 45	24.0
1	1	5	3	13 20	18.6	2	1	1	4	0 40	0.1
2	$\frac{1}{2}$	1	3	1 20	0.7	2	1	4	2	4 0	11.0
2	2	3	3	7 48	15.0	1	6	2	6	4 0	23.5
2	2	4	3	10 40	24.7	1	5	3	5	7 30	51.8
3	3	3	4	10 30	54.0	2	3	5	3	6 $25\frac{5}{7}$	52.8

As the given quantities are here arranged, the approximate values of V are all too great; but if the columns n and n' and the columns D and D' were interchanged, and V calculated, the approximate values of V would be just as much too small, the column of errors remaining the same.

83. Problem. *To measure the distance across a river on a given straight line.*

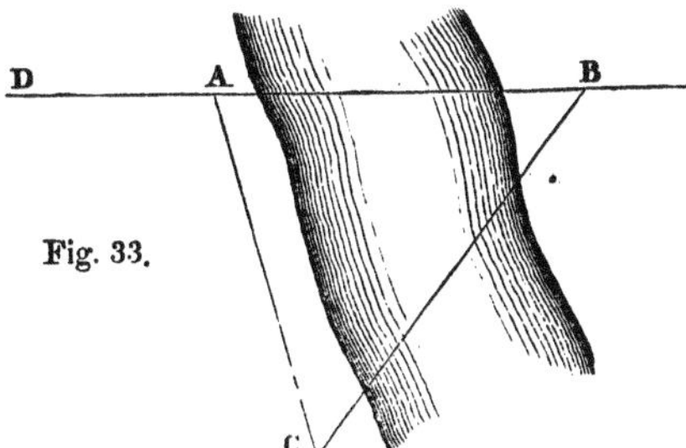

Fig. 33.

Solution. First Method. Let $A B$ (fig. 33) be the required distance. Measure a line $A C$ along the bank, and take the angles $B A C$ and $A C B$. Then in the triangle $A B C$ we have one side and two angles to find $A B$.

If $A C$ is of such a length that an angle $A C B = \frac{1}{2} D A C$ can be laid off to a point on the farther side, we have $A B C = \frac{1}{2} D A C = A C B$. Therefore, without calculation, $A B = A C$.

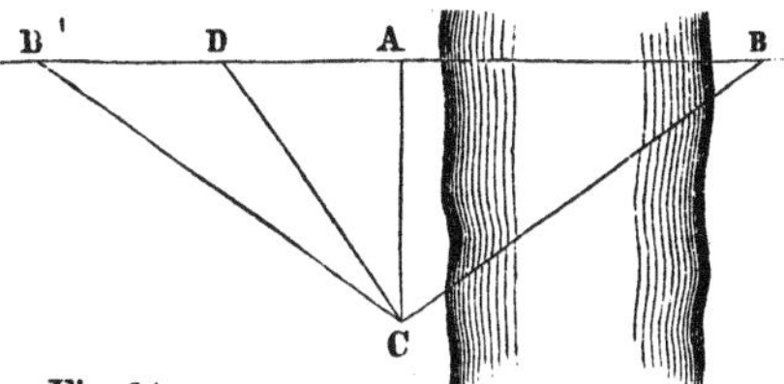

Fig. 34.

Second Method. Lay off $A C$ (fig. 34) perpendicular to $A B$. Measure $A C$, and at C lay off $C D$ perpendicular to the direction $C B$, and meeting the line of $A B$ in D. Measure $A D$. Then the triangles $A C D$ and $A B C$ are similar, and give $A D : A C = A C : A B$. Therefore, $A B = \frac{A C^2}{A D}$.

If from C, determined as before, the angle $A C B'$ be laid off equal to $A C B$, we have, without calculation, $A B = A B'$.

Third Method. Measure a line $A D$ (fig. 35) in an oblique direction from the bank, and fix its middle point C. From any convenient point E in the line of $A B$, measure the distance $E C$, and produce

$E\,C$ until $C\,F = E\,C$. Then, since the triangles $A\,C\,E$ and $D\,C\,F$ are similar by construction, we see that $D\,F$ is parallel to $E\,B$. Find

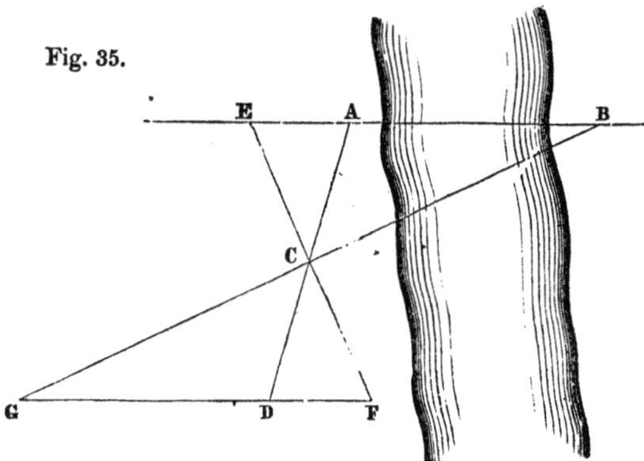

Fig. 35.

now a point G, that shall be at the same time in the line of $C\,B$ and of $D\,F$, and measure $G\,D$. Then the triangles $A\,B\,C$ and $D\,G\,C$ are equal, and $G\,D$ is equal to the required distance $A\,B$.

As the object of drawing $E\,F$ is to obtain a line parallel to $A\,B$, this line may be dispensed with, if by any other means a line $G\,F$ be drawn through D parallel to $A\,B$. A point G being found on this parallel in the line of $C\,B$, we have, as before, $G\,D = A\,B$.

CHAPTER II.

PARABOLIC CURVES.

ARTICLE I. — LOCATING PARABOLIC CURVES.

84. LET $A E B$ (fig. 36) be a parabola, $A C$ and $B C$ its tangents, and $A B$ the chord uniting the tangent points. Bisect $A B$ in D, and join $C D$. Then, according to Analytical Geometry, —

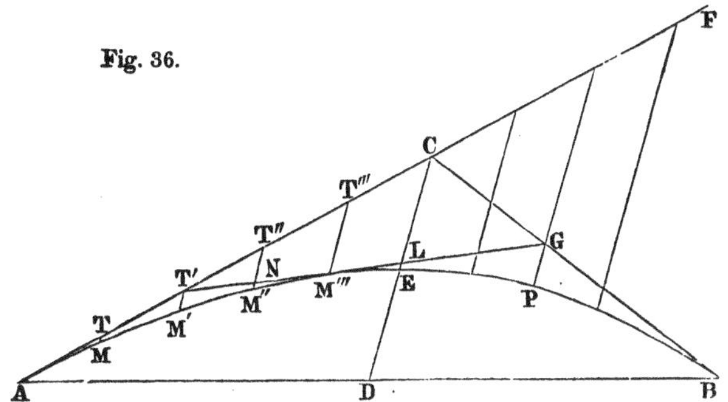

Fig. 36.

I. $C D$ is a *diameter* of the parabola, and the curve bisects $C D$ in E

II. If from any points T, T', T'', &c., on a tangent $A F$, lines be drawn to the curve *parallel to the diameter*, these lines $T M$, $T' M'$, $T'' M''$, &c., called *tangent deflections*, will be to each other as the squares of the distances $A T$, $A T'$, $A T''$, &c. from the tangent point A.

III. A line $E D$ (fig. 37), drawn from the middle of a chord $A B$ to the curve, and parallel to the diameter, may be called the *middle ordinate* of that chord; and if the secondary chords $A E$ and $B E$ be drawn, the middle ordinates of these chords, $K G$ and $L H$, are each equal to $\frac{1}{4} E D$. In like manner, if the chords $A K$, $K E$, $E L$, and $L B$ be drawn, their middle ordinates will be equal to $\frac{1}{4} K G$ or $\frac{1}{4} L H$.

IV. A tangent to the curve at the extremity of a middle ordinate, is parallel to the chord of that ordinate. Thus $M F$, tangent to the curve at E, is parallel to $A B$.

V. If any two tangents, as $A\,C$ and $B\,C$, be bisected in M and F the line $M\,F$, joining the points of bisection, will be a new tangent, its middle point E being the point of tangency.

85. **Problem.** *Given the tangents $A\,C$ and $B\,C$, equal or unequal (fig. 36,) and the chord $A\,B$, to lay out a parabola by tangent deflections.*

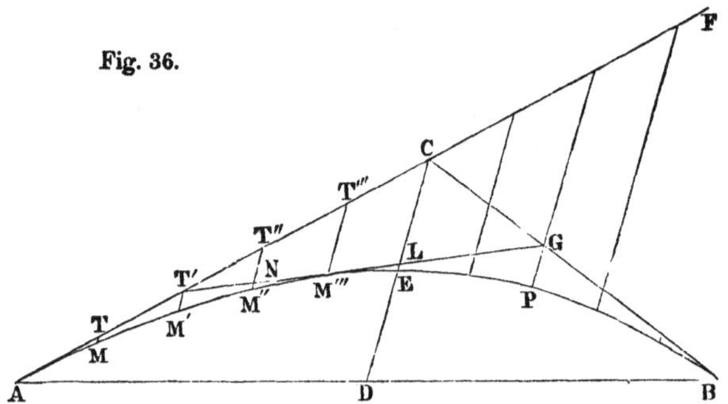

Solution. Bisect $A\,B$ in D, and measure $C\,D$ and the angle $A\,C\,D$; or calculate $C\,D$* and $A\,C\,D$ from the original data. Divide the tangent $A\,C$ into any number n of equal parts, and call the deflection $T\,M$ for the first point a. Then (§ 84, II.) the deflection for the second point will be $T'\,M' = 4\,a$, for the third point $T''\,M'' = 9\,a$, and so on to the nth point or C, where it will be $n^2\,a$. But the deflection at this last point is $C\,E = \frac{1}{2}\,C\,D$ (§ 84, I.). Therefore, $n^2\,a = C\,E$, and

$$a = \frac{C\,E}{n^2}.$$

Having thus found a, we have also the succeeding deflections $4\,a$, $9\,a$, $16\,a$, &c. Then laying off at T, T', &c. the angles $A\,T\,M$, $A\,T'\,M'$, &c. each equal to $A\,C\,D$, and measuring down the proper deflections, just found, the points M, M', &c. of the curve will be determined.

The curve may be finished by laying off on $A\,C$ produced n parts equal to those on $A\,C$, and the proper deflections will be, as before, a multiplied by the square of the number of parts from A. But an

* Since $C\,D$ is drawn to the middle of the base of the triangle $A\,B\,C$, we have, by Geometry, $C\,D^2 = \frac{1}{2}\,(A\,C^2 + B\,C^2) - A\,D^2$.

easier way generally of finding points beyond E is to divide the second tangent $B\,C$ into equal parts, and proceed as in the case of $A\,C$. If the number of parts on $B\,C$ be made the same as on $A\,C$, it is obvious that the deflections from both tangents will be of the same length for corresponding points. The angles to be laid off from $B\,C$ must, of course, be equal to $B\,C\,D$.

The points or stations thus found, though corresponding to equal distances on the tangents, are not themselves equidistant. The length of the curve is obtained by actual measurement.

86. **Problem.** *Given the tangents $A\,C$ and $B\,C$, equal or unequal, (fig. 37,) and the chord $A\,B$, to lay out a parabola by middle ordinates.*

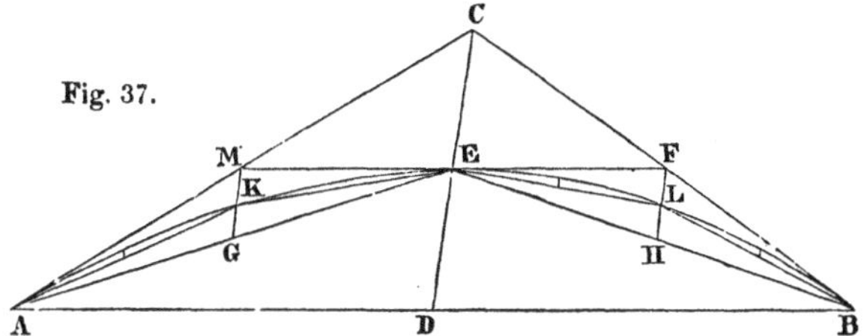

Solution. Bisect $A\,B$ in D, draw $C\,D$, and its middle point E will be a point on the curve (§ 84, I.). $D\,E$ is the first middle ordinate, and its length may be measured or calculated. To the point E draw the chords $A\,E$ and $B\,E$, lay off the second middle ordinates $G\,K$ and $H\,L$, each equal to $\frac{1}{4}\,D\,E$ (§ 84, III), and K and L are points on the curve. Draw the chords $A\,K$, $K\,E$, $E\,L$, and $L\,B$, and lay off third middle ordinates, each equal to one fourth the second middle ordinates, and four additional points on the curve will be determined. Continue this process, until a sufficient number of points is obtained

87. **Problem.** *To draw a tangent to a parabola at any station.*

Solution. I. If the curve has been laid out by tangent deflections (§ 85), let M''' (fig. 36) be the station, at which the tangent is to be drawn. From the preceding or succeeding station, lay off, parallel to $C\,D$, a distance $M''\,N$ or $E\,L$ equal to a, the first tangent deflection (§ 85), and $M'''\,N$ or $M'''\,L$ will be the required tangent. The same thing may be done by laying off from the second station a distance $M'\,T' = 4\,a$, or at the third station a distance $G\,P = 9\,a$; for the required tangent will then pass through T' or G. It will be seen, also, that the tangent at M''' passes through a point on the tangent at A corresponding to half the number of stations from A to M'''; that is, M''' is *four* stations from A, and the tangent passes through T', the *second* point on the tangent $A\,C$. In like manner, M''' is *six* stations from B, and the tangent passes through G, the *third* point on the tangent $B\,C$.

II. If the curve has been laid out by middle ordinates (§ 86), the tangent deflection for one station is equal to the last middle ordinate made use of in laying out the curve. For if the tangent $A\,C$ (fig. 37) were divided into four equal parts corresponding to the number of stations from A to E, the method of tangent deflections would give the same points on the curve, as were obtained by the method of § 86. In this case, the tangent deflection for one station would be $a = \frac{1}{16}\,C\,E = \frac{1}{16}\,D\,E$; but the last middle ordinate was made equal to $\frac{1}{4}\,G\,K$ or $\frac{1}{16}\,D\,E$. Therefore, a is equal to the last middle ordinate, and a tangent may be drawn at any station by the first method of this section.

A tangent may also be drawn at the extremity of any middle ordinate, by drawing a line through this extremity, parallel to the chord of that ordinate (§ 84, IV.).

88. In laying out a parabola by the method in § 85, it may sometimes be impossible or inconvenient to lay off all the points from the original tangents. A new tangent may then be drawn by § 87 to any station already found, as at M''' (fig. 36), and the tangent deflections a, $4\,a$, $9\,a$, &c. may be laid off from this tangent, precisely as from the first tangent. These deflections must be parallel to $C\,D$, and the distances on the new tangent must be equal to $T'\,N$ or $N\,M'''$, which may be measured.

89. **Problem.** *Given the tangents $A\,C$ and $B\,C$, equal or unequal, (fig 38,) to lay out a parabola by bisecting tangents.*

Solution. Bisect $A\,C$ and $B\,C$ in D and F, join $D\,F$, and find E, the middle point of $D\,F$. E will be a point on the curve (§ 84, V.). We have now two pairs of what may be called second tangents, $A\,D$ and $D\,E$, and $E\,F$ and $F\,B$. Bisect $A\,D$ in G and $D\,E$ in H, join $G\,H$, and its middle point M will be a point on the curve. Bisect $E\,F$ and $F\,B$ in K and L, join $K\,L$, and its middle point N will be a point on the curve. We have now four pairs of third tangents, $A\,G$ and $G\,M$, $M\,H$ and $H\,E$, $E\,K$ and $K\,N$, and $N\,L$ and $L\,B$. Bisect each pair in turn, join the points of bisection, and the middle points of the joining lines will be four new points, M', M'', N'', and N'. The same method may be continued, until a sufficient number of points is obtained.

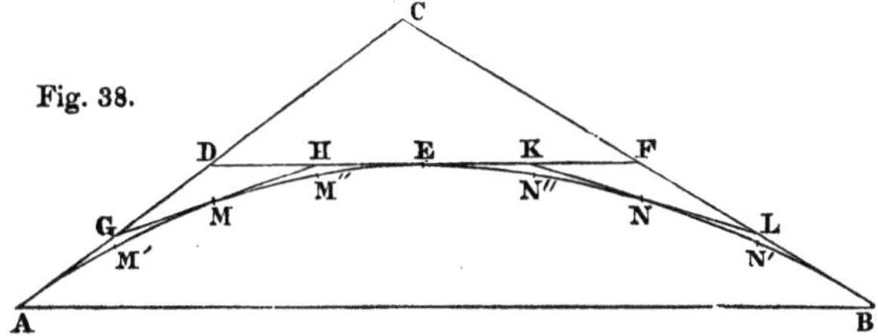

90. **Problem.** *Given the tangents A C and B C, equal or unequal, (fig. 39,) and the chord A B, to lay out a parabola by intersections.*

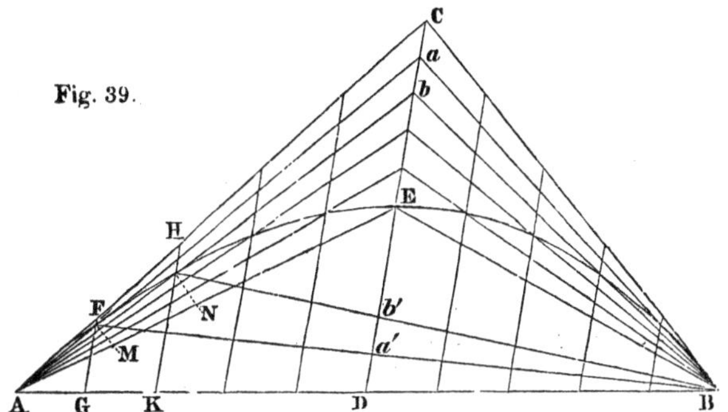

Solution. Bisect $A\,B$ in D, draw $C\,D$, and bisect it in E. Divide the tangents $A\,C$ and $B\,C$, the half-chords $A\,D$ and $D\,B$, and the line $C\,E$, into the same number of equal parts; five, for example. Then the intersection M of $A\,a$ and $F\,G$ will be a point on the curve. For $F\,M = \frac{1}{5}\,C\,a$, and $C\,a = \frac{1}{5}\,C\,E$. Therefore, $F\,M = \frac{1}{25}\,C\,E$, which is the proper deflection from the tangent at F to the curve (§ 85). In like manner, the intersection N of $A\,b$ and $H\,K$ may be shown to be a point on the curve, and the same is true of all the similar intersections indicated in the figure.

If the line $D\,E$ were also divided into five equal parts, the line $A\,a$ would be intersected in M on the curve by a line drawn from B through a', the line $A\,b$ would be intersected in N on the curve by a line drawn from B through b', and in general any two lines, drawn from A and B through two points on CD equally distant from the extremities C and D, will intersect on the curve. To show this for any point, as M, it is sufficient to show, that $B a'$ produced cuts $F G$ on the curve; for it has already been proved, that $A a$ cuts $F G$ on the curve. Now $D a' : M G = B D : B G = 5 : 9$, or $M G = \frac{9}{5} D a'$. But $D a' = \frac{1}{5} C E$. Therefore, $M G = \frac{9}{25} C E$. Again, $F G : C D = A G : A D = 1 : 5$. Therefore, $F G = \frac{1}{5} C D = \frac{2}{5} C E$. We have then $F M = F G - M G = \frac{2}{5} C E - \frac{9}{25} C E = \frac{1}{25} C E$. As this is the proper deflection from the tangent at F to the curve (§ 85), the intersection of $B a'$ with $F G$ is on the curve. This furnishes another method of laying out a parabola by intersections.

91. The following example is given in illustration of several of the preceding methods.

Example. Given $A C = B C = 832$ (fig. 40), and $A B = 1536$, to lay out a parabola $A E B$. We here find $C D = 320$. To begin with the method by tangent deflections (§ 85), divide the tangent $A C$ into eight equal parts. Then $a = \frac{C E}{n^2} = \frac{160}{64} = 2.5$. Lay off from the divisions on the tangent $F 1 = 2.5$, $G 2 = 4 \times 2.5 = 10$, $H 3 = 9 \times 2.5 = 22.5$, and $K 4 = 16 \times 2.5 = 40$. Suppose now that it is inconvenient to continue this method beyond K. In this case we may

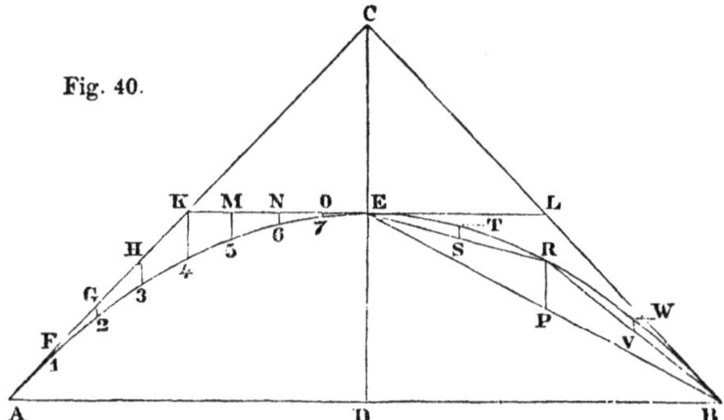

Fig. 40.

find a new tangent at E, by bisecting $A C$ and $B C$ (§ 89), and drawing $K L$ through the points of bisection. Divide the new tangent $K E = \frac{1}{2} A D = 384$ into four equal parts, and lay off from $K E$ the same tangent deflections as were laid off from AK, namely, $M5 = 22.5$, $N6 = 10$, and $O7 = 2.5$. To lay off the second half of the curve by middle ordinates (§ 86), measure $EB \doteq 784.49$. Bisect EB in P, and lay off the middle ordinate $PR = \frac{1}{4} DE = 40$. Measure $ER = 386.08$, and $BR = 402.31$, and lay off the middle ordinates ST and VW, each equal to $\frac{1}{4} PR = 10$. By measuring the chords ET, TR, RW, and WB, and laying off an ordinate from each, equal to 2.5, four additional points might be found.

Article II. — Radius of Curvature.

92. The curvature of circular arcs is always the same for the same arc, and in different arcs varies inversely as the radii of the arcs. Thus, the curvature of an arc of 1,000 feet radius is double that of an arc of 2,000 feet radius. The curvature of a parabola is continually changing. In fig. 39, for example, it is least at the tangent point A, the extremity of the longest tangent, and increases by a fixed law, until it becomes greatest at a point, called the vertex, where a tangent to the curve would be perpendicular to the diameter. From this point to B it decreases again by the same law. We may, therefore, consider a parabola to be made up of a succession of infinitely small circular arcs, the radii of which continually increase in going from the vertex to the extremities. The radius of the circular arc, corresponding to any part of a parabola, is called the *radius of curvature* at that point.

If a parabola forms part of the line of a railroad, it will be necessary, in order that the rails may be properly curved (§ 28), to know how the radius of curvature may be found. It will, in general, be necessary to find the radius of curvature at a few points only. In short curves it may be found at the two tangent points and at the middle station, and in longer curves at two or more intermediate points besides. The rails curved according to the radius at any point should be sufficient in number to reach, on each side of that point, half-way to the next point.

93. **Problem.** *To find the radius of curvature at certain stations on a parabola.*

Solution. Let AEB (fig. 41) be any parabola, and let it be required to find the radii of curvature at a certain number of stations from A to E. These stations must be selected at regular intervals from those determined by any of the preceding methods. Let n denote the number of parts into which $A\,E$ is divided, and divide $C\,D$ into the same number of equal parts. Draw lines from A to the points

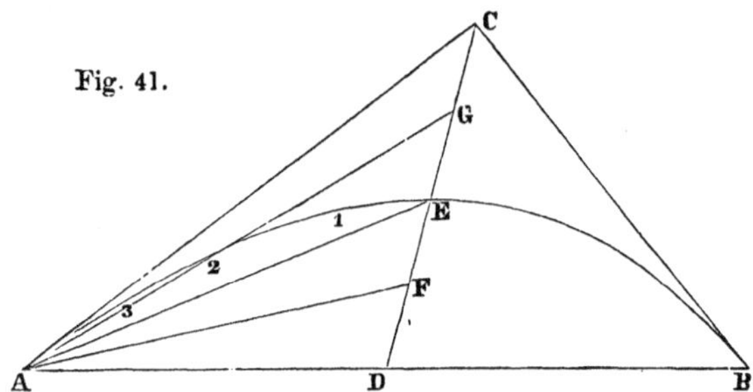

of division. Thus, if $n = 4$, as in the figure, divide $C\,D$ into four equal parts, and draw $A\,F$, $A\,E$, and $A\,G$. Let $A\,D = c$, $A\,F = c_1$, $A\,E = c_2$, $A\,G = c_3$, and $A\,C = T$. Denote, moreover, $C\,D$ by d, and the area of the triangle $A\,C\,B$ by A. Then the respective radii for the points E, 1, 2, 3, and A will be

$$R = \frac{c^3}{A}, \quad R_1 = \frac{c_1^{\,3}}{A}, \quad R_2 = \frac{c_2^{\,3}}{A}, \quad R_3 = \frac{c_3^{\,3}}{A}, \quad R_4 = \frac{T^3}{A}.$$

The area A may be found by form. 18, Tab. X.; c and T are known; and c_1, c_2, c_3 may be found approximately by measurement on a figure carefully constructed, or exactly by these general formulæ: —

$$c_1^{\,2} = c^2 + \frac{T^2 - c^2}{n} - \frac{(n-1)\,d^2}{n^2},$$

$$c_2^{\,2} = c_1^{\,2} + \frac{T^2 - c^2}{n} - \frac{(n-3)\,d^2}{n^2},$$

$$c_3^{\,2} = c_2^{\,2} + \frac{T^2 - c^2}{n} - \frac{(n-5)\,d^2}{n^2},$$

$$c_4^{\,2} = c_3^{\,2} + \frac{T^2 - c^2}{n} - \frac{(n-7)\,d^2}{n^2},$$

&c. &c.

It will be seen, that each of these values is formed from the preceding, by adding the same quantity $\frac{T^2 - c^2}{n}$, and subtracting $\frac{d^2}{n^2}$ multiplied in succession by $n - 1$, $n - 3$, $n - 5$, &c. Making $n = 4$, we have

$$c_1^2 = c^2 + \tfrac{1}{4}(T^2 - c^2) - \tfrac{3}{16}d^2,$$
$$c_2^2 = c_1^2 + \tfrac{1}{4}(T^2 - c^2) - \tfrac{1}{16}d^2,$$
$$c_3^2 = c_2^2 + \tfrac{1}{4}(T^2 - c^2) + \tfrac{1}{16}d^2.$$

All the quantities, which enter into the expressions for the radii, are now known, and the radii may, therefore, be determined. The same method will apply to the other half of the parabola.

The manner of obtaining the preceding formulæ is as follows. The
of curvature at any given point on a parabola is, by the Differ-
Calculus, $R = \frac{p}{2 \sin.^3 E}$, in which p represents the parameter of
arabola for rectangular coördinates, and E the angle made with
ameter by a tangent to the curve at the given point. First, let the
ddle station E (fig. 42) be the given point. Then the angle E is the

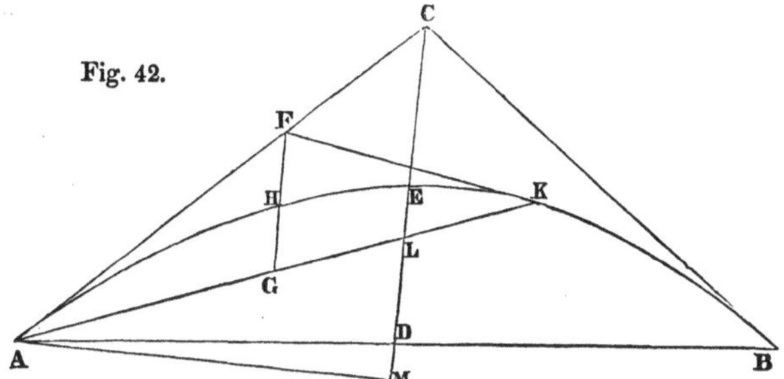

Fig. 42.

angle made with $E D$ by a tangent at E, or since $A B$ is parallel to the tangent at E (§ 84, IV.), sin. E = sin. $A D E$ = sin. $B D E$. Let p' be the parameter for the diameter $E D$. Then, by Analytical Geometry, $p = p' \sin.^2 E$. Therefore, at this point $R = \frac{p}{2 \sin.^3 E} = \frac{p' \sin.^2 E}{2 \sin.^3 E} = \frac{p'}{2 \sin. E}$. But $p' = \frac{A D^2}{E D} = \frac{c^2}{\frac{1}{2} d}$. Therefore, $R = \frac{c^2}{d \sin. E} = \frac{c^3}{c d \sin. E} = \frac{c^3}{A}$; since $A = c d \sin. E$ (Tab. X. 17).

Next, to find R_1, or the radius of curvature at H, the first station from E. Through H draw $F G$ parallel to $C D$, and from F draw the tangent $F K$. Join $A K$, cutting $C D$ in L. Then from what has just been proved for the radius of curvature at E, we have for the radius of curvature at H, $R_1 = \frac{A G^3}{A F K}$. Now $A G : A L = A F : A C =$

$n-1:n$, or $A\,G=\frac{n-1}{n}\times A\,L$. But $A\,L=c_1$. For, since $A\,F=\frac{n-1}{n}\times A\,C$, the tangent deflection $F\,H=\frac{(n-1)^2}{n^2}\cdot\frac{d}{2}$ (§ 84, II.), and $F\,G=2\,F\,H=\frac{(n-1)^2}{n^2}d$. Then, since $C\,L:F\,G=A\,C:A\,F=n:n-1$, $C\,L=\frac{n}{n-1}\times F\,G=\frac{n-1}{n}d$. Hence $L\,D=d-\frac{n-1}{n}d=\frac{1}{n}d$, that is, $A\,L=c_1$. Substituting this value in the expression for $A\,G$ above, we have $A\,G=\frac{n-1}{n}c_1$. Moreover, since $A\,F=\frac{n-1}{n}\times A\,C$, and because similar triangles are to each other as the squares of their homologous sides, we have the triangle $A\,F\,G=\frac{(n-1)^2}{n^2}\times A\,C\,L$. But $A\,C\,L:A\,C\,D=C\,L:C\,D=n-1:n$, or $A\,C\,L=\frac{n-1}{n}\times A\,C\,D$. Therefore, $A\,F\,G=\frac{(n-1)^3}{n^3}\times A\,C\,D$, and $A\,F\,K=2\,A\,F\,G=\frac{(n-1)^3}{n^3}\times A\,C\,B=\frac{(n-1)^3}{n^3}A$. Substituting these values of $A\,G$ and $A\,F\,K$ in the equation $R_1=\frac{A\,G^3}{A\,F\,K}$, and reducing, we find $R_1=\frac{c_1^3}{A}$. By similar reasoning we should find $R_2=\frac{c_2^3}{A}$, $R_3=\frac{c_3^3}{A}$, &c.

It remains to find the values of c_1, c_2, &c. Through A draw $A\,M$ perpendicular to $C\,D$, produced if necessary. Then, by Geometry, we have $A\,D^2=A\,L^2+L\,D^2-2\,L\,D\times L\,M$, and $A\,C^2=A\,L^2+C\,L^2+2\,C\,L\times L\,M$. Finding from each of these equations the value of $2\,L\,M$, and putting these values equal to each other, we have $\frac{A\,L^2+L\,D^2-A\,D^2}{L\,D}=\frac{A\,C^2-A\,L^2-C\,L^2}{C\,L}$. But $A\,L=c_1$, $L\,D=\frac{1}{n}d$, $A\,D=c$, $A\,C=T$, and $C\,L=\frac{n-1}{n}d$. Substituting these values in the last equation, and reducing, we find

$$c_1^2=\frac{T^2}{n}+\frac{(n-1)c^2}{n}-\frac{(n-1)d^2}{n^2}.$$

By similar reasoning we should find

$$c_2^2=\frac{2\,T^2}{n}+\frac{(n-2)c^2}{n}-\frac{2(n-2)d^2}{n^2},$$

$$c_3^2=\frac{3\,T^2}{n}+\frac{(n-3)c^2}{n}-\frac{3(n-3)d^2}{n^2},$$

&c. &c.

From these equations the values of c_1^2, c_2^2, c_3^2, &c. given on page 72 are readily obtained. That given for c_1^2 is obtained from the first of these equations by a simple reduction; that given for c_2^2 is obtained by subtracting the first of these equations from the second, and reducing; that given for c_3^2 is obtained by subtracting the second equation from the third, and reducing; and so on.

94. *Example.* Given (fig. 41) $A\,C = T = 600$, $B\,C = T' = 520$, and $A\,D = c = 550$, to find R, R_1, R_2, R_3, and R_4, the radii of curvature at E, 1, 2, 3, and A.

To find $C\,D = d$, we have, by Geometry, $d^2 = \frac{1}{2}(T^2 + T'^2) - c^2$, which gives $d^2 = 12700$.

To find the area of $A\,C\,B = A$, we have (Tab. X. 18) $A = \sqrt{s\,(s-a)\,(s-b)\,(s-c)}$.

$s = 1110$	3.045323
$s - a = 590$	2.770852
$s - b = 510$	2.707570
$s - c = 10$	1.000000
	2)9.523745
log. A	4.761872

Next $\frac{1}{n}(T^2 - c^2) = \frac{1}{4}(T + c)\,(T - c) = \frac{1150 \times 50}{4} = 14375$, and $\frac{d^2}{n^2} = \frac{12700}{16} = 793.75$. Then

$$c^2 = 550^2 = 302500$$

$$c_1^2 = 302500 + 14375 - 3 \times 793.75 = 314493.75$$

$$c_2^2 = 314493.75 + 14375 - 793.75 = 328075$$

$$c_3^2 = 328075 + 14375 + 793.75 = 343243.75$$

To find R, we have $R = \frac{c^3}{A}$, or log. $R = 3$ log. c — log. A.

$c = 550$	2.740363
c^3	8.221089
A	4.761872
$R = 2878.8$	3.459217

To find R_1, we have $R_1 = \frac{c_1^3}{A}$, or log. $R_1 = \frac{3}{2}$log c_1^2 — log. A.

$c_1^2 = 314493.75$	5.497612
c_1^3	8.246418
A	4.76 872
$R_1 = 3051.7$	3.484546

In the same way we should find $R_2 = 3251.5$, $R_3 = 3479.6$, $R_4 = 3737.5$.

To find the radii for the second part EB of the parabola, the same formulæ apply, except that T' takes the place of T. We have then $\frac{1}{n}(T'^2 - c^2) = \frac{1}{4}(T' + c)(T' - c) = \frac{1070 \times -30}{4} = -8025$

Hence

$$c_1^2 = 302500 - 8025 - 2381.25 = 292093.75$$

$$c_2^2 = 292093.75 - 8025 - 793.75 = 283275.$$

$$c_3^2 = 283275 - 8025 + 793.75 = 276043.75$$

To find R_1, we have $R_1 = \frac{c_1^3}{A}$, or $\log. R_1 = \frac{3}{2}\log. c_1^2 - \log. A$

$c_1^2 = 292093.75$	5.465523
c_1^3	8.198284
A	4.761872
$R_1 = 2731.6$	3.436412

In the same way we should find $R_2 = 2608.8$, $R_3 = 2509.5$, $R_4 = 2433$.

It will be seen, that the radii in this example decrease from one tangent point to the other, which shows that both tangent points lie on the same side of the vertex of the parabola (§ 92). This will be the case, whenever the angle BCD, adjacent to the shorter tangent, exceeds 90°, that is, whenever c^2 exceeds $T'^2 + d^2$. If $BCD = 90°$, the tangent point B falls on the vertex. If BCD is less than 90°, one tangent point falls on each side of the vertex, and the curvature will, therefore, decrease towards both extremities.

95. If the tangents T and T' are equal, the equations for c_1^2, c_2^2, &c. will be more simple; for in this case d is perpendicular to c, and $T^2 - c^2 = d^2$. Substituting this value, we get

$$c_1^2 = c^2 + \frac{d^2}{n^2},$$

$$c_2^2 = c_1^2 + \frac{3d^2}{n^2},$$

$$c_3^2 = c_2^2 + \frac{5d^2}{n^2},$$

&c. &c.

Example. Given, as in § 91, $T = T' = 832$, $c = 768$, and $d =$

320, to find the radii R, R_1, and R_2 at the points E, 4, and A (fig. 40). Here $A = c\,d = 245760$, $n = 2$, and $c_1{}^2 = c^2 + \frac{1}{4}d^2 = 615424$. Then $R = \frac{c^3}{c\,d} = \frac{c^2}{d} = \frac{768^2}{320} = 1843.2$, $R_1 = \frac{c_1{}^3}{c\,d}$, and $R_2 = \frac{T^3}{c\,d}$.

$c_1{}^2 = 615424$	5.789174
$c_1{}^3$	8.683761
$c\,d = 245760$	5.390511
$R_1 = 1964.5$	3.293250
$T = 832$	2.920123
T^3	8.760369
$c\,d = 245760$	5.390511
$R_2 = 2343.5$	3.369858

R is the radius at the point R also, and R_2 the radius at the point B

CHAPTER III.

LEVELLING.

ARTICLE I. — HEIGHTS AND SLOPE STAKES.

96. THE Level is an instrument consisting essentially of a telescope, supported on a tripod of convenient height, and capable of being so adjusted, that its line of sight shall be horizontal, and that the telescope itself may be turned in any direction on a vertical axis. The instrument when so adjusted is said to be *set*.

The line of sight, being a line of indefinite length, may be made to describe a horizontal plane of indefinite extent, called the *plane of the level.*

The levelling rod is used for measuring the vertical distance of any point, on which it may be placed, below the plane of the level. This distance is called the *sight* on that point.

97. **Problem.** *To find the difference of level of two points, as A and B (fig.* 43).

Solution. Set the level between the two points,* and take sights on both points. Subtract the less of these sights from the greater, and the difference will be the difference of level required. For if FP represent the plane of the level, and AG be drawn through A parallel to FP, AF will be the sight on A, and BP the sight on B. Then the required difference of level $BG = BP - PG = BP - AF$.

If the distance between the points, or the nature of the ground, makes it necessary to set the level more than once, set down all the backward sights in one column and all the forward sights in another. Add up these columns, and take the less of the two sums from the greater, and the difference will be the difference of level required. Thus, to find the difference of level between A and D (fig. 43), the level is first set between A and B, and sights are taken on A and B; the level is then set between B and C, and sights are taken on B and

* The level should be placed midway between the two points, when practicable, in order to neutralize the effect of inaccuracy in the adjustment of the instrument, and for the reason given in § 105.

C; lastly, the level is set between *C* and *D*, and sights are taken on *C* and *D*. Then the difference of level between *A* and *D* is $ED = (BP + KC + OD) - (AF + BI + NC)$. For $ED = HC - LC = HM + MC - LC$. But $HM = bG = BP - AF$, $MC = KC - BI$, and $LC = NC - OD$. Substituting these values, we have $ED = BP - AF + KC - BI - NC + OD = (BP + KC + OD) - (AF + BI + NC)$.

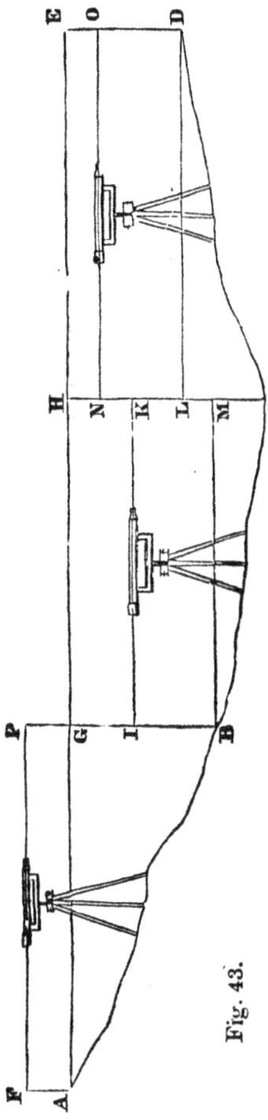

Fig. 43.

98. It is often convenient to refer all heights to an imaginary level plane called the *datum plane*. This plane may be assumed at starting to pass through, or at some fixed distance above or below, any permanent object, called a *bench-mark*, or simply a *bench*. It is most convenient, in order to avoid minus heights, to assume the datum plane at such a distance below the bench-mark, that it will pass below all the points on the line to be levelled. Thus if *A B* (fig. 44) were part of the line to be levelled, and if *A* were the starting point, we should assume the datum plane *C D* at such a distance below some permanent object near *A*, as would make it pass below all the points on the line. If, for instance, we had reason to believe that no point on this line was more than 15 or 20 feet below A, we might safely assume *C D* to be 25 feet below the bench near *A*, in which case all the distances from the line to the datum plane would be positive. Lines before being levelled are usually divided into regular stations, the height of each of which above the datum plane is required.

99. **Problem.** *To find the heights above a datum plane of the several stations on a given line.*

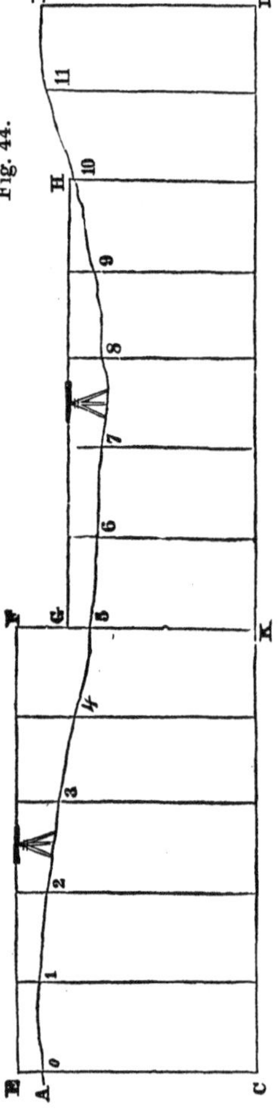

Solution. Let $A\,B$ (fig. 44) represent a portion of the line, divided into regular stations, marked 0, 1, 2, 3, 4, 5, &c., and let $C\,D$ represent the datum plane, assumed to be 25 feet below a bench-mark near A. Suppose the level to be set first between stations 2 and 3, and a sight upon the bench-mark to be taken, and found to be 3.125. Now as this sight shows that the plane of the level $E\,F$ is 3.125 feet above the bench-mark and as the datum plane is 25 feet below this mark, we shall find the height of the plane of the level above the datum plane by adding these heights, which gives for the height of $E\,F$ $25 + 3.125 = 28.125$ feet This height may for brevity's sake be called the *height of the instrument*, meaning by this the height of the line of sight of the instrument.

If now a sight be taken on station 0, we shall obtain the height of this station above the datum plane, by subtracting this sight from the height of the instrument; for the height of this station is $0\,C$ and $0\,C = E\,C - E\,0$. Thus if $E\,0 = 3.413$, $0\,C = 28.125 - 3.413 = 24.712$. In like manner, the heights of stations 1, 2, 3, 4, and 5 may be found, by taking sights on them in succession, and subtracting these sights from the height of the instrument. Suppose these sights to be respectively 3.102, 3.827, 4.816, 6.952, and 9.016, and we have

height of station 0 = 28.125 — 3.413 = 24.712,
" " " 1 = 28.125 — 3.102 = 25.023,

height of station 2 = 28.125 — 3.827 = 24.298,
" " " 3 = 28.125 — 4.816 = 23.309,
" " " 4 = 28.125 — 6.952 = 21.173,
" " " 5 = 28.125 — 9.016 = 19.109.

Next, set the level between stations 7 and 8, and as the height of station 5 is known, take a sight upon this point. This sight, being added to the height of station 5, will give the height of the instrument in its new position; for $GK = G5 + 5K$. Suppose this sight to be $G5 = 2.740$, and we have $GK = 19.109 + 2.740 = 21.849$. A point like station 5, which is used to get the height of the instrument after resetting, is called a *turning point.* The height of the instrument being found, sights are taken on stations 6, 7, 8, 9, and 10, and the heights of these stations found by subtracting these sights from the height of the instrument. Suppose these sights to be respectively 3.311, 4.027, 3.824, 2.516, and 0.314, and we have

height of station 6 = 21.849 — 3.311 = 18.538,
" " " 7 = 21.849 — 4.027 = 17.822,
" " " 8 = 21.849 — 3.824 = 18.025,
" " " 9 = 21.849 — 2.516 = 19.333,
" " " 10 = 21.849 — 0.314 = 21.535.

The instrument is now again carried forward and reset, station 10 is used as a turning point to find the height of the instrument, and every thing proceeds as before.

At convenient distances along the line, permanent objects are selected, and their heights obtained and preserved, to be used as starting points in any further operations. These are also called benches. Let us suppose, that a bench has been thus selected near station 9, and that the sight upon it from the instrument, when set between stations 7 and 8, is 2.635. Then the height of this bench will be 21.849 — 2.635 = 19 214.

100. From what has been shown above, it appears that the first thing to be done, after setting the level, is to take a sight upon some point of known height, and that this sight is always to be *added* to the known height, in order to get the height of the instrument. This first sight may therefore be called a *plus* sight. The next thing to be done is to take sights on those points whose heights are required, and to *subtract* these sights from the height of the instrument, in order to get the required heights. These last sights may therefore be called *minus* sights.

101. The field notes are kept in the following form. The *first* column in the table contains the *stations*, and also the benches marked B., and the turning points marked *t. p.*, except when coincident with a station. The *second* column contains the *plus sights*; the *third* column shows the *height of the instrument*; the *fourth* contains the *minus sights*; and the *fifth* contains the *heights* of the points in the first column.

Station	+S.	H. I.	—S.	H.
B.	3.125			25.000
0		28.125	3.413	24.712
1			3.102	25.023
2			3 827	24.298
3			4.816	23.309
4			6.952	21.173
5	2.740		9.016	19.109
6		21.849	3.311	18.538
7			4.027	17.822
8			3.824	18.025
9			2.516	19.333
B.			2.635	19.214
10			0.314	21.535

The height of the bench is set down as assumed above, namely, 25 feet; the first plus sight is set opposite B., on which point it was taken, and, being added to the height in the same line, gives the height of the instrument, which is set opposite 0; the minus sights are set opposite the points on which they are taken, and, being subtracted from the height of the instrument, give the heights of these points, as set down in the fifth column. The minus sights are subtracted from the same height of the instrument, as far as the turning point at station 5, inclusive. The plus sight on station 5 is set opposite this station, and a new height obtained for the instrument by adding the plus sight to the height of the turning point. This new height of the instrument is set opposite station 6, where the minus sights to be subtracted from it commence. These sights are again set opposite the points on which they were taken, and, being subtracted from the new height of the instrument, give the heights in the last column.

102. **Problem.** *To set slope stakes for excavations and embankments.*

Solution. Let $A\ B\ H\ K\ C$ (fig. 45) be a cross-section of a proposed excavation, and let the centre cut $A\ M = c$, and the width of the road-bed $HK = b$. The slope of the sides BH or CK is usually given by the ratio of the base KN to the height EN. Suppose, in the present case, that $KN : EN = 3 : 2$, and we have the slope $= \frac{3}{2}$. Then if the ground were level, as DAE, it is evident that the distance from

Fig. 45.

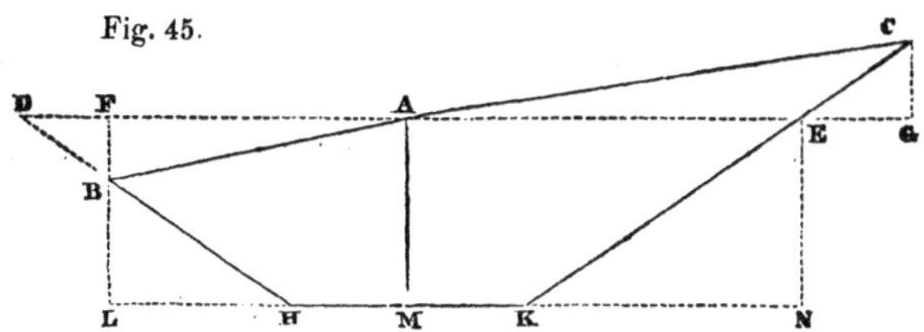

the centre A to the slope stakes at D and E would be $AD = AE = MK + KN = \frac{1}{2}b + \frac{3}{2}c$. But as the ground rises from A to C through a height $CG = g$, the slope stake must be set farther out a distance $EG = \frac{3}{2}g$; and as the ground falls from A to B through a height $BF = g$, the slope stake must be set farther in a distance $DF = \frac{3}{2}g$.

To find B and C, set the level, if possible, in a convenient position for sighting on the points A, B, and C. From the known cut at the centre find the value of $AE = \frac{1}{2}b + \frac{3}{2}c$. Estimate by the eye the rise from the centre to where the slope stake is to be set, and take this as the probable value of g. To AE add $\frac{3}{2}g$, as thus estimated, and measure from the centre a distance out, equal to the sum. Obtain now by the level the rise from the centre to this point, and if it agrees with the estimated rise, the distance out is correct. But if the estimated rise prove too great or too small, assume a new value for g, measure a corresponding distance out, and test the accuracy of the estimate by the level, as before. These trials must be continued, until the estimated rise agrees sufficiently well with the rise found by the level at the corresponding distance out. The *distance out* will then be $\frac{1}{2}b + \frac{3}{2}c + \frac{3}{2}g$. The same course is to be pursued, when the ground falls from the centre, as at B; but as g here becomes *minus*, the distance out, when the true value of g is found, will be $AF = AD - DF = \frac{1}{2}b + \frac{3}{2}c - \frac{3}{2}g$.

For embankment, the process of setting slope stakes is the same as for excavation except that a *rise* in the ground from the centre on embankments corresponds to a *fall* on excavations, and *vice versâ*. This will be evident by inverting figure 45, which will then represent an embankment. What was before a *fall* to B, becomes now a *rise*. and what was before a *rise* to C, becomes now a *fall*.

When the section is partly in excavation and partly in embankment, the method above applies directly only to the side which is in excavation at the same time that the centre of the road-bed is in excavation, or in embankment at the same time that the centre is in embankment. On the opposite side, however, it is only necessary to make c in the expressions above *minus*, because its effect here is to diminish the distance out. The formula for this distance out will, therefore, become $\frac{1}{2} b - \frac{3}{2} c + \frac{3}{2} g$.

Article II. — Correction for the Earth's Curvature and for Refraction.

103. Let $A\,C$ (fig. 46) represent a portion of the earth's surface. Then, if a level be set at A, the *line of sight* of the level will be the tangent $A\,D$, while the *true level* will be $A\,C$. The difference $D\,C$ between the line of sight and the true level is the correction for the earth's curvature for the distance $A\,D$.

104. A correction in the opposite direction arises from refraction. Refraction is the change of direction which light undergoes in passing from one medium into another of different density. As the atmosphere increases in density the nearer it lies to the earth's surface, light, passing from a point B to a lower point A, enters continually air of greater and greater density, and its path is in consequence a curve concave towards the earth. Near the earth's surface this path may be taken as the arc of a circle whose radius is seven times the radius of the earth.* Now a level at A, having its line of sight in the direction $A\,D$, tangent to the curve $A\,B$, is in the proper position to receive the light from an object at B; so that this object appears to the observer to be at D. The effect of refraction, therefore, is to make an object appear higher than its true position. Then, since the correction for the earth's curvature $D\,C$ and the correction for refraction $D\,B$ are in opposite directions, the correction for both will be $B\,C = D\,C - D\,B$.

* Peirce's Spherical Astronomy, Chap. X., § 125 It should be observed, however, that the effect of refraction is very uncertain, varying with the state of the atmosphere. Sometimes the path of a ray is even made convex towards the earth, and sometimes the rays are refracted horizontally as well as vertically.

This correction must be *added* to the height of any object as determined by the level.

105. **Problem.** *Given the distance $A D = D$ (fig. 46), the radius of the earth $A E = R$, and the radius of the arc of refracted light $= 7 R$, to find the correction $B C = d$ for the earth's curvature and for refraction.*

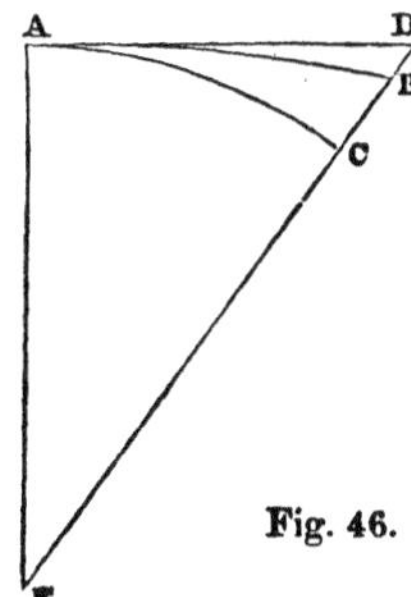

Fig. 46.

Solution. To find the correction for the earth's curvature $D C$, we have, by Geometry, $D C (D C + 2 E C) = A D^2$, or $D C (D C + 2 R) = D^2$. But as $D C$ is always very small compared with the diameter of the earth, it may be dropped from the parenthesis, and we have $D C \times 2 R = D^2$, or $D C = \frac{D^2}{2 R}$. The correction for refraction $D B$ may be found by the method just used for finding $D C$, merely changing R into $7 R$. Hence $D B = \frac{D^2}{14 R}$. We have then $d = B C = D C - D B = \frac{D^2}{2 R} - \frac{D^2}{14 R}$, or

☞ $$d = \frac{3 D^2}{7 R}.$$

By this formula Table III. is calculated, taking $R = 20{,}911{,}790$ ft., as given by Bowditch. The necessity for this correction may be avoided, whenever it is possible to set the level midway between the points whose height is required. In this case, as the distance on each side of the level is the same, the corrections will be equal, and will destroy each other.

ARTICLE III. — VERTICAL CURVES.

106. VERTICAL curves are used to round off the angles formed by the meeting of two grades. Let $A\,C$ and $C\,B$ (fig. 47) be two grades meeting at C. These grades are supposed to be given by the *rise* per station in going in some particular direction. Thus, starting from A, the grades of $A\,C$ and $C\,B$ may be denoted respectively by g and g'; that is, g denotes what is added to the height at every station on $A\,C$, and g' denotes what is added to the height at every station on $C\,B$; but since $C\,B$ is a descending grade, the quantity added is a minus quantity, and g' will therefore be negative. The parabola furnishes a very simple method of putting in a vertical curve.

107. **Problem.** *Given the grade g of $A\,C$ (fig. 47), the grade g of $C\,B$, and the number of stations n on each side of C to the tangent points A and B, to unite these points by a parabolic vertical curve.*

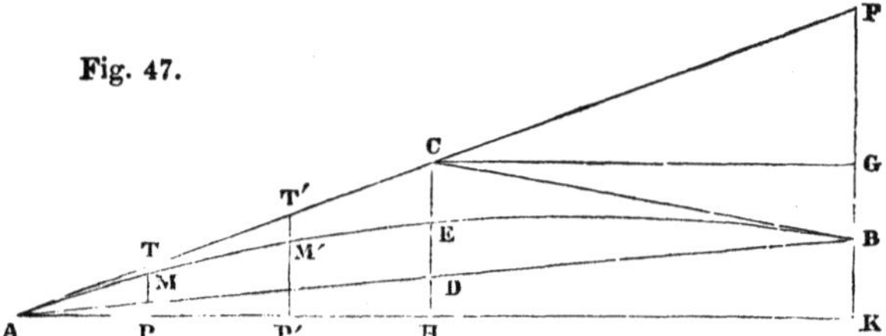

Fig. 47.

Solution. Let $A\,E\,B$ be the required parabola. Through B and C draw the vertical lines $F\,K$ and $C\,H$, and produce $A\,C$ to meet $F\,K$ in F. Through A draw the horizontal line $A\,K$, and join $A\,B$, cutting $C\,H$ in D. Then, since the distance from C to A and B is measured horizontally, we have $A\,H = H\,K$, and consequently $A\,D = D\,B$. The vertical line $C\,D$ is, therefore, a diameter of the parabola (§ 84, I.), and the distances of the curve in a vertical direction from the stations on the tangent $A\,F$ are to each other as the squares of the number of stations from A (§ 84, II.). Thus, if a represent this distance at the first station from A, the distance at the second station would be $4\,a$, at the third station $9\,a$, and at B, which is $2\,n$ stations from A, it would be $4\,n^2 a$; that is, $F\,B = 4\,n^2 a$, or $a = \frac{F\,B}{4\,n^2}$. To find a, it will then be necessary to find $F\,B$ first. Through C draw the horizontal line $C\,G$, and we have, from the equal triangles $C\,F\,G$ and

$A\,C\,H$, $F\,G = C\,H$. But $C\,H$ is the rise of the first grade g in the n stations from A to C; that is, $C\,H = n\,g$, or $F\,G = n\,g$. $G\,B$ is also the rise of the second grade g' in n stations, but since g' is negative (§ 106), we must put $G\,B = -n\,g'$. Therefore, $F\,B = F\,G + G\,B = n\,g - n\,g'$. Substituting this value of $F\,B$ in the equation for a we have $a = \frac{ng - ng'}{4\,n^2}$, or

$$a = \frac{g - g'}{4\,n}.$$

The value of a being thus determined, all the distances of the curve from the tangent $A\,F$, viz. a, $4\,a$, $9\,a$, $16\,a$, &c., are known. Now if T and T' be the first and second stations on the tangent, and vertical lines $T\,P$ and $T'\,P'$ be drawn to the horizontal line $A\,K$, the height $T\,P$ of the first station above A will be g, the height $T'\,P'$ of the second station above A will be $2\,g$, and in like manner for succeeding stations we should find the heights $3\,g$, $4\,g$, &c As we have already found $T\,M = a$, $T'\,M' = 4\,a$, &c., we shall have for the heights of the curve above the level of A, $M\,P = T\,P - T\,M = g - a$, $M'\,P' = T'\,P' - T'\,M' = 2\,g - 4\,a$, and in like manner for the succeeding heights $3\,g - 9\,a$, $4\,g - 16\,a$, &c. Then to find the grades for the curve at the successive stations from A, that is, the rise of each height over the preceding height, we must subtract each height from the next following height, thus: $(g - a) - 0 = g - a$, $(2\,g - 4\,a) - (g - a) = g - 3\,a$, $(3\,g - 9\,a) - (2\,g - 4\,a) = g - 5\,a$, $(4\,g - 16\,a) - (3\,g - 9\,a) = g - 7\,a$, &c. *The successive grades for the vertical curve are, therefore,*

$$g - a,\quad g - 3\,a,\quad g - 5\,a,\quad g - 7\,a,\ \text{\&c.}$$

In finding these grades, strict regard must be paid to the algebraic signs. The results are then general; though the figure represents but one of the six cases that may arise from various combinations of ascending and descending grades. If proper figures were drawn to represent the remaining cases, the above solution, with due attention to the signs, would apply to them all, and lead to precisely the same formulæ.

168. *Examples.* Let the number of stations on each side of C be 3, and let $A\,C$ ascend .9 per station, and $C\,B$ descend .6 per station. Here $n = 3$, $g = .9$, and $g' = -.6$. Then $a = \frac{g - g'}{4\,n} = \frac{.9 - (-.6)}{4 \times 3} = \frac{1.5}{12} = .125$, and the grades from A to B will be

$$g - a = .9 - .125 = .775,$$
$$g - 3a = .9 - .375 = .525,$$
$$g - 5a = .9 - .625 = .275,$$
$$g - 7a = .9 - .875 = .025,$$
$$g - 9a = .9 - 1.125 = -.225,$$
$$g - 11a = .9 - 1.375 = -.475.$$

As a second example, let the first of two grades descend .8 per station, and the second ascend .4 per station, and assume two stations on each side of C as the extent of the curve. Here $g = -.8$, $g' = .4$, and $n = 2$. Then $a = \frac{-.8 - .4}{4 \times 2} = \frac{-1.2}{8} = -.15$, and the four grades required will be

$$g - a = -.8 - (-.15) = -.8 + .15 = -.65,$$
$$g - 3a = -.8 - (-.45) = -.8 + .45 = -.35,$$
$$g - 5a = -.8 - (-.75) = -.8 + .75 = -.05,$$
$$g - 7a = -.8 - (-1.05) = -.8 + 1.05 = +.25.$$

It will be seen, that, after finding the first grade, the remaining grades may be found by the continual subtraction of $2a$. Thus, in the first example, each grade after the first is .25 less than the preceding grade, and in the second example, a being here negative, each grade after the first is .3 greater than the preceding grade.

109. The grades calculated for the whole stations, as in the foregoing examples, are sufficient for all purposes except for laying the track. The grade stakes being then usually only 20 feet apart, it will be necessary to ascertain the proper grades on a vertical curve for these sub-stations. To do this, nothing more is necessary than to let g and g' represent the given grades for a sub-station of 20 feet, and n the number of sub-stations on each side of the intersection, and to apply the preceding formulæ. In the last example, for instance, the first grade descends .8 per station, or .16 every 20 feet, the second grade ascends .4 per station, or .08 every 20 feet, and the number of sub-stations in 200 feet is 10. We have then $g = -.16$, $g' = .08$, and $n = 10$ Hence $a = \frac{-.16 - .08}{4 \times 10} = \frac{-.24}{40} = -.006$. The first grade is, therefore, $g - a = -.16 + .006 = -.154$, and as each subsequent grade increases .012 (§ 108), the whole may be written down without farther trouble, thus: $-.154$, $-.142$, $-.130$, $-.118$, $-.106$, $-.094$, $-.082$, $-.070$, $-.058$, $-.046$, $-.034$, $-.022$, $-.010$, $+.002$, $+.014$, $+.026$, $+.038$ $+.050$, $+.062$, $+.074$.

ARTICLE IV. — ELEVATION OF THE OUTER RAIL ON CURVES.

110. **Problem.** *Given the radius of a curve R, the gauge of the track g, and the velocity of a car per second v, to determine the proper elevation e of the outer rail of the curve.*

Solution. A car moving on a curve of radius R, with a velocity per second $= v$, has, by Mechanics, a centrifugal force $= \frac{v^2}{R}$. To counteract this force, the outer rail on a curve is raised above the level of the inner rail, so that the car may rest on an inclined plane. This elevation must be such, that the action of gravity in forcing the car down the inclined plane shall be just equal to the centrifugal force, which impels it in the opposite direction. Now the action of gravity on a body resting on an inclined plane is equal to 32.2 multiplied by the ratio of the height to the length of the plane. But the height of the plane is the elevation e, and its length the gauge of the track g. This action of gravity, which is to counteract the centrifugal force is, therefore, $= \frac{32.2\,e}{g}$. Putting this equal to the centrifugal force, we have $\frac{32.2\,e}{g} = \frac{v^2}{R}$. Hence

☞ $$e = \frac{g\,v^2}{32.2\,R}.$$

If we substitute for R its value (§ 10) $R = \frac{50}{\sin. D}$, we have $e = \frac{g\,v^2 \sin. D}{50 \times 32.2} = .00062112\,g\,v^2 \sin. D$. If the velocity is given in miles per hour, represent this velocity by M, and we have $v = \frac{M \times 5280}{60 \times 60}$. Substituting this value of v, we find $e = .0013361\,g\,M^2 \sin. D$. When $g = 4.7$, this becomes $e = .00627966\,M^2 \sin. D$ By this formula Table IV. is calculated. In determining the proper elevation in any given case, the usual practice is to adopt the highest customary speed of passenger trains as the value of M.

111. Still the outer rail of a curve, though elevated according to the preceding formula, is generally found to be much more worn than the inner rail On this account some are led to distrust the formula, and to give an increased elevation to the rail. So far, however, as the centrifugal force is concerned, the formula is undoubtedly correct, and the evil in question must arise from other causes, — causes which are not counteracted by an additional elevation of the outer rail. The principal of these causes is probably improper "coning" of the wheels. Two wheels, immovable on an axle, and of the same radius, must, if

no slip is allowed, pass over equal spaces in a given number of revolutions. Now as the outer rail of a curve is longer than the inner rail, the outer wheel of such a pair must on a curve fall behind the inner wheel. The first effect of this is to bring the flange of the outer wheel against the rail, and to keep it there. The second is a strain on the axle consequent upon a slip of the wheels equal in amount to the difference in length of the two rails of the curve. To remedy this, coning of the wheels was introduced, by means of which the radius of the outer wheel is in effect increased, the nearer its flange approaches the rail, and this wheel is thus enabled to traverse a greater distance than the inner wheel.

To find the amount of coning for a play of the wheels of one inch, let r and r' represent the proper radii of the inner and outer wheels respectively, when the flange of the outer wheel touches the rail. Then $r' - r$ will be the coning for one inch in breadth of the tire. To enable the wheels to keep pace with each other in traversing a curve, their radii must be proportional to the lengths of the two rails of the curve, or, which is the same thing, proportional to the radii of these rails. If R be taken as the radius of the inner rail, the radius of the outer rail will be $R + g$, and we shall have $r : r' = R : R + g$. Therefore, $r R + r g = r' R$, or

$$r' - r = \frac{r g}{R}.$$

As an example, let $R = 600$, $r = 1.4$, and $g = 4.7$. Then we have $r' - r = \frac{1.4 \times 4.7}{600} = 011$ ft. For a tire 3.5 in. wide, the coning would be $3.5 \times .011 = .0385$ ft., or nearly half an inch. Wheels coned to this amount would accommodate themselves to any curves of not less than 600 feet radius. On a straight line the flanges of the two wheels would be equally distant from the rails, making both wheels of the same diameter. On a curve of say 2400 feet radius, the flange of the outer wheel would assume a position one fourth of an inch nearer to the rail than the flange of the inner wheel, which would increase the radius of the outer wheel just one fourth of the necessary increase on a curve of 600 feet. Should the flange of the outer wheel get too near the rail, the disproportionate increase of the radius of this wheel would make it get the start of the inner wheel, and cause the flange to recede from the rail again. If the shortest radius were taken as 900 feet, r and g remaining the same, we should have $r' - r = \frac{1.4 \times 4.7}{900}$

= .0073, and for the coning of the whole tire 3.5 × .0073 = .0256 ft., or about three tenths of an inch. Wheels coned to this amount would accommodate themselves to any curve of not less than 900 feet radius. If the wheels are larger, the coning must be greater, or if the gauge of the track is wider, the coning must be greater. If the play of the wheels is greater, the coning may be diminished. Hence it might be advisable to increase the play of the wheels on short curves, by a slight increase of the gauge of the track.

Two distinct things, therefore, claim attention in regard to the motion of cars on a curve. The first is the centrifugal force, which is generated in all cases, when a body is constrained to move in a curvilinear path, and which may be effectually counteracted for any given velocity by elevating the outer rail. The second is the unequal length of the two rails of a curve, in consequence of which two wheels fixed on an axle cannot traverse a curve properly, unless some provision is made for increasing the diameter of the outer wheel. Coning of the wheels seems to be the only thing yet devised for obtaining this increase of diameter. At present, however, there is little regularity either in the coning itself, or in the distance between the flanges of wheels for tracks of the same gauge. The tendency has been to diminish the coning,* without substituting any thing in its place. If the wheels could be made to turn independently of each other, the whole difficulty would vanish ; but if this is thought to be impracticable, the present method ought at least to be reduced to some system.

---

* Bush and Lobdell, extensive wheel-makers, say, in a note published in Appletons' Mechanic's Magazine for August, 1852, that wheels made by them for the New York and Erie road have a coning of but one sixteenth of an inch. This coning on a track of six feet gauge with the other data as given above, would suit no curve of less than a mile radius.

CHAPTER IV.

EARTH-WORK.

Article I. — Prismoidal Formula.

112. Earth-work includes the regular excavation and embankment on the line of a road, borrow-pits, or such additional excavations as are made necessary when the embankment exceeds the regular excavation, and, in general, any transfers of earth that require calculation. We begin with the prismoidal formula, as this formula is frequently used in calculating cubical contents both of earth and masonry.

A *prismoid* is a solid having two parallel faces, and composed of prisms, wedges, and pyramids, whose common altitude is the perpendicular distance between the parallel faces.

113. **Problem.** *Given the areas of the parallel faces B and B', the middle area M, and the altitude a of a prismoid, to find its solidity S.*

Solution. The middle area of a prismoid is the area of a section midway between the parallel faces and parallel to them, and the altitude is the perpendicular distance between the parallel faces. If now b represents the base of any prism of altitude a, its solidity is $a\,b$. If b represents the base of a regular wedge or half-parallelopipedon of altitude a, its solidity is $\frac{1}{2} a b$. If b represents the base of a pyramid of altitude a, its solidity is $\frac{1}{3} a\,b$. The solidity of these three bodies admits of a common expression, which may be found thus. Let m represent the middle area of either of these bodies, that is, the area of a section parallel to the base and midway between the base and top. In the prism, $m = b$, in the regular wedge, $m = \frac{1}{2} b$, and in the pyramid, $m = \frac{1}{4} b$. Moreover, the upper base of the prism $= b$, and the upper base of the wedge or pyramid $= 0$. Then the expressions $a\,b$, $\frac{1}{2} a\,b$, and $\frac{1}{3} a b$ may be thus transformed. Solidity of

$$\text{prism} \quad = a\,b = \frac{a}{6} \times 6\,b = \frac{a}{6}(b + b + 4\,b) = \frac{a}{6}(b + b + 4\,m),$$

$$\text{wedge} \quad = \tfrac{1}{2}\,a\,b = \frac{a}{6} \times 3\,b = \frac{a}{6}(0 + b + 2\,b) = \frac{a}{6}(0 + b + 4\,m),$$

$$\text{pyramid} = \tfrac{1}{3}\,a\,b = \frac{a}{6} \times 2\,b = \frac{a}{6}(0 + b + b) \quad = \frac{a}{6}(0 + b + 4\,m).$$

Hence, the solidity of either of these bodies is found by adding together the area of the upper base, the area of the lower base, and four times the middle area, and multiplying the sum by one sixth of the altitude. Irregular wedges, or those not half-parallelopipedons, may be measured by the same rule, since they are the sum or difference of a regular wedge and a pyramid of common altitude, and as the rule applies to both these bodies, it applies to their sum or difference.

Now a prismoid, being made up of prisms, wedges, and pyramids of common altitude with itself, will have for its solidity the sum of the solidities of the combined solids. But the sum of the areas of the upper and lower bases of the combined solids is equal to $B + B'$, the sum of the areas of the parallel faces of the prismoid; and the sum of the middle areas of the combined solids is equal to M, the middle area of the prismoid. Therefore

☞ $$S = \frac{a}{6}(B + B' + 4M).$$

Article II. — Borrow-Pits.

114. For the measurement of small excavations, such as borrow-pits, &c., the usual method of preparing the ground is to divide the surface into parallelograms * or triangles, small enough to be considered planes, laid off from a base line, that will remain untouched by the excavation. A convenient bench-mark is then selected, and levels taken at all the angles of the subdivisions. After the excavation is made, the same subdivisions are laid off from the base line upon the bottom of the excavation, and levels referred to the same bench-mark are taken at all the angles.

This method divides the excavation into a series of vertical prisms, generally truncated at top and bottom. The vertical edges of these prisms are known, since they are the differences of the levels at the top and bottom of the excavation. The horizontal section of the prisms is also known, because the parallelograms or triangles, into which the surface is divided, are always measured horizontally.

115. **Problem.** *Given the edges h, h_1, and h_2, to find the solidity*

---

* If the ground is divided into rectangles, as is generally done, and one side be made 27 feet, or some multiple of 27 feet, the contents may be obtained at once in cubic yards, by merely omitting the factor 27 in the calculation.

S of a vertical prism, whether truncated or not, whose horizontal section is a triangle of given area A.

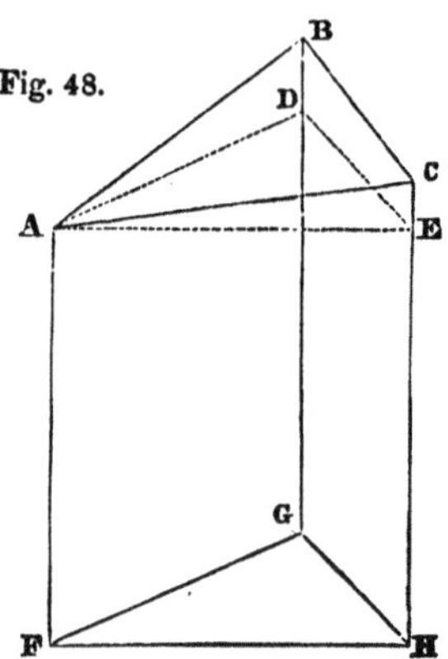

Fig. 48.

Solution. When the prism is not truncated, we have $h = h_1 = h_2$. The ordinary rule for the solidity of a prism gives, therefore, $S = A h = A \times \frac{1}{3}(h + h_1 + h_2)$. When the prism is truncated, let $A B C \cdot F G H$ (fig. 48) represent such a prism, truncated at the top. Through the lowest point A of the upper face draw a horizontal plane $A D E$ cutting off a pyramid, of which the base is the trapezoid $B D E C$, and the altitude a perpendicular let fall from A on $D E$. Represent this perpendicular by p, and we have (Tab. X. 52) the solidity of the pyramid $= \frac{1}{3} p \times B D E C = \frac{1}{3} p \times D E \times \frac{1}{2}(B D + C E) = \frac{1}{2} p \times D E \times \frac{1}{3}(B D + C E) = A \times \frac{1}{3}(B D + C E)$, since $\frac{1}{2} p \times D E = A D E = A$. But $\frac{1}{3}(B D + C E)$ is the mean height of the vertical edges of the truncated portion, the height at A being 0. Hence the formula already found for a prism not truncated, will apply to the portion above the plane $A D E$, as well as to that below. The same reasoning would apply, if the lower end also were truncated. Hence, for the solidity of the whole prism, whether truncated or not, we have

$$S = A \times \tfrac{1}{3}(h + h_1 + h_2).$$

116. **Problem.** *Given the edges h, h_1, h_2, and h_3, to find the solidity S of a vertical prism, whether truncated or not, whose horizontal section is a parallelogram of given area A.*

Solution. Let $B\,H$ (fig. 49) represent such a prism, whether truncated or not, and let the plane $B\,F\,H\,D$ divide it into two triangular

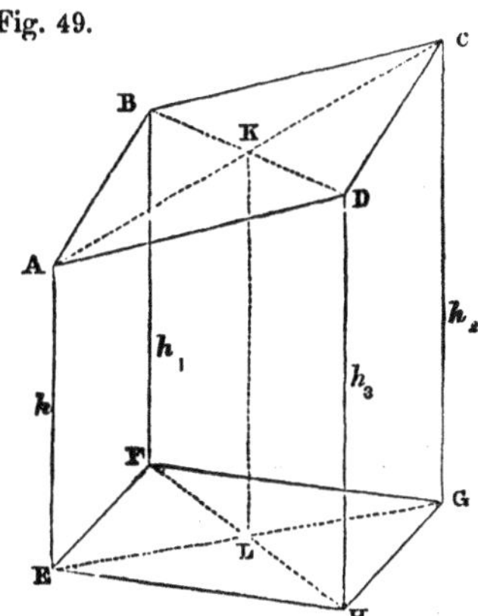

Fig. 49.

prisms $A\,F\,H$ and $C\,F\,H$. The horizontal section of each of these prisms will be $\frac{1}{2}A$, and if h, h_1, h_2, and h_3 represent the edges to which they are attached in the figure, we have for their solidity (§ 115), $A\,F\,H = \frac{1}{2}A \times \frac{1}{3}(h + h_1 + h_3)$, and $C\,F\,H = \frac{1}{2}A \times \frac{1}{3}(h_1 + h_2 + h_3)$. Therefore, the whole prism will have for its solidity $S = \frac{1}{2}A \times \frac{1}{3}(h + 2\,h_1 + h_2 + 2\,h_3)$. Let the whole prism be again divided by the plane $A\,E\,G\,C$ into two triangular prisms $B\,E\,G$ and $D\,E\,G$ Then we have for these prisms, $B\,E\,G = \frac{1}{2}A \times \frac{1}{3}(h + h_1 + h_2)$, and $D\,E\,G = \frac{1}{2}A \times \frac{1}{3}(h + h_2 + h_3)$, and for the whole prism, $S = \frac{1}{2}A \times \frac{1}{3}(2\,h + h_1 + 2\,h_2 + h_3)$. Adding the two expressions found for S, we have $2\,S = \frac{1}{2}A\,(h + h_1 + h_2 + h_3)$, or

☞ $$S = A \times \tfrac{1}{4}(h + h_1 + h_2 + h_3).$$

It will be seen by the figure, that $\frac{1}{2}(h + h_2) = K\,L = \frac{1}{2}(h_1 + h_3)$, or $h + h_2 = h_1 + h_3$. The expression for S might, therefore, be reduced to $S = A \times \frac{1}{2}(h + h_2)$, or $S = A \times \frac{1}{2}(h_1 + h_3)$. But as the ground surfaces $A\,B\,C\,D$ and $E\,F\,G\,H$ are seldom perfect planes, it is considered better to use the mean of the four heights, instead of the mean of two diagonally opposite.

117. **Corollary.** When all the prisms of an excavation have the same horizontal section A, the calculation of any number of them

may be performed by one operation. Let figure 50 be a plan of such an excavation, the heights at the angles being denoted by a, a_1, a_2, b

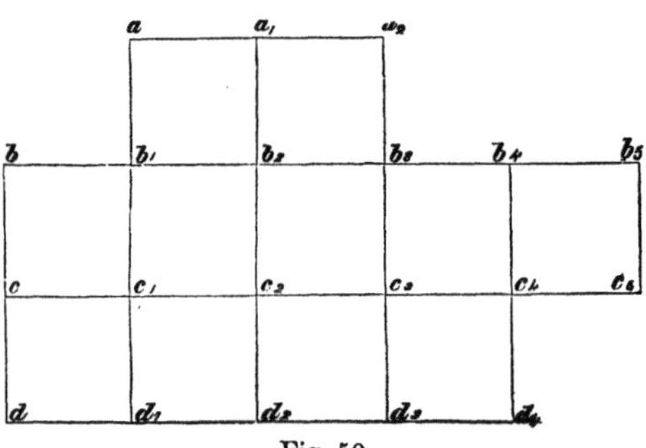

Fig. 50.

b_1, &c. Then the solidity of the whole will be equal to $\frac{1}{4} A$ multiplied by the sum of the heights of the several prisms (§ 116). Into this sum the corner heights a, a_2, b, b_5, c_5, d, and d_4 will enter but *once*, each being found in but one prism; the heights a_1, b_4, c, d_1, d_2, and d_3 will enter *twice*, each being common to two prisms; the heights b_1, b_3, and c_4 will enter *three* times, each being common to three prisms, and the heights b_2, c_1, c_2, and c_3 will enter *four* times, each being common to four prisms. If, therefore, the sum of the first set of heights is represented by s_1, the sum of the second by s_2, of the third by s_3, and of the fourth by s_4, we shall have for the solidity of all the prisms

☞ $$S = \frac{1}{4} A \,(s_1 + 2\,s_2 + 3\,s_3 + 4\,s_4).$$

Article III. — Excavation and Embankment.

118. As embankments have the same general shape as excavations, it will be necessary to consider excavations only. The simplest case is when the ground is considered level on each side of the centre line. Figure 51 represents the mass of earth between two stations in an excavation of this kind. The trapezoid $G\,B\,F\,H$ is a section of the mass at the first station, and $G_1\,B_1\,F_1\,H_1$ a section at the second station; $A\,E$ is the centre height at the first station, and $A_1\,E_1$ the centre height at the second station; $H\,H_1\,F_1\,F$ is the road-bed, $G\,G_1\,B_1\,B$ the

surface of the ground, and $G\,G_1\,H_1\,H$ and $B\,B_1\,F_1\,F$ the planes forming the side slopes. This solid is a prismoid, and might be calculated by the prismoidal formula (§ 113). The following method gives the same result.

A. *Centre Heights alone given.*

119. **Problem.** *Given the centre heights c and c_1, the width of the road-bed b, the slope of the sides s, and the length of the section l, to find the solidity S of the excavation.*

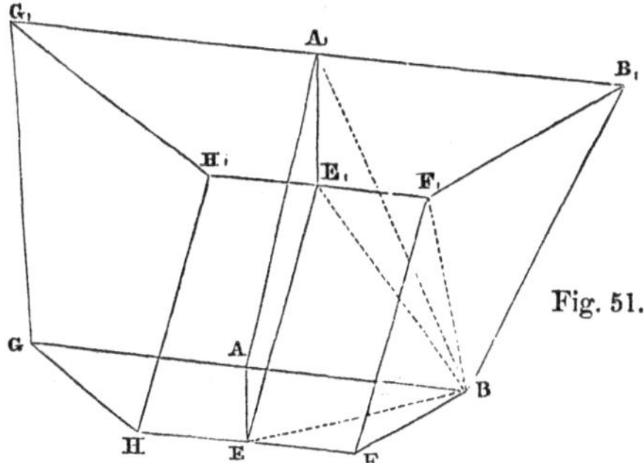

Fig. 51.

Solution. Let c be the centre height at A (fig. 51) and c_1 the height at A_1. The slope s is the ratio of the base of the slope to its perpendicular height (§ 102). We have then the distance out $A\,B = \frac{1}{2}b + s\,c$, and the distance out $A_1\,B_1 = \frac{1}{2}\,b + s\,c_1$ (§ 102). Divide the whole mass into two equal parts by a vertical plane $A\,A_1\,E_1\,E$ drawn through the centre line, and let us find first the solidity of the right-hand half. Through B draw the planes $B\,E\,E_1$, $B\,A_1\,E_1$, and $B\,E_1\,F_1$, dividing the half-section into three quadrangular pyramids, having for their common vertex the point B, and for their bases the planes $A\,A_1\,E_1\,E$, $E\,E_1\,F_1\,F$, and $A_1\,B_1\,F_1\,E_1$. For the areas of these bases we have

$$\text{Area of } A\,A_1\,E_1\,E = \tfrac{1}{2}E\,E_1 \times (A\,E + A_1\,E_1) = \tfrac{1}{2}\,l\,(c + c_1),$$
$$\text{`` `` } E\,E_1\,F_1\,F = E\,F \times E\,E_1 = \tfrac{1}{2}\,b\,l,$$
$$\text{`` `` } A_1\,B_1\,F_1\,E_1 = \tfrac{1}{2}\,A_1\,E_1 \times (E_1\,F_1 + A_1\,B_1) = \tfrac{1}{2}\,(b\,c_1 + s\,c_1^2);$$

and for the perpendiculars from the vertex B on these bases, produced when necessary,

Perpendicular on $A A_1 E_1 E \;\; = A B \;\; = \frac{1}{2} b + s c$,
" " $E E_1 F_1 F \;\; = A E \;\; = c$,
" " $A_1 B_1 F_1 E_1 = E E_1 = l$.

Then (Tab. X. 52) the solidities of the three pyramids are

$$B - A A_1 E_1 E \;\; = \tfrac{1}{3} (\tfrac{1}{2} b + s c) \times \tfrac{1}{2} l (c + c_1) = \tfrac{1}{6} l (\tfrac{1}{2} b c + \tfrac{1}{2} b c_1 + s c^2 + s c c_1)$$

$$B - E E_1 F_1 F \;\; = \tfrac{1}{3} c \times \tfrac{1}{2} b l \qquad = \tfrac{1}{6} l b c,$$

$$B - A_1 B_1 F_1 E_1 = \tfrac{1}{3} l \times \tfrac{1}{2} (b c_1 + s c_1^2) \qquad = \tfrac{1}{6} l (b c_1 + s c_1^2).$$

Their sum, or the solidity of the half-section, is

$$\tfrac{1}{2} S = \tfrac{1}{6} l \, [\tfrac{3}{2} b \, (c + c_1) + s \, (c^2 + c_1^2 + c c_1)].$$

Therefore the solidity of the whole section is

$$S = \tfrac{1}{3} l \, [\tfrac{3}{2} b \, (c + c_1) + s \, (c^2 + c_1^2 + c c_1)],$$

or

☞ $$S = \tfrac{1}{2} l \, [b \, (c + c_1) + \tfrac{2}{3} s \, (c^2 + c_1^2 + c c_1)].$$

When the slope is $1\frac{1}{2}$ to 1, $s = \frac{3}{2}$, and the factor $\frac{2}{3} s = 1$ may be dropped.

120. **Problem.** *To find the solidity S of any number n of successive sections of equal length.*

Solution. Let c, c_1, c_2, c_3, &c. denote the centre heights at the successive stations. Then we have (§ 119)

Solidity of first section $= \frac{1}{2} l \, [b \, (c + c_1) + \frac{2}{3} s \, (c^2 + c_1^2 + c \, c_1)]$,
" " second section $= \frac{1}{2} l \, [b \, (c_1 + c_2) + \frac{2}{3} s \, (c_1^2 + c_2^2 + c_1 c_2)]$,
" " third section $= \frac{1}{2} l \, [b \, (c_2 + c_3) + \frac{2}{3} s \, (c_2^2 + c_3^2 + c_2 c_3)]$
&c. &c.

For the solidity of any number n of sections, we should have $\frac{1}{2} l$ multiplied by the sum of the quantities in n parentheses formed as those just given. The last centre height, according to the notation adopted will be represented by c_n, and the next to the last by c_{n-1}. Collecting the terms multiplied by b into one line, the squares multiplied by $\frac{2}{3} s$ into a second line, and the remaining terms into a third line, we have for the solidity of n sections

☞ $$S = \tfrac{1}{2} l \left| \begin{array}{l} b \, (c + 2 c_1 + 2 c_2 + 2 c_3 \ldots\ldots + 2 c_{n-1} + c_n) \\ + \tfrac{2}{3} s \, (c^2 + 2 c_1^2 + 2 c_2^2 + 2 c_3^2 \ldots\ldots + 2 c_{n-1}^2 + c_n^2) \\ + \tfrac{2}{3} s \, (c \, c_1 + c_1 c_2 + c_2 c_3 + c_3 c_4 \ldots\ldots + c_{n-1} \, c_n). \end{array} \right.$$

When $s = \frac{3}{2}$, the factor $\frac{2}{3} s = 1$ may be dropped.

Example. Given $l = 100$, $b = 28$, $s = \frac{3}{2}$, and the stations and centre heights as set down in the first and second columns of the annexed table. The calculation is thus performed. Square the heights, and set the squares in the third column. Form the successive products $c c_1$, $c_1 c_2$, &c., and place them in the fourth column. Add up the last three columns. To the sum of the second column add the sum itself, minus the first and the last height, and to the sum of the third column add the sum itself, minus the first and the last square. Then 86 is the multiplier of b in the first line of the formula, 592 is the second line, since $\frac{2}{3} s$ is here 1, and 274 is the third line. The product of 86 by $b = 28$ is 2408, and the sum of 274, 592, and 2408 is 3274. This multiplied by $\frac{1}{2} l = 50$ gives for the solidity 163,700 cubic feet.

Station.	c.	c^2.	$c c_1$.
0	2	4	
1	4	16	8
2	7	49	28
3	6	36	42
4	10	100	60
5	7	49	70
6	6	36	42
7	4	16	24
	46	306	274
	40	286	592
	86	592	2408
	28		2)3274
	2408		163700.

B. *Centre and Side Heights given.*

121 When greater accuracy is required than can be attained by the preceding method, the side heights and the distances out (§ **102**) are introduced. Let figure 52 represent the right-hand side of an excavation between two stations. $A A_1 B_1 B$ is the ground surface; $A E = c$ and $A_1 E_1 = c_1$ are the centre heights; $B G = h$ and $B_1 G_1 = h_1$, the side heights; and d and d_1, the distances out, or the horizontal distances of B and B_1 from the centre line. The whole ground surface may sometimes be taken as a plane, and sometimes the part on each side of the centre line may be so taken; * but neither of these suppo-

* It is easy in any given case to ascertain whether a surface like $A A_1 B_1 B$ is a

sitions is sufficiently accurate to serve as the basis of a general method. In most cases, however, we may consider the surface on each side of the centre line to be divided into two triangular planes by a diagonal passing from one of the centre heights to one of the side heights. A ridge or depression will, in general, determine which diagonal ought to be taken as the dividing line, and this diagonal must be noted in the field. Thus, in the figure a ridge is supposed to run from B to A_1, from which the ground slopes downward on each side to A and B_1. Instead of this, a depression might run from A to B_1, and the ground rise each way to A_1 and B. If the ridge or depression is very marked, and does not cross the centre or side lines at the regular stations, intermediate stations must be introduced to make the triangular planes conform better to the nature of the ground. If the surface happens to be a plane, or nearly so, the diagonal may be taken in either direction. It will be seen, therefore, that the following method is applicable to all ordinary ground. When, however, the ground is very irregular, the method of § 127 is to be used.

122. **Problem.** *Given the centre heights c and c_1, the side heights on the right h and h_1, on the left h' and h'_1, the distances out on the right d and d_1, on the left d' and d'_1, the width of the road-bed b, the length of the section l, and the direction of the diagonals, to find the solidity S of the excavation.*

Solution. Let figure 52 represent the right-hand side of the excavation, and let us suppose first, that the diagonal runs, as shown in the figure, from B to A_1. Through B draw the planes $B E E_1$, $B A_1 E_1$, and $B E_1 F_1$, dividing the half-section into three quadrangular pyramids, having for their common vertex the point B, and for their bases the planes $A A_1 E_1 E$, $E E_1 F_1 F$, and $A_1 B_1 F_1 E_1$. For the areas of these bases we have

$$\text{Area of } A A_1 E_1 E = \tfrac{1}{2} E E_1 \times (A E + A_1 E_1) = \tfrac{1}{2} l (c + c_1),$$
$$\text{" " } E E_1 F_1 F = E F \times E E_1 = \tfrac{1}{2} b l,$$
$$\text{" " } A_1 B_1 F_1 E_1 = \tfrac{1}{2} A_1 E_1 \times d_1 + \tfrac{1}{2} E_1 F_1 \times h_1 = \tfrac{1}{2} d_1 c_1 + \tfrac{1}{4} b h_1,$$

and for the perpendiculars from the vertex B on these bases, produced when necessary,

---

plane; for if it is a plane, the descent from A to B will be to the descent from A_1 to B_1, as the distance out at the first station is to the distance out at the second station, that is, $c - h : c_1 - h_1 = d : d_1$. If we had $c = 9$, $h = 6$, $c_1 = 12$, $h_1 = 8$, $d = 24$, and $d_1 = 27$, the formula would give $3 : 4 = 24 : 27$, which shows that the surface is not a plane.

Perpendicular on $A\,A_1\,E_1\,E \;= E\,G \;= d$,
" " $E\,E_1\,F_1\,F \;= B\,G \;= h$,
" " $A_1\,B_1\,F_1\,E_1 = E\,E_1 = l$.

Fig. 52.

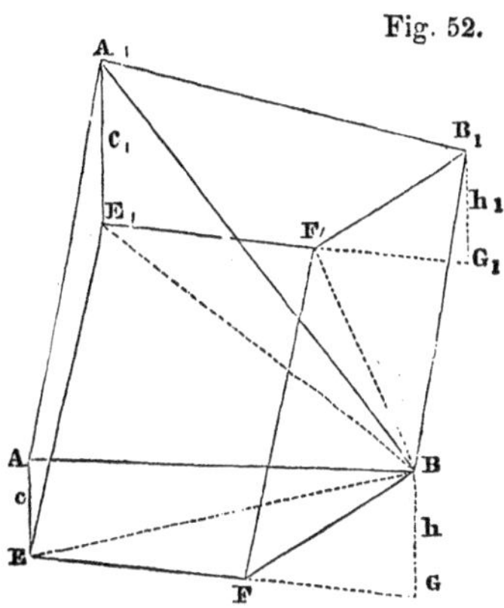

Then (Tab. X. 52) the solidities of the three pyramids are

$$B - A\,A_1\,E_1\,E \;= \tfrac{1}{3}d \times \tfrac{1}{2}l\,(c + c_1) \qquad = \tfrac{1}{6}l\,(d\,c + d\,c_1),$$
$$B - E\,E_1\,F_1\,F \;= \tfrac{1}{3}h \times \tfrac{1}{2}b\,l \qquad = \tfrac{1}{6}l\,b\,h,$$
$$B - A_1\,B_1\,F_1\,E_1 = \tfrac{1}{3}l \times \tfrac{1}{2}(d_1\,c_1 + \tfrac{1}{2}b\,h_1) = \tfrac{1}{6}l\,(d_1\,c_1 + \tfrac{1}{2}b\,h_1).$$

Their sum, or the solidity of the half-section, is

$$\tfrac{1}{6}l\,(d\,c + d_1\,c_1 + d\,c_1 + b\,h + \tfrac{1}{2}b\,h_1). \qquad (1)$$

Next, suppose that the diagonal runs from A to B_1. In this case, hrough B_1 draw the planes $B_1\,E_1\,E$, $B_1\,A\,E$, and $B_1\,E\,F$ (not rep·'esented in the figure), dividing the half-section again into three [uadrangular pyramids, having for their common vertex the point 3$_1$, and for their bases the planes $A\,A_1\,E_1\,E$, $E\,E_1\,F_1\,F$, and $A\,B\,F\,E$. ?or the areas of these bases we have

Area of $A\,A_1\,E_1\,E = \tfrac{1}{2}E\,E_1 \times (A\,E + A_1\,E_1) = \tfrac{1}{2}l\,(c + c_1)$,
" " $E\,E_1\,F_1\,F \;= E\,F \times E\,E_1 \qquad = \tfrac{1}{2}b\,l$,
" " $A\,B\,F\,E \;= \tfrac{1}{2}A\,E \times d + \tfrac{1}{2}E\,F \times h \;= \tfrac{1}{2}d\,c + \tfrac{1}{4}b\,h$;

nd for the perpendiculars from B_1 on these bases, produced when ecessary,

$$\begin{aligned} \text{Perpendicular on } A A_1 E_1 E &= E_1 G_1 = d_1, \\ \text{" \quad " } E E_1 F_1 F &= B_1 G_1 = h_1, \\ \text{" \quad " } A B F E &= E E_1 = l. \end{aligned}$$

Then (Tab. X. 52) the solidities of the three pyramids are

$$\begin{aligned} B_1 - A A_1 E_1 E &= \tfrac{1}{3} d_1 \times \tfrac{1}{2} l (c + c_1) &&= \tfrac{1}{6} l (d_1 c + d_1 c_1), \\ B_1 - E E_1 F_1 F &= \tfrac{1}{3} h_1 \times \tfrac{1}{2} b l &&= \tfrac{1}{6} l b h_1, \\ B_1 - A B F E &= \tfrac{1}{3} l \times \tfrac{1}{2} (d c + \tfrac{1}{2} b h) &&= \tfrac{1}{6} l (d c + \tfrac{1}{2} b h). \end{aligned}$$

Their sum, or the solidity of the half-section, is

$$\tfrac{1}{6} l (d c + d_1 c_1 + d_1 c + b h_1 + \tfrac{1}{2} b h). \qquad (2)$$

We have thus found the solidity of the half-section for both directions of the diagonal. Let us now compare the results (1) and (2), and express them, if possible, by one formula. For this purpose let (1) be put under the form

$$\tfrac{1}{6} l [d c + d_1 c_1 + d c_1 + \tfrac{1}{2} b (h + h_1 + h)],$$

and (2) under the form

$$\tfrac{1}{6} l [d c + d_1 c_1 + d_1 c + \tfrac{1}{2} b (h + h_1 + h_1)].$$

The only difference in these two expressions is, that $d c_1$ and the last h in the first, become $d_1 c$ and h_1 in the second. But in the first case, c_1 and h are the heights at the extremities of the diagonal, and d is the distance out corresponding to h; and in the second case, c and h_1 are the heights at the extremities of the diagonal, and d_1 is the distance out corresponding to h_1. *Denote the centre height touched by the diagonal by C, the side height touched by the diagonal by H, and the distance out corresponding to the side height H by D.* We may then express both $d c_1$ and $d_1 c$ by $D C$, and both h and h_1 by H; so that the solidity of the half-section on the right of the centre line, whichever way the diagonal runs, may be expressed by

$$\tfrac{1}{6} l [d c + d_1 c_1 + D C + \tfrac{1}{2} b (h + h_1 + H)]. \qquad (3)$$

To obtain the contents of the portion on the left of the centre line, we designate the quantities on the left by the same letters used for corresponding quantities on the right, merely attaching a ($'$) to them to distinguish them. Thus the side heights are h' and h'_1, and the distances out d' and d'_1, while D, C, and H become D', C', and H'. The solidity of the half-section on the left may therefore be taken directly from (3), which will become

$$\tfrac{1}{6} l \left[d' c + d'_1 c_1 + D' C' + \tfrac{1}{2} b (h' + h'_1 + H')\right]. \quad (4)$$

Finally, by uniting (3) and (4), we obtain the following formula for the solidity of the whole section between two stations

$$S = \tfrac{1}{6} l \left[(d + d') c + (d_1 + d'_1) c_1 + D C + D' C' + \tfrac{1}{2} b (h + h_1 + H + h' + h'_1 + H')\right].$$

Example. Given $l = 100$, $b = 18$, and the remaining data, as arranged in the first six columns of the following table. The first column gives the stations; the fourth gives the centre heights, namely, $c = 13.6$ and $c_1 = 8$; the two columns on the left of the centre heights give the side heights and distances out on the left of the centre line of the road, and the two columns on the right of the centre heights give the side heights and distances out on the right. The direction of the diagonals is marked by the oblique lines drawn from $h' = 8$ to $c_1 = 8$ and from $c = 13.6$ to $h_1 = 12$.

Sta.	d'.	h'.	c.	h.	d.	$d+d'$.	$(d+d')c$.	$D' C'$	$D C$.
0	21	8 ╲	13.6 ╲	10	24	45	612		
1	15	4	╲ 8.0	╲ 12	27	42	336	168	367.2
		12		12			168		
				20			367.2		
				54 × 9 =			486		
							6)1969.20		
							32820.		

To apply the formula, the distances out at each station are added together, and their sum placed in the seventh column; these sums, multiplied by the respective centre heights, are placed in the eighth column; the product of $d' = 21$ (which is the distance out corresponding to the side height touched by the left-hand diagonal) by $c_1 = 8$ (which is the centre height touched by the same diagonal) is placed in the ninth column, and the similar product of $d_1 = 27$ by $c = 13.6$ is placed in the last column. The terms in the formula multiplied by $\tfrac{1}{2} b$ are all the side heights, and in addition all the side heights touched by diagonals, or $8 + 4 + 10 + 12 + 8 + 12 = 54$. Then by substitution in the formula, we have $S = \tfrac{1}{6} \times 100\ (612 + 336 + 168 + 367.2 + 9 \times 54) = 32{,}820$ cubic feet.*

* The example here given is the same as that calculated in Mr. Borden's "Sys-

By applying the rule given in the note to § 121, we see that the surface on the left of the centre line in the preceding example is a plane; since $13.6 - 8 : 8 - 4 = 21 : 15$. The diagonal on that side might, therefore, be taken either way, and the same solidity would be obtained. This may be easily seen by reversing the diagonal in this example, and calculating the solidity anew. The only parts of the formula affected by the change are $D' C'$ and $\frac{1}{2} b H'$. In the one case the sum of these terms is $21 \times 8 + 9 \times 8$, and in the other $15 \times 13.6 + 9 \times 4$, both of which are equal to 240.

123 **Problem.** *To find the solidity S of any number n of successive sections of equal length.*

Solution. Let c, c_1, c_2, c_3, &c. be the centre heights at the successive stations; h, h_1, h_2, h_3, &c. the right-hand side heights; h', h'_1, h'_2, h'_3, &c. the left-hand side heights; d, d_1, d_2, d_3, &c. the distances out on the right; and d', d'_1, d'_2, d'_3, &c. the distances out on the left. Then the formula for the solidity of one section (§ 122) gives for the solidities of the successive sections

$$\tfrac{1}{6} l \left[(d + d') c + (d_1 + d'_1) c_1 + D C + D' C' + \tfrac{1}{2} b (h + h_1 + H + h' + h'_1 + H')\right],$$

$$\tfrac{1}{6} l \left[(d_1 + d'_1) c_1 + (d_2 + d'_2) c_2 + D_1 C_1 + D'_1 C'_1 + \tfrac{1}{2} b (h_1 + h_2 + H_1 + h'_1 + h'_2 + H'_1)\right],$$

$$\tfrac{1}{6} l \left[(d_2 + d'_2) c_2 + (d_3 + d'_3) c_3 + D_2 C_2 + D'_2 C'_2 + \tfrac{1}{2} b (h_2 + h_3 + H_2 + h'_2 + h'_3 + H'_2)\right],$$

and so on, for any number of sections. For the solidity of any number n of sections, we should have $\frac{1}{6} l$ multiplied by the sum of n parentheses formed as those just given. Hence

$$☞ \quad S = \tfrac{1}{6} l \left| \begin{array}{l} (d + d') c + 2 (d_1 + d'_1) c_1 + 2 (d_2 + d'_2) c_2 \ldots + (d_n + d'_n) c_n \\ + DC + D'C' + D_1 C_1 + D'_1 C'_1 + D_2 C_2 + D'_2 C'_2 + \text{\&c.} \\ + \tfrac{1}{2} b \left| \begin{array}{l} h + 2 h_1 + 2 h_2 \ldots\ldots + h_n + H + H_1 + H_2 + \text{\&c.} \\ + h' + 2 h'_1 + 2 h'_2 \ldots + h'_n + H' + H'_1 + H'_2 + \text{\&c.} \end{array} \right. \end{array} \right.$$

tem of Useful Formulæ, &c ," page 187. It will be seen, that his calculation makes the solidity 32,460 cubic feet, which is 360 cubic feet less than the result above. This difference is owing to the omission, by Mr. Borden's method, of a pyramid inclosed by the four pyramids, into which the upper portion of the right-hand half section is by that method divided.

Example. Given $l = 100$, $b = 28$, and the remaining data as given in the first six columns of the following table.

Sta.	d'.	h'.	c.	h.	d.	$d + d'$.	$(d + d')\,c$.	$D'C'$.	$D\,C$.
0	17	2	2	2	17	34	68		
1	18.5	3	4	5	21.5	40	160	68	43
2	20	4	5	6	23	43	215	80	92
3	23	6	6	8	26	49	294	115	130
4	21.5	5	6	7	24.5	46	276	129	147
5	20	4	6	4	20	40	240	120	147
6	15.5	1	4	3	18.5	34	136	93	80
		25		35			1389	605	639
		22		30			1185		
		22		37			605		
		69		102			639		
		102					2394		
		171 × 14 = 2394					6)6212		
							103533 cubic feet.		

The data in this table are arranged precisely as in the example for calculating one section (§ 122), and the remaining columns are calculated as there shown. Then, to obtain the first line of the formula, add all the numbers in the column headed $(d + d')\,c$, making 1389, and afterwards all the numbers except the first and the last, making 1185. The next line of the formula is the sum of the columns $D'\,C'$ and $D\,C$, which give respectively 605 and 639. To obtain the first line of the quantities multiplied by $\frac{1}{2}b$, add all the numbers in column h, making 35, next all the numbers except the first and the last, making 30, and lastly all the numbers touched by diagonals (doubling any one touched by two diagonals), making 37. The second line of the quantities multiplied by $\frac{1}{2}b$ is obtained in the same way from the column marked h'. The sum of these numbers is 171, and this multiplied by $\frac{1}{2}b = 14$ gives 2394. We have now for the first line of the formula $1389 + 1185$, for the second $605 + 639$, and for the remainder 2394. By adding these together, and multiplying the sum by $\frac{1}{6}l = \frac{100}{6}$, we get the contents of the six sections in feet.

124. When the section is partly in excavation and partly in embankment, the preceding formulæ are still applicable; but as this application introduces minus quantities into the calculation, the following method, similar in principle, is preferable.

125. **Problem.** *Given the widths of an excavation at the road-bed*

$AF = w$ and $A_1 F_1 = w_1$ (fig. 53), the side heights h and h_1, the length of the section l, and the direction of the diagonal, to find the solidity S of the excavation, when the section is partly in excavation and partly in embankment.

Fig. 53.

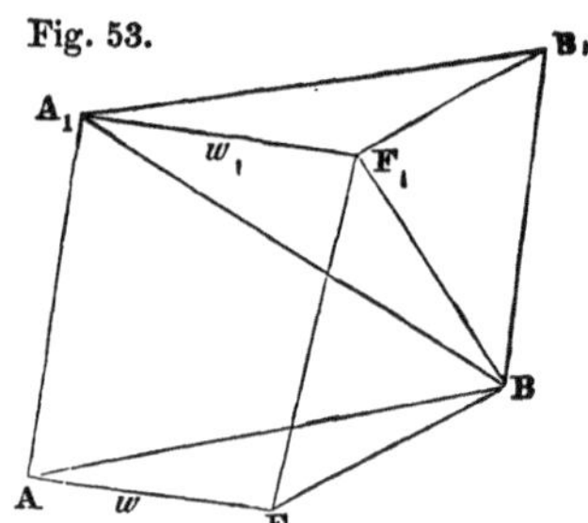

Solution. Suppose, first, that the surface is divided into two triangles by the diagonal BA_1. Through B draw the plane BA_1F_1, dividing that part of the section which is in excavation into two pyramids $B-AA_1F_1F$ and $B-A_1B_1F_1$, the solidities of which are

$$B-AA_1F_1F = \tfrac{1}{3}h \times \tfrac{1}{2}l(w+w_1) = \tfrac{1}{6}l(wh + w_1 h),$$
$$B-A_1B_1F_1 = \tfrac{1}{3}l \times \tfrac{1}{2}w_1 h_1 \qquad = \tfrac{1}{6}l w_1 h_1.$$

The whole solidity is, therefore,

$$S = \tfrac{1}{6}l(wh + w_1 h_1 + w_1 h).$$

Next, suppose the dividing diagonal to run from A to B_1. Through B_1 draw a plane B_1AF (not represented in the figure), dividing the excavation again into two pyramids, of which the solidities are

$$B_1-AA_1F_1F = \tfrac{1}{3}h_1 \times \tfrac{1}{2}l(w+w_1) = \tfrac{1}{6}l(wh_1 + w_1 h_1),$$
$$B_1-ABF \qquad = \tfrac{1}{3}l \times \tfrac{1}{2}wh \qquad = \tfrac{1}{6}lwh.$$

The whole solidity is, therefore,

$$S = \tfrac{1}{6}l(wh + w_1 h_1 + wh_1).$$

The only difference in these two expressions is, that $w_1 h$ in the first becomes wh_1 in the second. But in the first case the diagonal touches w_1 and h, and in the second case it touches w and h_1. If, then, we designate the width touched by the diagonal by W, and the height touched by the diagonal by H, we may express both $w_1 h$ and wh_1 by WH; so that the solidity in either case may be expressed by

☞ $$S = \tfrac{1}{6} l \,(w\,h + w_1 h_1 + W H).$$

Corollary. When several sections of equal length succeed one another, the whole may be calculated together. For this purpose, the preceding formula gives for the solidities of the successive sections

$$\tfrac{1}{6} l \,(w\,h \;\; + w_1 h_1 + W H),$$
$$\tfrac{1}{6} l \,(w_1 h_1 + w_2 h_2 + W_1 H_1),$$
$$\tfrac{1}{6} l \,(w_2 h_2 + w_3 h_3 + W_2 H_2),$$

and so on for any number of sections. Hence for the solidity of any number n of sections we should have

☞ $$S = \tfrac{1}{6} l \,(w\,h + 2\,w_1 h_1 + 2\,w_2 h_2 \ldots\ldots + w_n h_n + WH + W_1 H_1 + W_2 H_2 + \&c.)$$

Example. Given $l = 100$, and the remaining data as given in the first three columns of the following table.

Station.	w.	h.	$w\,h$.	WH.
0	2	1	2	
1	8	6	48	8
2	10	7	70	56
3	13	7	91	70
4	9	4	36	52
			247	186
			209	
			186	
			6)642	
			10700.	

The fourth column contains the products of the several widths by the corresponding heights, and the next column the products of those widths and heights touched by diagonals. The sum of the products in the fourth column is 247, the sum of all but the first and the last is 209, and the sum of the products in the fifth column is 186. These three sums are added together, multiplied by 100, and divided by 6, according to the formula. This gives the solidity of the four sections = 10700 cubic feet.

126. When the excavation does not begin on a line at right angles to the centre line, intermediate stations are taken where the excavation begins on each side of the road-bed, and the section may be calcu-

lated as a pyramid, having its vertex at the first of these points, and for its base the cross-section at the second. The preceding method gives the same result, since w and h in this case become 0, and reduce the formula to $S = \frac{1}{6} l w_1 h_1$. The same remarks apply to the end of an excavation.

C. *Ground very Irregular.*

127. **Problem.** *To find the solidity of a section, when the ground is very irregular.*

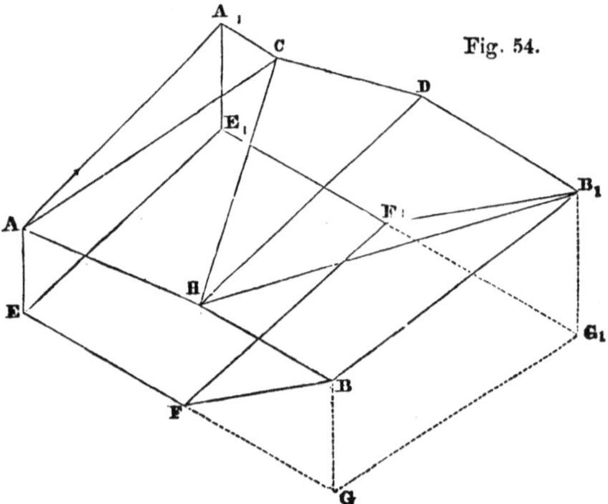

Fig. 54.

Solution. Let $A\,H\,B\,F\,E - A_1\ C\,D\,B_1\,F_1\,E_1$ (fig. 54) represent one side of a section, the surface of which is too irregular to be divided into two planes. Suppose, for instance, that the ground changes at H, C, and D, making it necessary to divide the surface into five triangles running from station to station.* Let heights be taken at H, C, and D, and let the distances out of these points be measured. If now we suppose the earth to be excavated vertically downward through the side line $B\,B_1$ to the plane of the road-bed, we may form as many vertical triangular prisms as there are triangles on the surface. This will be made evident by drawing vertical planes through the sides

* It will often be necessary to introduce intermediate stations, in order to make the subdivision into triangles more conveniently and accurately.

$A\,C$, $H\,C$, $H\,D$, and $H\,B_1$. *Then the solidity of the half-section will be equal to the sum of these prisms, minus the triangular mass* $B\,F\,G - B_1\,F_1\,G_1$.

The horizontal section of the prisms may be found from the distances out and the length of the section, and the vertical edges or heights are all known. Hence the solidities of these prisms may be calculated by § 115.

To find the solidity of the portion $B\,F\,G - B_1\,F_1\,G_1$, which is to be deducted, represent the slope of the sides by s (§ 102), the heights at B and B_1 by h and h_1, and the length of the section by l. Then we have $F\,G = s\,h$, and $F_1\,G_1 = s\,h_1$. Moreover, the area of $B\,F\,G = \frac{1}{2}\,s\,h^2$, and that of $B_1\,F_1\,G_1 = \frac{1}{2}\,s\,h_1^2$. Now as the triangles $B\,F\,G$ and $B_1\,F_1\,G_1$ are similar, the mass required is the frustum of a pyramid, and the mean area is $\sqrt{\frac{1}{2}\,s\,h^2 \times \frac{1}{2}\,s\,h_1^2} = \frac{1}{2}\,s\,h\,h_1$. Then (Tab. X 53) the solidity is $B\,F\,G - B_1\,F_1\,G_1 = \frac{1}{6}\,l\,s\,(h^2 + h_1^2 + h\,h_1)$.

Example. Given $l = 50$, $b = 18$, $s = \frac{3}{2}$, the heights at A, H, and B respectively 4, 7, and 6, the distances $A\,H = 9$ and $HB = 9$, the heights at A_1, C, D, and B_1 respectively 6, 7, 9, and 8, and the distances $A_1\,C = 4$, $C\,D = 5$, and $D\,B_1 = 12$ Then the horizontal section of the first prism adjoining the centre line is $\frac{1}{2}\,l \times A_1\,C$, since the distance $A_1\,C$ is measured horizontally; and the mean of the three heights is $\frac{1}{3}\,(4 + 6 + 7) = \frac{1}{3} \times 17$. The solidity of this prism is therefore $\frac{1}{2}\,l \times A_1\,C \times \frac{1}{3} \times 17 = \frac{1}{6}\,l \times 4 \times 17$, that is, equal to $\frac{1}{6}\,l$ multiplied by the base of the triangle and by the sum of the heights. In this way we should find for the solidity of the five prisms

$$\tfrac{1}{6}\,l\,(4 \times 17 + 9 \times 18 + 5 \times 23 + 12 \times 24 + 9 \times 21) = \tfrac{1}{6}\,l \times 822.$$

For the frustum to be deducted, we have

$$\tfrac{1}{6}\,l \times \tfrac{3}{2}\,(6^2 + 8^2 + 6 \times 8) = \tfrac{1}{6}\,l \times 222.$$

Hence the solidity of the half-section is

$$\tfrac{1}{6}\,l\,(822 - 222) = \tfrac{1}{6} \times 50 \times 600 = 5000 \text{ cubic feet.}$$

128. Let us now examine the usual method of calculating excavation, when the cross-section of the ground is not level. This method consists, first, in finding the area of a cross-section at each end of the mass; secondly, in finding the height of a section, *level at the top*, equivalent in area to each of these end sections; thirdly, in finding from the average of these two heights the middle area of the mass;

and, lastly, in applying the prismoidal formula to find the contents The heights of the equivalent sections level at the top may be found approximately by Trautwine's Diagrams,* or exactly by the following method. Let A represent the area of an irregular cross-section, b the width of the road-bed, and s the slope of the sides. Let x be the required height of an equivalent section level at the top. The bottom of the equivalent section will be b, the top $b + 2\,s\,x$, and the area will be the sum of the top and bottom lines multiplied by half the height or $\frac{1}{2}x\,(2\,b + 2\,s\,x) = s\,x^2 + b\,x$. But this area is to be equal to A Therefore, $s\,x^2 + b\,x = A$, and from this equation the value of x may be found in any given case.

According to this method, the contents of the section already calculated in § 122 will be found thus. Calculating the end areas, we find the first end area to be 387 and the second to be 240. Then as s is here $\frac{3}{2}$ and $b = 18$, the equations for finding the heights of the equivalent end sections will be $\frac{3}{2}x^2 + 18\,x = 387$, and $\frac{3}{2}x^2 + 18\,x = 240$. Solving these equations, we have for the height at the first station $x = 11.146$, and at the second, $x = 8$. The middle area will, therefore, have the height $\frac{1}{2}\,(11.146 + 8) = 9.573$, and from this height the middle area is found to be 309.78. Then by the prismoidal formula (§ 113) the solidity will be $S = \frac{1}{6} \times 100\,(387 + 240 + 4 \times 309.78) = 31102$ cubic feet.

But the true solidity of this section was found to be 32820 cubic feet, a difference of 1718 feet. The error, of course, is not in the prismoidal formula, but in assuming that, if the earth were levelled at the ends to the height of the equivalent end sections, the intervening earth might be so disposed as to form a plane between these level ends, thus reducing the mass to a prismoid. This supposition, however, may sometimes be very far from correct, as has just been shown. If the diagonal on the right-hand side in this example were reversed, that is, if the dividing line were formed by a depression, the true solidity found by § 122 would be 29600 feet; whereas the method by equivalent sections would give the same contents as before, or 1502 feet too much.

D. *Correction in Excavation on Curves*

129. In excavations on curves the ends of a section are not parallel

---

* A New Method of Calculating the Cubic Contents of Excavations and Embankments by the aid of Diagrams. By John C. Trautwine.

to each other, but converge towards the centre of the curve. A section between two stations 100 feet apart on the centre line will, therefore, measure less than 100 feet on the side nearest to the centre of the curve, and more than 100 feet on the side farthest from that centre. Now in calculating the contents of an excavation, it is assumed that the ends of a section are parallel, both being perpendicular to the chord of the curve. Thus, let figure 55 represent the plan of two sections of

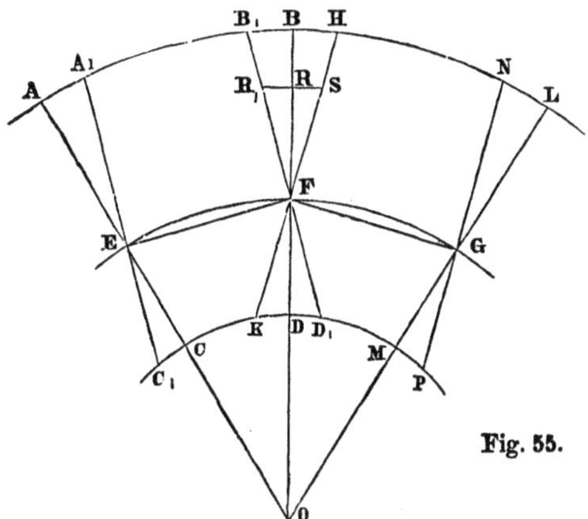

Fig. 55.

an excavation, $E\ F\ G$ being the centre line, $A\ L$ and $C\ M$ the extreme side lines, and O the centre of the curve. Then the calculation of the first section would include all between the lines $A_1\ C_1$ and $B_1\ D_1$; while the true section lies between $A\ C$ and $B\ D$. In like manner, the calculation of the second section would include all between $H\ K$ and $N\ P$; while the true section lies between $B\ D$ and $L\ M$. It is evident, therefore, that at each station on the curve, as at F, the calculation is too great by the wedge-shaped mass represented by $K\ F\ D_1$, and too

Fig. 56.

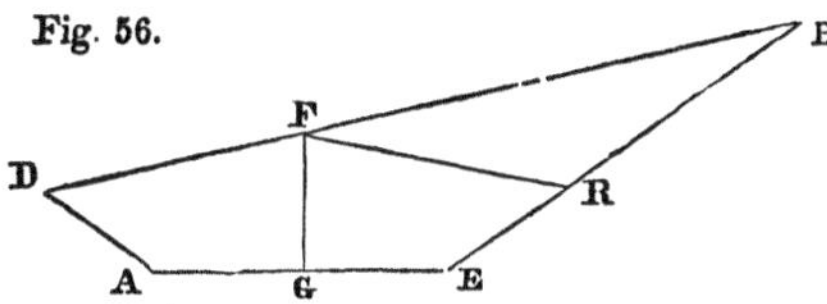

small by the mass represented by $B_1\ F\ H$. These masses balance each other, when the distances out on each side of the centre line are equal, that is, when the cross-section may be represented by $ADFRE$ (fig. 56). But if the excavation is on the side of a hill, so that the distances out differ very much, and the cross-section is of the shape $ADFBE$, the difference of the wedge-shaped masses may require consideration.

130. **Problem.** *Given the centre height c, the greatest side height h, the least side height h', the greatest distance out d, the least distance out d', and the width of the road-bed b, to find the correction in excavation C, at any station on a curve of radius R or deflection angle D.*

Solution. The correction, from what has been said above, is a triangular prism of which BFR (fig. 56) is a cross-section. The height of this prism at B (fig. 55) is $B_1 H$, the height at R is $R_1 S$, and the height at F is 0. $B_1 H$ and $R_1 S$, being very short, are here considered straight lines. Now we have the cross-section $BFR = FBEG - FREG = (\frac{1}{2}cd + \frac{1}{4}bh) - (\frac{1}{2}cd' + \frac{1}{4}bh') = \frac{1}{2}c(d - d') + \frac{1}{4}b(h - h')$. To find the height $B_1 H$, we have the angle $BFH = BFB_1 = D$, and therefore $B_1 H = 2HF \sin. D = 2d \sin. D$. In like manner, $R_1 S = KD_1 = 2KF \sin. D = 2d' \sin. D$. Then since the height at F is 0, one third of the sum of the heights of the prism will be $\frac{2}{3}(d + d') \sin. D$, and the correction, or the solidity of the prism, will be (§ 115)

☞ $$C = [\tfrac{1}{2}c(d - d') + \tfrac{1}{4}b(h - h')] \times \tfrac{2}{3}(d + d') \sin. D.$$

When R is given, and not D, substitute for sin. D its value (§ 9) $\sin. D = \frac{50}{R}$. The correction then becomes

☞ $$C = [\tfrac{1}{2}c(d - d') + \tfrac{1}{4}b(h - h')] \times \frac{100(d + d')}{3R}.$$

This correction is to be *added*, when the highest ground is on the convex side of the curve, and *subtracted*, when the highest ground is on the concave side. At a tangent point, it is evident, from figure 55, that the correction will be just half of that given above.

Example. Given $c = 28$, $h = 40$, $h' = 16$, $d = 74$, $d' = 38$, $b = 28$, and $R = 1400$, to find C. Here the area of the cross-section $BFR = \frac{28}{2}(74 - 38) + \frac{28}{4}(40 - 16) = 672$, and one third of the sum of the heights of the prism is $\frac{100(74 + 38)}{3 \times 1400} = \frac{8}{3}$. Hence $C = 672 \times \frac{8}{3} = 1792$ cubic feet.

131. When the section is partly in excavation and partly in embankment, the cross-section of the excavation is a triangle lying wholly on one side of the centre line, or partly on one side and partly on the other. The surface of the ground, instead of extending from B to D (fig. 56), will extend from B to a point between G and E, or to a point between A and G. In the first case, the correction will be a triangular prism lying between the lines $B_1 F$ and HF (fig. 55), but not extending below the point F. In the second case, the excavation extends below F, and the correction, as in § 129, is the difference between the masses above and below F. This difference may be obtained in a very simple manner, by regarding the mass on both sides of F as one triangular prism the bases of which intersect on the line GF (fig. 56), in which case the height of the prism at the edge below F must be considered to be *minus*, since the direction of this edge, referred to either of the bases, is contrary to that of the two others. The solidity of this prism will then be the difference required.

132. **Problem.** *Given the width of the excavation at the road-bed w, the width of the road-bed b, the distance out d, and the side height h, to find the correction in excavation C, at any station on a curve of radius R or deflection angle D, when the section is partly in excavation and partly in embankment.*

Solution. When the excavation lies wholly on one side of the centre line, the correction is a triangular prism having for its cross-section the cross-section of the excavation. Its area is, therefore, $\frac{1}{2} w h$. The height of this prism at B (fig. 56) is (§ 130) $B_1 H = 2 HF \sin. D = 2 d \sin. D$. In a similar manner, the height at E will be $2 GE \sin. D = b \sin. D$, and at the point intermediate between G and E, the distance of which from the centre line is $\frac{1}{2} b - w$, the height will be $2 (\frac{1}{2} b - w) \sin. D = (b - 2 w) \sin. D$. Hence, the correction, or the solidity of the prism, will be (§ 115) $C = \frac{1}{2} w h \times \frac{1}{3} (2 d + b + b - 2 w) \sin. D = \frac{1}{2} w h \times \frac{2}{3} (d + b - w) \sin. D$.

When the excavation lies on both sides of the centre line, the correction, from what has been said above, is a triangular prism having also for its cross-section the cross-section of the excavation. Its area will, therefore, be $\frac{1}{2} w h$. The height of this prism at B is also $2 d \sin. D$, and the height at E, $b \sin. D$; but at the point intermediate between A and G, the distance of which from the centre line is $w - \frac{1}{2} b$, the height will be $2 (w - \frac{1}{2} b) \sin. D = (2 w - b) \sin D$. As this height is to be considered *minus*, it must be subtracted from the others, and the correction required will be $C = \frac{1}{2} w h \times \frac{1}{3} (2 d + b - 2 w + b) \sin. D$

$= \frac{1}{2} w h \times \frac{2}{3} (d + b - w) \sin. D$. Hence, in all cases, when the section is partly in excavation and partly in embankment, we have the formula

$$C = \tfrac{1}{2} w h \times \tfrac{2}{3} (d + b - w) \sin. D.$$

When R is given, and not D, substitute for sin. D its value (§ 9) $\sin. D = \frac{50}{R}$. The correction then becomes

$$C = \tfrac{1}{2} w h \times \frac{100\,(d + b - w)}{3\,R}.$$

This correction is to be *added*, when the highest ground is on the convex side of the curve, and *subtracted* when the highest ground is on the concave side. At a tangent point the correction will be just half of that given above.

Example. Given $w = 17$, $b = 30$, $d = 51$, $h = 24$, and $R = 1600$, to find C. Here the area of the cross-section is $\frac{1}{2} w h = 17 \times 12 = 204$, and one third of the sum of the heights of the prism is $\frac{100\,(d + b - w)}{3\,R} = \frac{100\,(51 + 30 - 17)}{3 \times 1600} = \frac{4}{3}$. Hence $C = 204 \times \frac{4}{3} = 272$ cubic feet.

133. The preceding corrections (§ 130 and § 132) suppose the length of the sections to be 100 feet. If the sections are shorter, the angle $B F H$ (fig. 55) may be regarded as the same part of D that $F G$ is of 100 feet, and $B_1 F B$ as the same part of D that $E F$ is of 100 feet. The true correction may then be taken as the same part of C that the sum of the lengths of the two adjoining sections is of 200 feet.

TABLE I.

RADII, ORDINATES, DEFLECTIONS,

AND

ORDINATES FOR CURVING RAILS.

Formula for Radii, § 10; for Ordinates, § 25; for Deflections, § 19
for Curving Rails, § 29.

Degree.	Radii.	Ordinates.				Tangent Deflection.	Chord Deflection.	Ordinates for Rails.	
		12½.	25.	37½.	50.			18.	20.
° ′									
0 5	68754.94	.008	.014	.017	.018	.073	.145	.001	.001
10	34377.48	.016	.027	.034	.036	.145	.291	.001	.001
15	22918 33	.024	.041	.051	.055	.218	.436	.002	.002
20	17188.76	.032	.055	.068	.073	.291	.582	.002	.003
25	13751.02	.040	.068	.085	.091	.364	.727	.003	.004
30	11459.19	.048	.082	.102	.109	.436	.873	.004	.004
35	9822.18	.056	.095	.119	.127	.509	1.018	.004	.005
40	8594.41	.064	.109	.136	.145	.582	1.164	.005	.006
45	7639.49	.072	.123	.153	.164	.654	1.309	.005	.007
50	6875.55	.080	.136	.170	.182	.727	1.454	.006	.007
55	6250.51	.087	.150	.187	.200	.800	1.600	.006	.008
1 0	5729.65	.095	.164	.205	.218	.873	1.745	.007	.009
5	5288.92	.103	.177	.222	.236	.945	1.891	.008	.009
10	4911 15	.111	.191	.239	.255	1.018	2.036	.008	.010
15	4583.75	.119	.205	.256	.273	1.091	2.182	.009	.011
20	4297.28	.127	.218	.273	.291	1.164	2.327	.009	.012
25	4044.51	.135	.232	.290	.309	1.236	2.472	.010	.012
30	3819.83	.143	.245	.307	.327	1.309	2.618	.011	.013
35	3618.80	.151	.259	.324	.345	1 382	2.763	.011	.014
40	3437.87	.159	.273	.341	.364	1.454	2.909	.012	.015
45	3274.17	.167	.286	.358	.382	1.527	3.054	.012	.015
50	3125.36	.175	.300	.375	.400	1.600	3.200	.013	.016
55	2989.48	.183	.314	.392	.418	1.673	3.345	.014	017
2 0	2864.93	.191	.327	.409	.436	1.745	3.490	.014	.017
5	2750.35	.199	.341	.426	.455	1.818	3.636	.015	.018
10	2644.58	.207	.355	.443	.473	1.891	3.781	.015	.019
15	2546.64	.215	.368	.460	.491	1.963	3.927	.016	.020
20	2455.70	.223	.382	.477	.509	2.036	4.072	.016	.020
25	2371.04	.231	.395	.494	.527	2.109	4.218	.017	.021
30	2292.01	.239	.409	.511	.545	2.181	4.363	.018	.022
35	2218.09	.247	.423	.528	.564	2.254	4.508	.018	.023
40	2148.79	.255	.436	.545	.582	2.327	4.654	.019	.023
45	2083.68	.263	.450	.562	.600	2.400	4.799	.019	.024
50	2022.41	.270	.464	.580	.618	2.472	4.945	.020	.025
55	1964 64	.278	.477	.597	.636	2.545	5.090	.021	.025
3 0	1910.08	.286	.491	.614	.655	2.618	5.235	.021	.026
5	1858.47	.294	.505	.631	.673	2.690	5.381	.022	.027
10	1809.57	.302	.518	.648	.691	2.763	5.526	.022	.028
15	1763 18	.310	.532	.665	.709	2.836	5.672	.023	.028
20	1719.12	.318	.545	.682	.727	2.908	5.817	.024	.029
25	1677 20	.326	.559	.699	.745	2.981	5.962	.024	.030
30	1637.28	.334	.573	.716	.764	3.054	6.108	.025	.031
35	1599.21	.342	.586	.733	.782	3.127	6.253	.025	.031
40	1562.88	.350	.600	.750	.800	3.199	6.398	.026	.032
45	1528.16	.358	.614	.767	.818	3.272	6.544	.027	.033
50	1494.95	.366	.627	.784	.836	3.345	6.689	.027	.033
55	1463.16	.374	.641	.801	.855	3 417	6.835	.028	.034
4 0	1432.69	.382	.655	.818	.873	3.490	6.980	.028	.035
5	1403 46	.390	.668	.835	.891	3.563	7.125	.029	.036
10	1375.40	.398	.682	.852	.909	3.635	7.271	.029	.036
15	1348.45	.406	.695	.869	.927	3.708	7.416	.030	.037
20	1322.53	.414	.709	.886	.945	3.781	7.561	.031	.038
25	1297.58	.422	.723	.903	.964	3.853	7.707	.031	.039
30	1273.57	.430	.736	.921	.982	3.926	7.852	.032	.039
35	1250 42	.438	.750	.938	1.000	3.999	7.997	.032	.040
40	1228.11	.446	.764	.955	1.018	4.071	8.143	.033	.041
45	1206.57	.454	.777	.972	1.036	4.144	8.288	.034	.041
50	1185.78	.462	.791	.989	1.055	4.217	8.433	.034	.042
55	1165.70	.469	.805	1.006	1.073	4.289	8.579	.035	.043
5 0	1146.28	.477	.818	1.023	1.091	4.362	8.724	.035	.044

Degree.	Radii.	Ordinates.				Tangent Deflection.	Chord Deflection.	Ordinates for Rails.	
		12½.	25.	37½.	50.			18.	20.
° ′									
5 5	1127.50	.485	.832	1.040	1.109	4.435	8.869	.036	.044
10	1109.33	.493	.846	1.057	1.127	4.507	9.014	.037	.045
15	1091 73	.501	.859	1.074	1.146	4.580	9.160	.037	.046
20	1074.68	.509	.873	1.091	1.164	4.653	9.305	.038	.047
25	1058.16	.517	.887	1.108	1.182	4.725	9.450	.038	.047
30	1042.14	.525	.900	1.125	1 200	4.798	9.596	.039	.048
35	1026.60	.533	.914	1.142	1.218	4.870	9.741	.039	.049
40	1011.51	.541	.928	1.159	1.237	4.943	9.886	.040	.049
45	996.87	.549	.941	1.176	1.255	5.016	10.031	.041	.050
50	982.64	.557	.955	1.193	1.273	5.088	10.177	.041	.051
55	968.81	.565	.968	1.210	1.291	5.161	10.322	.042	.052
6 0	955 37	.573	.982	1.228	1.309	5.234	10.467	.042	.052
5	942.29	.581	.996	1.245	1.327	5.306	10.612	.043	.053
10	929 57	.589	1.009	1.262	1.346	5.379	10.758	.044	.054
15	917.19	.597	1.023	1.279	1.364	5.451	10.903	.044	.055
20	905.13	.605	1.037	1.296	1.382	5.524	11.048	.045	.055
25	893.39	.613	1.050	1.313	1.400	5.597	11.193	.045	.056
30	881.95	.621	1.064	1.330	1.418	5.669	11.339	.046	.057
35	870.79	.629	1.078	1.347	1.437	5.742	11.484	.047	.057
40	859.92	.637	1.091	1.364	1.455	5.814	11.629	.047	.058
45	849.32	.645	1.105	1.381	1.473	5.887	11.774	.048	.059
50	838.97	.653	1.118	1.398	1.491	5.960	11.919	.048	.060
55	828.88	.661	1.132	1.415	1.510	6.032	12.065	.049	.060
7 0	819.02	.669	1.146	1.432	1.528	6.105	12.210	.049	061
5	809.40	.677	1.159	1.449	1.546	6.177	12.355	.050	.062
10	800.00	.685	1.173	1.466	1.564	6.250	12.500	.051	.063
15	790.81	.693	1.187	1.483	1.582	6.323	12.645	.051	.063
20	781.84	.701	1.200	1.501	1.600	6.395	12.790	.052	.064
25	773.07	.709	1.214	1.517	1.619	6.468	12.936	.052	.065
30	764.49	.717	1.228	1.535	1.637	6.540	13.081	.053	.065
35	756.10	.725	1.242	1.552	1.655	6.613	13.226	.054	.066
40	747.89	.733	1.255	1.569	1.673	6.685	13.371	.054	.067
45	739.86	.740	1.269	1.586	1.691	6.758	13.516	.055	.068
50	732.01	.748	1.283	1.603	1.710	6.831	13.661	.055	.068
55	724.31	.756	1.296	1.620	1.728	6.903	13.806	.056	.069
8 0	716.78	.764	1.310	1.637	1.746	6.976	13.951	.057	070
5	709.40	.772	1.324	1.654	1.764	7.048	14.096	.057	.070
10	702.18	.780	1.337	1.671	1.782	7.121	14.241	.058	.071
15	695.09	.788	1.351	1.688	1.801	7.193	14.387	.058	.072
20	688.16	.796	1.365	1.705	1.819	7.266	14.532	.059	.073
25	681 35	.804	1.378	1.722	1.837	7.338	14.677	.059	.073
30	674.69	.812	1.392	1.739	1.855	7.411	14.822	.060	.074
35	668.15	.820	1.406	1.757	1.873	7.483	14.967	.061	.075
40	661.74	.828	1.419	1.774	1.892	7.556	15.112	.061	.076
45	655.45	.836	1.433	1.791	1.910	7.628	15.257	.062	.076
50	649.27	.844	1.447	1.808	1.928	7.701	15.402	.062	.077
55	643.22	.852	1.460	1.825	1.946	7.773	15.547	.063	.078
9 0	637.27	.860	1.474	1.842	1.965	7.846	15.692	.064	.078
5	631.44	.868	1.488	1.859	1.983	7 918	15.837	.064	.079
10	625 71	.876	1.501	1.876	2.001	7.991	15.982	.065	.080
15	620.09	.884	1.515	1.893	2.019	8.063	16.127	.065	.081
20	614.56	.892	1.529	1.910	2.037	8.136	16.272	.066	.081
25	609.14	.900	1.542	1.927	2.056	8.208	16.417	.066	.082
30	603.80	.908	1.556	1.944	2.074	8.281	16.562	.067	.083
35	598 57	.916	1.570	1.961	2.092	8.353	16.707	.068	.084
40	593.42	.924	1.583	1.979	2.110	8.426	16.852	.068	.084
45	588.36	.932	1.597	1.996	2.128	8.498	16.996	.069	.085
50	583.38	.940	1.611	2.013	2.147	8.571	17.141	.069	.086
55	578.49	.948	1.624	2.030	2.165	8.643	17.286	.070	.086
10 0	573.69	.956	1.638	2.047	2.183	8.716	17.431	.071	.087

Degree.	Radii.	Ordinates.				Tangent Deflection.	Chord Deflection.	Ordinates for Rails.	
		12½.	25.	37½.	50.			18.	20.
10° 10′	564.31	.972	1.665	2.081	2.219	8.860	17.721	.072	.089
20	555.23	.988	1.693	2.115	2.256	9.005	18.011	.073	.090
30	546.44	1.004	1.720	2.149	2.292	9.150	18.300	.074	.092
40	537.92	1.020	1.748	2.184	2.329	9.295	18.590	.075	.093
50	529.67	1.036	1.775	2.218	2.365	9.440	18.880	.076	.094
11 0	521.67	1.052	1.802	2.252	2.402	9.585	19.169	.078	.096
10	513.91	1.068	1.830	2.286	2.438	9.729	19.459	.079	.097
20	506.38	1.084	1.857	2.320	2.475	9.874	19.748	.080	.099
30	499.06	1.100	1.884	2.354	2.511	10.019	20.038	.081	.100
40	491.96	1.116	1.912	2.389	2.547	10.164	20.327	.082	.102
50	485.05	1.132	1.938	2.423	2.584	10.308	20.616	.084	.103
12 0	478.34	1.148	1.967	2.457	2.620	10.453	20.906	.085	.105
10	471.81	1.164	1.994	2.491	2.657	10.597	21.195	.086	.106
20	465.46	1.180	2.021	2.525	2.693	10.742	21.484	.087	.107
30	459.28	1.196	2.049	2.560	2.730	10.887	21.773	.088	.109
40	453.26	1.212	2.076	2.594	2.766	11.031	22.063	.089	.110
50	447.40	1.228	2.104	2.628	2.803	11.176	22.352	.091	.112
13 0	441.68	1.244	2.131	2.662	2.839	11.320	22.641	.092	.113
10	436.12	1.260	2.159	2.697	2.876	11.465	22.930	.093	.115
20	430.69	1.277	2.186	2.731	2.912	11.609	23.219	.094	.116
30	425.40	1.293	2.213	2.765	2.949	11.754	23.507	.095	.118
40	420.23	1.309	2.241	2.799	2.985	11.898	23.796	.096	.119
50	415.19	1.325	2.268	2.833	3.022	12.043	24.085	.098	.120
14 0	410.28	1.341	2.296	2.868	3.058	12.187	24.374	.099	.122
10	405.47	1.357	2.323	2.902	3.095	12.331	24.663	.100	.123
20	400.78	1.373	2.351	2.936	3.131	12.476	24.951	.101	.125
30	396.20	1.389	2.378	2.970	3.168	12.620	25.240	.102	.126
40	391.72	1.405	2.406	3.005	3.204	12.764	25.528	.103	.128
50	387.34	1.421	2.433	3.039	3.241	12.908	25.817	.105	.129
15 0	383.06	1.437	2.461	3.073	3.277	13.053	26.105	.106	.131
10	378.88	1.453	2.488	3.107	3.314	13.197	26.394	.107	.132
20	374.79	1.469	2.515	3.142	3.350	13.341	26.682	.108	.133
30	370.78	1.486	2.543	3.176	3.387	13.485	26.970	.109	.135
40	366.86	1.502	2.570	3.210	3.423	13.629	27.258	.110	.136
50	363.02	1.518	2.598	3.245	3.460	13.773	27.547	.112	.138
16 0	359.26	1.534	2.625	3.279	3.496	13.917	27.835	.113	.139
10	355.59	1.550	2.653	3.313	3.533	14.061	28.123	.114	.141
20	351.98	1.566	2.680	3.347	3.569	14.205	28.411	.115	.142
30	348.45	1.582	2.708	3.382	3.606	14.349	28.699	.116	.143
40	344.99	1.598	2.736	3.416	3.643	14.493	28.986	.117	.145
50	341.60	1.615	2.763	3.450	3.679	14.637	29.274	.119	.146
17 0	338.27	1.631	2.791	3.485	3.716	14.781	29.562	.120	.148
10	335.01	1.647	2.818	3.519	3.752	14.925	29.850	.121	.149
20	331.82	1.663	2.846	3.553	3.789	15.069	30.137	.122	.151
30	328.68	1.679	2.873	3.588	3.825	15.212	30 425	.123	.152
40	325.60	1.695	2.901	3.622	3 862	15.356	30.712	.124	.154
50	322.59	1.711	2.928	3.656	3.898	15.500	31.000	.126	.155
18 0	319.62	1.728	2.956	3.691	3.935	15.643	31.287	.127	.156
10	316.71	1.744	2.983	3.725	3.972	15.787	31.574	.128	.158
20	313.86	1.760	3.011	3.759	4.008	15.931	31.861	.129	.159
30	311.06	1.776	3.039	3.794	4.045	16.074	32.149	.130	.161
40	308.30	1.792	3.066	3.828	4.081	16.218	32.436	.131	.162
50	305.60	1.809	3.094	3.862	4.118	16.361	32.723	.133	.164
19 0	302.94	1.825	3.121	3.897	4.155	16.505	33.010	.134	.165
10	300.33	1.841	3.149	3.931	4.191	16.648	33.296	.135	.166
20	297.77	1.857	3.177	3.965	4.228	16.792	33.583	.136	.168
30	295.25	1.873	3.204	4.000	4.265	16.935	33.870	.137	.169
40	292.77	1.890	3.232	4.034	4.301	17.078	34.157	.138	.171
50	290.33	1.906	3.259	4.069	4.338	17.222	34.443	.140	.172
20 0	287.94	1.922	3.287	4.103	4.374	17.365	34.730	.141	.174

TABLE II.

LONG CHORDS. § 69.

Degree of Curve.	2 Stations.	3 Stations.	4 Stations.	5 Stations	6 Stations.
° ′ 0 10	200.000	299.999	399.998	499.996	599.993
20	199.999	.997	.992	.983	.970
30	.998	.992	.981	.962	.933
40	.997	.986	.966	.932	.882
50	.995	.979	.947	.894	.815
1 0	199.992	299.970	399.924	499.848	599.733
10	.990	.959	.896	.793	.637
20	.986	.946	.865	.729	.526
30	.983	.932	.829	.657	.401
40	.979	.915	.789	.577	.260
50	.974	.898	.744	.488	.105
2 0	199.970	299.878	399.695	499.391	598.934
10	.964	.857	.643	.285	.750
20	.959	.834	.586	.171	.550
30	.952	.810	.524	·049	.336
40	.916	.783	.459	498.918	.106
50	.939	.756	.389	.778	597.862
3 0	199.931	299.726	399.315	498.630	597.604
10	.924	.695	.237	.474	.331
20	.915	.662	.154	.309	.043
30	.907	.627	.068	.136	596.740
40	.898	.591	398.977	497.955	.423
50	.888	.553	.882	.765	.091
4 0	199.878	299.513	398.782	497.566	595.744
10	.868	.471	.679	.360	.383
20	.857	.428	.571	.145	.007
30	.846	.383	.459	496.921	594.617
40	.834	.337	.343	.689	.212
50	.822	.289	.223	.449	593.792
5 0	199.810	299.239	398.099	496.200	593.358
10	.797	.187	397.970	495.944	592.909
20	.783	.134	.837	.678	446
30	.770	.079	.700	.405	591.968
40	.756	.023	.559	.123	.476
50	.741	298.964	.413	494.832	590.970
6 0	199.726	298.904	397.264	494.534	590.449
10	.710	.843	.110	.227	589.913
20	.695	.779	396.952	493.912	.364
30	.678	.714	.790	.588	588.800
40	.662	.648	.623	257	.221
50	.644	.579	453	492.917	587.628
7 0	199.627	298.509	396.278	492.568	587.021
10	.609	.438	.099	.212	586.400
20	.591	.364	395.916	491.847	585.765
30	.572	.289	.729	.474	.115
40	·553	.212	.538	.093	584.451
50	.533	.134	.342	490.704	583.773
8 0	.513	298.054	395.142	490.306	583.081

TABLE III.

CORRECTION FOR THE EARTH'S CURVATURE AND FOR REFRACTION. § 105.

D.	*d.*	*D.*	*d.*	*D.*	*d.*	*D.*	*d.*
300	.002	1800	.066	3300	.223	4800	.472
400	.003	1900	.074	3400	.237	4900	.492
500	.005	2000	.082	3500	.251	5000	.512
600	.007	2100	.090	3600	.266	5100	.533
700	.010	2200	.099	3700	.281	5200	.554
800	.013	2300	.108	3800	.296	1 mile	.571
900	.017	2400	.118	3900	.312	2 "	2.285
1000	.020	2500	.128	4000	.328	3 "	5.142
1100	.025	2600	.139	4100	.345	4 "	9.142
1200	.030	2700	.149	4200	.362	5 "	14.284
1300	.035	2800	.161	4300	.379	6 "	20.568
1400	.040	2900	.172	4400	.397	7 "	27.996
1500	.046	3000	.184	4500	.415	8 "	36.566
1600	.052	3100	.197	4600	.434	9 "	46.279
1700	.059	3200	.210	4700	.453	10 "	57.135

TABLE IV.

ELEVATION OF THE OUTER RAIL ON CURVES. § 110.

Degree.	*M* = 15.	*M* = 20.	*M* = 25.	*M* = 30.	*M* = 40.	*M* = 50.
1°	.012	.022	.034	.049	.088	.137
2	.025	.044	.068	.099	.175	.274
3	.037	.066	.103	.148	.263	.411
4	.049	.088	.137	.197	.351	.548
5	.062	.110	.171	.247	.438	.685
6	.074	.131	.205	.296	.526	.822
7	.086	.153	.240	.345	.613	.958
8	.099	.175	.274	.394	.701	1.095
9	.111	.197	.308	.443	.788	1.232
10	.123	.219	.342	.493	.876	1.368

TABLE V.

FROG ANGLES, CHORDS, AND ORDINATES FOR TURNOUTS.

This table is calculated for $g = 4.7$, $d = .42$, and $S = 1° 20'$. Formula for frog angle F, and chord BF, § 50; for m, the middle ordinate of BF, § 25; for m', the middle ordinate for curving an 18 ft. rail, § 29.

R.	F. ° ′ ″	B F.	m.	m′.	R.	F. ° ′ ″	B F.	m.	m′.
1000	5 27 44	72.22	.651	.041	600	6 57 48	59.17	.727	.068
975	5 31 39	71.53	.655	.042	575	7 6 26	58.16	.733	.070
950	5 35 44	70.83	.659	.043	550	7 15 40	57.12	.739	.074
925	5 39 59	70.11	.663	.044	525	7 25 33	56.05	.745	.077
900	5 44 24	69.38	.667	.045	500	7 36 10	54.94	.752	.081
875	5 49 1	68.64	.671	.046	475	7 47 37	53.79	.758	.085
850	5 53 50	67.88	.676	.048	450	8 0 1	52.61	.765	.090
825	5 58 52	67.10	.680	.049	425	8 13 30	51.37	.773	.095
800	6 4 9	66.30	.685	.051	400	8 28 14	50.09	.780	.101
775	6 9 41	65.49	.690	.052	375	8 44 26	48.75	.788	.108
750	6 15 30	64.65	.695	.054	350	9 2 20	47.35	.796	.116
725	6 21 37	63.80	.700	.056	325	9 22 16	45.88	.805	.125
700	6 28 4	62.92	.705	.058	300	9 44 39	44.34	.814	.135
675	6 34 52	62.02	.710	.060	275	10 10 1	42.72	.824	.147
650	6 42 4	61.09	.716	.062	250	10 39 6	41.00	.834	.162
625	6 49 42	60.14	.721	.065	225	11 12 55	39.16	.845	.180

TABLE VI.

LENGTH OF CIRCULAR ARCS IN PARTS OF RADIUS.

°				′				″			
1	.01745	32925	19943	1	.00029	08882	08666	1	.00000	48481	36811
2	.03490	65850	39887	2	.00058	17764	17331	2	.00000	96962	73622
3	.05235	98775	59830	3	.00087	26646	25997	3	.00001	45444	10433
4	.06981	31700	79773	4	.00116	35528	34663	4	.00001	93925	47244
5	.08726	64625	99716	5	.00145	44410	43329	5	.00002	42406	84055
6	.10471	97551	19660	6	.00174	53292	51994	6	.00002	90888	20867
7	.12217	30476	39603	7	.00203	62174	60660	7	.00003	39369	57678
8	.13962	63401	59546	8	.00232	71056	69326	8	.00003	87850	94489
9	.15707	96326	79490	9	.00261	79938	77991	9	.00004	36332	31300

TABLE VII.

EXPANSION BY HEAT.

Bodies.	32° to 212°.	1°.	Authority.
Platina,	.0008842	.000004912	Hassler.
Gold,	.001466	.000008141	"
Silver,	.001909	.000010605	"
Mercury,	.018018	.0001001	"
Brass,	.00189163	.000010509	"
Iron,	.00125344	.000006964	"
Water,	.0466	not uniform.	"
Granite,	.00086850	.000004825	Prof. Bartlett.
Marble,	.00102024	.000005668	"
Sandstone,	.00171576	.000009532	"

TABLE VIII.

PROPERTIES OF MATERIALS.

The authorities referred to by the capital letters in the table are: —

B Barlow, *On the Strength of Materials.*
Be. Bevan.
Br. Lieut. Brown.
C. Couch.
F. Franklin Institute, *Report on Steam Boilers.*
G. Gordon, *Eng. Translation of Weisbach.*
H. Hodgkinson, *Reports to Brit. Association.*
Ha. Hassler, *Tables.*
L. Lamé.
M. Musschenbroek, *Int. to Nat Phil.*
R. Rennie, *Phil. Trans.*
Ro. Rondelet, *L'Art de Batir.*
T. Telford.
Ta. Taylor, *Statistics of Coal.*
W. Weisbach, *Mech. of Machinery and Engineering.*

The numbers without letters are taken from Prof. Moseley's *Engineering and Architecture*

In finding the weights, a cubic foot of water has, for convenience, been taken at 62.5 lbs.

The numbers for compression taken from Hodgkinson were obtained by him from prisms high enough to allow the wedge of rupture to slide freely off. He shows that this is essential in experiments on compression.

The modulus of rupture S is the breaking weight of a prism 1 in. broad, 1 in. deep, and 1 in. between the supports, the weight being applied in the middle. To find the corresponding breaking weight W of a rectangular beam of any other size, let l = its length, b = its breadth, and d = its depth, all in inches. Then $W = \frac{2bd^2}{3l} \times S$.

The numbers in the last three columns express absolute strength For safety, a certain proportion only of these numbers is taken. The divisors for wood may be from 6 to 10, for metal from 3 to 6, for stone 10, and for ropes 3.

When double numbers are used in the column headed "Crushing Force per Square Inch in lbs.," the first applies to specimens moderately dry, the second to specimens turned and kept dry in a warm place two months longer. In the case of American Birch, Elm, and Teak, the numbers apply to seasoned specimens.

Materials.	Specific Gravity.	Weight per Cubic Foot in lbs.	Tensile Strength per Square Inch in lbs.	Crushing Force per Square Inch in lbs.	Modulus of Rupture S in lbs.
Metals.					
Brass, cast,	8.399	524.94	17963 R.		
Copper, cast,	8.607	537.94	19072		
" rolled,	8.864 F.	554.00	32826 F.		
" wire-drawn,	8.878	554.87	61228		
Gold,	19.258 Ha. 19.361 Ha.	1203.62 1210.06			
Iron, cast,					
Carron No. 2, cold blast,	7.066 H	441.62	16683 H.	106375 H.	38556 H.
" " hot "	7.046 H.	440.37	13505 H.	108540 H.	37503 H.
Devon No. 3, cold "	7.295 H.	455.94			36288 H.
" " hot "	7.229 H.	451.81	21907 H.	145435 H.	43497 H.
Buffery No. 1, cold "	7.079 H.	442.44	17466 H.	93385 H.	37503 H.
" " hot "	6.998 H.	437.37	13434 H.	86397 H.	35316 H.
Iron, wrought,					
English bar,	7.700	481.25	57120 L.	56000 ? G.	54000 G.
Welsh "			64960 T.		
Swedish "			64960 T.		
" "	7.478 F.	467.37	58184 F.		
Lancaster Co., Pa., bars,	7.740 F.	483.75	58661 F.		
Tennessee "	7.805 F.	487.81	52099 F.		
Missouri "	7.722 F.	482.62	47909 F.		
Iron wire,					
English, diam = [illegible] in.			80214 T.		
Phillipsb'g, Pa. " [illegible] "	7.[illegible]27 F.	482.94	84186 F.		
" " .190 "			73888 F.		
" " .156 "			89162 F.		
Lead, cast,	11.446 M.	715.37	1824 R.		
Lead wire,	[illegible]	707.31	2581 M.		
Mercury,	13.598 W.	849.87			
Platina,	19.500 Ha. 22.669 Ha.	1218.75 1416.81			
Silver,	10.474 Ha.	654.62	40902 M.		
Steel, soft,	7.780	486.25	120000		
" razor-tempered,	7.840	490.00	150000		
Tin, cast,	7.291	455.69	[illegible]322 M.		
Zinc, fused,	7.050 W.	4[illegible]0.[illegible]2			
" rolled,	7.540 W.	471.25			
Woods.					
Ash, English,	.760 B	47.50	[illegible] B.	8683 H. 9363 H.	12156 B
Birch, English,	.792 B.	49.50	[illegible]	3297 H. 6402 H.	10920 B.
" American,	.648 B.	40.50		11663 H.	9624 B
Box,	.960 B.	60.00	20000 B.	9771 H.	
Cedar, Canadian,	.909 C.	56.81	11400 Bo.	[illegible] H. 5863 M.	
Chestnut,	.657 Ro.	41.06	13300 Ro.		
Deal, Christiania middle,	.698 B.	43.62	12400		9864 B.
" Memel "	.590 B.	36.87			10386 B.
" Norway Spruce,	.340	21.25	17600		
" English,	.470	29.37	7000		
Elm, seasoned,	.553 B.	34.56	13489 M.	10331 H.	6078 B.
Fir, New England,	.553 B.	34.56			6612 B.
" Riga,	.753 B.	47.06	12000 B.	5748 H. 6586 H.	[illegible] B.
Lignum-vitæ,	1.220	76.25	11800 M.		
Mahogany, Spanish,	.800	50.00	16500	8198 H. 8198 H.	

Materials.	Specific Gravity.	Weight per Cubic Foot in lbs.	Tensile Strength per Square Inch in lbs.	Crushing Force per Square Inch in lbs.	Modulus of Rupture S in lbs.
Woods.					
Oak, English, ...	.934 **B.**	58.37	10000 **B.**	6184 **H.** 10058 **H.**	10032 **B.**
" Canadian, ...	.872 **B.**	54.50	10253	4231 **H.** 5982 **H.**	10596 **B.**
Pine, pitch, ... : .	.660 **B.**	41.25	7818 **M.**	6790 **H.** 6790 **H.**	9792 **B.**
" red, ...	.657 **B.**	41.06		5395 **H.** 7518 **H.**	8046 **B.**
" American, white, . .	.455 **Br.**	28.44			7829 **Br.**
" " Southern,	.872 **Br.**	54.50			13937 **Br.**
Poplar, ...	.333 **M.**	23.94	7200 **Be.**	3107 **H.** 5124 **H.**	
Teak, ...	.745 **B.**	46.56	15000 **B.**	12101 **H.**	14772 **B.**
Other Materials.					
Brick, red, ...	2.168 **R.**	135.50	280	808 **R.**	340 **W.**
" pale red, ...	2.085 **R.**	130.31	300	562 **R.**	180 **W.**
Chalk, ...	2.784 1.869	174.00 116.81		501 **R.**	
Coal, Penn. anthracite, .	1.327 **Ta.** 1.700 **Ta.**	82.94 106.25			
" " semi-bituminous,	1.453 **Ta.**	90.81			
" Md. "	1.552 **Ta.**	97.00			
" Penn. bituminous, .	1.312 **Ta.**	82.00			
" Ohio " .	1.270 **Ta.**	79.37			
" English " .	1.259 **Ta.**	78.69			
Earth,					
loamy hard-stamped, fresh,	2.060 **W.**	128.75			
" " dry,	1.930 **W.**	120.62			
garden, fresh, ...	2.050 **W.**	128.12			
" dry, ...	1.630 **W.**	101.87			
dry, poor, ...	1.340 **W.**	83.75			
Glass, plate, ...	2.453	153.31	9420		
Gravel, ...	1.920	120.00			
Granite, Aberdeen, . . .	2.625 **R.**	164.06		10914 **R.**	
Ivory, ...	1.826	114.12	16626		
Limestone, ...	2.400 **W.** 2.860 **W.**	150.00 178.75		1500 **W.** 6000 **W.**	700 **W.** 1700 **W**
Marble, white Italian, . .	2.638 **H.**	164.87		9583 **G.**	1062
" black Galway, . .	2.695 **H.**	168.44			2664
Masonry, quarry stone, dry,	2.400 **W.**	150.00			
" sandstone, "	2.050 **W.**	128.12			
" brick, dry, .	1.470 **W.** 1.590 **W.**	91.87 99.37			
Ropes,					
hemp, under 1 inch diam.,			9280 **W.**		
" from 1 to 3 in. "			7218 **W.**		
" over 3 inches "			5156 **W.**		
Sand, river, ...	1.886	117.87			
Sandstone, ...	1.900 **W.** 2.700 **W.**	118.75 168.75		1400 **W.** 13000 **W.**	600 **W.** 800 **W.**
" Dundee, ...	2.530 **R.**	158.12		6630 **R.**	
" Derby, red and friable,	2.316 **R.**	144.75		3142 **R.**	
Slate, Welsh, ...	2.888	180.50	12800		
" Scotch, ...			9600		

TABLE IX.

MAGNETIC VARIATION.

THE following table has been made up from various sources, prin cipally, however, from the results of the United States Coast Survey kindly furnished in manuscript by the Superintendent, Prof. A. D Bache. "These results," he remarks in an accompanying note, "ar from preliminary computations, and may be somewhat changed by th final ones." Among the other sources may be mentioned the Smith sonian Contributions for 1852, Trans. Am. Phil. Soc. for 1846, Lond Phil. Trans. for 1849, Silliman's Journal for 1838, 1840, 1846, an 1852, and the various American, British, and Russian Governmen Observations. The latitudes and longitudes here given are not alway to be relied on as minutely correct. Many of them, for places in th Western States, were confessedly taken from maps and other uncer tain sources. Those of the Coast Survey Stations, however, as wel as those of American and foreign Government Observatories and Sta tions, are presumed to be accurate.

It will be seen that the variation of the magnetic needle in th United States is in some places west and in others east. *The line of n variation* begins in the northwest part of Lake Huron, and runs throug the middle of Lake Erie, the southwest corner of Pennsylvania, th central parts of Virginia, and through North Carolina to the coast All places on the east of this line have the variation of the needl west, — all places on the west of this line have the variation of th needle east; and, as a general rule, the farther a place lies from thi line, the greater is the variation. The position of the line of no varia tion given above is the position assigned to it by Professor Loomis fo the year 1840. But this line has for many years been moving slowl westward, and this motion still continues. Hence places whose varia tion is west are every year farther and farther from this line, so th the variation west is constantly increasing. On the contrary, place whose variation is east are every year nearer and nearer to this line so that the variation east is constantly decreasing. The rate of thi increase or decrease, as the case may be, is said to average about 2′ fo the Southern States, 4′ for the Middle and Western States, and 6′ fo the New England States.* The increase in Washington in 1840 – was 3′ 44.2″; in Toronto in 1841 – 2 it was 4′ 46 2″. The changes i

* Prof. Loomis in Silliman's Journal, Vol. XXXIX., 1840.

Cambridge, Mass. may be seen from the following determinations of the variation, taken from the Memoirs of the American Academy for 1846.

		° ′			° ′
Cambridge,	1708,	9 0	Cambridge,	1788,	6 38
"	1742,	8 0	Boston,	1793,	6 30
"	1757,	7 20	Salem,	1805,	5 57
"	1761,	7 14	"	1808,	5 20
"	1763,	7 0	"	1810,	6 22
"	1780,	7 2	Cambridge,	1810,	7 30
"	1782,	6 46	"	1835,	8 51
"	1783,	6 52	"	1840,	9 18

But besides this change in the variation, which may be called secular, there is an annual and a diurnal change, and very frequently there are irregular changes of considerable amount. With respect to the annual change, the variation west in the Northern hemisphere is generally found to be somewhat greater, and the variation east somewhat less, in the summer than in the winter months. The amount of this change is different in different places, but it is ordinarily too small to be of any practical importance. The diurnal change is well determined. At Washington in 1840 – 2, the mean diurnal change in the variation was,* —

Summer, 10′ 4″.1 Autumn, 6′ 21″.2 Winter, 5′ 9″.1 Spring, 8′ 10″.7

At Toronto the means were, † —

	1841.	1843.	1845.	1847.	1849.	1850.	1851.
	′	′	′	′	′	′	′
Winter,	6.67	5.64	5.73	7.28	8.25	8.01	7.01
Spring and Autumn,	9.46	9.36	9.15	10.08	12.25	10.90	10.82
Summer,	12.38	11.70	13.36	13.84	14.80	13.74	12.61

The diurnal change in the variation is such that the north end of the needle in the Northern hemisphere attains its extreme westerly position about 2 o'clock, P. M., and its extreme easterly position about 8 o'clock, A. M. In places, therefore, whose variation is west, the maximum variation occurs about 2 P. M., while in places whose variation is east, the maximum variation occurs about 8 A. M. In Washington, according to the report of Lieutenant Gilliss, the maximum variation, taking the mean of two years' observations, occurs at $1^{h.}$ $33^{m.}$ P. M., the minimum at $8^{h.}$ $6^{m.}$ A. M.

The determinations of the Coast Survey are distinguished by the letters C. S. attached to the name of the observer. In some instances the name of the nearest town has been added to the name of the Coast Survey station.

* Lieut. Gilliss's Report, Senate Document 172, 1845

† London Philosophical Transactions, 1852

Place.	Latitude.	Longitude.	Authority.	Date.	Variation.
Maine.	° ′	° ′			° ′
Agamenticus,	43 13.4	70 41.2	T. J. Lee, C. S.	Sept., 1847	10 10.0 W
Bethel,	44 28.0	70 51.0	J. Locke,	June, 1845	11 50.0 "
Bowdoin Hill, Portland,	43 38.8	70 16.2	J. E. Hilgard, C S.	Aug., 1851	11 41.1 "
Cape Neddick, York	43 11.6	70 36.1	J. E. Hilgard, C. S.	Aug., 1851	11 9.0 "
Cape Small,	43 46.7	69 50.4	G. W. Dean, C. S.	Oct., 1851	12 5.5 "
Kennebunkport,	43 21.4	70 27.8	J. E. Hilgard, C. S.	Aug., 1851	11 23.6 "
Kittery Point,	43 4.8	70 43.3	J. E. Hilgard, C. S.	Sept., 1850	10 30.2 "
Mt. Pleasant,	44 1.6	70 49.0	G. W. Dean, C. S.	Aug., 1851	14 32.0 "
Portland,	43 41.0	70 20.5	J. Locke,	June, 1845	11 28.3 "
Richmond Island,	43 32.4	70 14.0	J. E. Hilgard, C. S.	Sept., 1850	12 17.9 "
New Hampshire.					
Fabyan's Hotel,	44 16.0	71 29.0	J. Locke,	June, 1845	11 32.0 W.
Hanover,	43 42.0	72 10.0	Prof Young,	1839	9 15.0 "
Isle of Shoals,	42 59.2	70 36.5	T. J. Lee, C. S.	Aug, 1847	10 3.4 "
Patuccawa,	43 7.2	71 11.5	G. W. Dean, C. S.	Aug., 1849	10 42.9 "
Unkonoonuc,	42 59.0	71 35.0	J. S. Ruth, C. S.	Oct, 1848	9 5.6 "
Vermont.					
Burlington,	44 27.0	73 10.0	J. Locke,	June, 1845	9 22.0 W.
Massachusetts.					
Annis-squam,	42 39.4	70 40.3	G. W. Keely, C. S.	Aug., 1849	11 36.7 W.
Baker's Island,	42 32.2	70 46.8	G. W. Keely, C. S.	Sept., 1849	12 17.0 "
Blue Hill, Milton,	42 12.7	71 6.5	T. J. Lee, C. S.	Sept. and Oct., 1845	9 13.8 "
Cambridge,	42 22.9	71 7.2	W. C. Bond,	1852	10 8.0 "
Chappaquidick, Edgartown,	41 22.7	70 28.7	T. J. Lee, C. S.	July, 1846	8 47.7 "
Coddon's Hill, Marblehead,	42 31.0	70 50.9	G. W. Keely, C. S.	Sept, 1849	11 49.8 "
Copecut Hill,	41 43.3	71 3.3	T. J. Lee, C. S.	Sept and Oct, 1844	9 12.1 "
Dorchester,	42 19.0	71 4.0	W. C. Bond,	1839	9 2.0 "
Fort Lee, Salem,	42 31.9	70 52.1	G. W. Keely, C. S.	Aug., 1849	10 14.5 "
Hyannis,	41 38.0	70 18.0	T. J Lee, C. S.	Aug., 1846	9 22.0 "
Indian Hill,	41 25.7	70 40.3	T. J. Lee, C. S.	Aug., 1846	8 49.3 "
Little Nahant,	42 26.2	70 55.5	G. W. Keely, C. S.	Aug., 1849	9 40.9 "
Nantasket,	42 18.2	70 54.0	T. J. Lee, C. S.	Sept., 1847	9 33.5 "
Nantucket,	41 17.0	70 6.0	T. J. Lee, C. S.	July, 1846	9 14.0 "
New Bedford,	41 38.0	70 54.0	T. J. Lee, C. S.	Oct., 1845	8 54.6 "
Shootflying Hill, Barnstable,	41 41.1	70 20.5	T. J. Lee, C. S.	Aug., 1846	9 40.1 "
Tarpaulin Cove,	41 28.1	70 45.1	T. J. Lee, C. S.	Aug., 1846	9 10.1 "
Rhode Island.					
Beacon-pole Hill,	41 59.7	71 26.7	T. J Lee, C. S.	Oct. and Nov., 1844	9 29.8 W.
McSparran Hill,	41 29.7	71 27.1	T. J. Lee, C. S.	July, 1844	8 53.3 "
Point Judith,	41 21.9	71 28.9	R.H. Fauntleroy, C.S.	Sept, 1847	8 59.4 "
Spencer Hill,	41 40.7	71 29.3	T. J. Lee, C. S.	July and Aug. 1844	9 11.9 "
Connecticut.					
Black Rock, Fairfield,	41 8.6	73 12.6	J. Renwick, C. S.	Sept., 1845	6 53.5 W.
Bridgeport,	41 10.0	73 11.0	J. Renwick, C. S.	Sept., 1845	6 19.3 "
Fort Wooster,	41 16.9	72 53.2	J. S. Ruth, C. S.	Aug., 1848	7 26.4 "
Groton Point, New London,	41 18.0	72 0.0	J. Renwick, C. S.	Aug., 1845	7 29.5 "

Place.	Latitude.	Longitude.	Authority.	Date.	Variation.
	° ′	° ′			° ′
Milford,	41 16.0	73 1.0	J. Renwick, C S.	Sept., 1845	6 38.3 W.
New Haven, Pavilion,	41 18.5	72 55.4	J. S. Ruth, C. S	Aug., 1848	6 37.5 "
New Haven, Yale College,	41 18.5	72 55.4	J. Renwick, C. S.	Sept., 1845	6 17.3 "
Norwalk,	41 7.1	73 24.2	J. Renwick, C. S.	Sept., 1844	6 46.3 "
Oyster Point, New Haven,	41 17.0	72 55.4	J. S. Ruth, C. S.	Aug., 1848	6 32.3 "
Sachem's Head, Guilford,	41 17.0	72 43.0	J. Renwick, C. S.	Aug., 1845	6 15.2 "
Sawpits,	40 59.5	73 39.4	J. Renwick, C. S.	Sept., 1844	6 1.6 "
Saybrook,	41 16.0	72 20.0	J. Renwick, C. S.	Aug., 1845	6 49.9 "
Stamford,	41 3.5	73 32.0	J. Renwick, C. S.	Sept., 1844	6 40.4 "
Stonington,	41 20.0	71 54.0	J. Renwick, C. S.	Aug., 1845	7 38.2 "
New York.					
Albany,	42 39.0	73 44.0	Regents' Report,	1836	6 47.0 W.
Bloomingdale Asylum,	40 48.8	73 57.4	J. Locke, C. S.	April, 1846	5 10.9 "
Cole, Staten Island,	40 31.8	74 13.8	J. Locke, C. S.	April, 1846	5 33.8 "
Drowned Meadow, L. I.,	40 56.1	73 3.5	J. Renwick, C. S.	Sept., 1845	6 3.6 "
Flatbush, L. I.,	40 40.2	73 57.7	J. Locke, C. S.	April, 1846	5 54.6 "
Greenport, L. I.,	41 6.0	72 21.0	J. Renwick, C. S.	Aug., 1845	7 14.6 "
Leggett,	40 48.9	73 53.0	R.H. Fauntleroy, C.S.	Oct., 1847	5 40.6 "
Lloyd's Harbor, L. I.,	40 55.6	73 24.8	J. Renwick, C. S.	Sept., 1844	6 12.5 "
New Rochelle,	40 52.5	73 47.0	J. Renwick, C. S.	Sept., 1844	5 31.5 "
New York,	40 42.7	74 0.1	J. Renwick, C. S.	Sept., 1845	6 25.3 "
Oyster Bay, L. I.,	40 52.3	73 31.3	J. Renwick, C. S.	Sept., 1844	6 53.6 "
Rouse's Point,	45 0.0	73 21.0	Boundary Survey,	Oct., 1845	11 28.0 "
Sands Lighthouse, L. I.,	40 51.9	73 43.5	R.H. Fauntleroy, C.S.	Oct., 1847	6 9.7 "
Sands Point, L. I.,	40 52.0	73 43.0	J. Renwick, C. S.	Sept., 1845	7 14.6 "
Watchhill, Fire Island,	40 41.4	72 58.9	R.H. Fauntleroy, C.S.	Oct., 1847	7 33.5 "
West Point	41 25.0	73 56.0	Prof. Davies,	Sept., 1835	6 32.0 "
New Jersey.					
Cape May Lighthouse,	38 55.8	74 57.6	J. Locke, C. S.	June, 1846	3 3.2 W.
Chew,	39 48.2	75 9.7	J. Locke, C. S.	July, 1846	3 20.4 "
Church Landing,	39 40.9	75 30.3	J. Locke, C. S.	June, 1846	*5 45.8 "
Egg Island,	39 10.4	75 7.8	J. Locke, C. S.	June, 1846	3 18.2 "
Hawkins,	39 25.5	75 17.1	J. Locke, C. S.	June, 1846	2 58.7 "
Mt. Rose, Princeton,	40 22.2	74 42.9	J. E. Hilgard, C. S.	Aug., 1852	5 31.8 "
Newark,	40 44.8	74 7.0	J. Locke, C. S.	April, 1846	5 32.7 "
Pine Mountain,	39 25.0	75 19.9	J. Locke, C. S.	June, 1846	2 52.0 "
Port Norris.	39 14.5	75 1.9	J. Locke, C. S.	June, 1846	3 6.5 "
Sandy Hook,	40 28.0	73 59.8	J. Renwick, C. S	Aug., 1844	5 54.0 "
Town Bank, Cape May,	38 58.6	74 57.4	J. Locke, C. S.	June, 1846	3 3.2 "
Tucker's Island,	39 30.8	74 16.9	T. J. Lee, C. S.	Nov., 1846	4 23.8 "
White Hill, Bordentown,	40 8.3	74 43.8	J. Locke, C. S.	April, 1846	4 22.5 "
Pennsylvania.					
Girard College, Philadelphia,	39 58.4	75 9.9	J. Locke, C. S.	May, 1846	3 50.7 W.
Pittsburg,	40 26.0	79 58.0	J. Locke,	May, 1845	0 33.1 "
Vanuxem, Bristol,	40 5.9	74 52.7	J. Locke, C. S.	July, 1846	4 20.5 "

* Local attraction exists here, according to Prof. Locke.

Place.	Latitude.	Longitude.	Authority.	Date.	Variation.
Delaware.					
Bombay Hook Lighthouse,	39° 21.8′	75° 30.3′	J. Locke, C. S.	June, 1846	3° 17.9′ W.
Fort Delaware, Delaware River,	39 35.3	75 33.8	J. Locke, C. S.	June, 1846	3 16.0 "
Lewes Landing,	38 48.8	75 11.5	J. Locke, C. S.	July, 1846	2 47.7 "
Pilot Town,	38 47.1	75 9.2	J. Locke, C. S.	July, 1846	2 42.2 "
Sawyer,	39 42.0	75 33.5	J. Locke, C. S.	June, 1846	2 47.8 "
Wilmington,	39 44.9	75 33.6	J. Locke, C. S.	May, 1846	2 31.8 "
Maryland.					
Annapolis,	38 56.0	76 35.0	T. J. Lee, C. S.	June, 1845	2 14.0 W.
Bodkin,	39 8.0	76 25.2	T. J. Lee, C. S.	April, 1847	2 2.6 "
Finlay,	39 24.4	76 31.2	J. Locke, C. S.	April, 1846	2 19.5 "
Fort McHenry, Baltimore,	39 15.7	76 34.5	T. J. Lee, C. S.	April, 1847	2 13.0 "
Hill,	38 53.9	76 52.5	G. W. Dean, C. S.	Sept., 1850	2 15.4 "
Kent Island,	39 1.8	76 18.8	J. Heuston, C. S.	July, 1849	2 30.5 "
Marriott's,	38 52.4	76 36.3	T. J Lee, C. S.	June, 1849	2 5.2 "
North Point,	39 11.7	76 26.3	T. J. Lee, C. S.	July, 1846	1 42.1 "
Osborne's Ruin,	39 27.9	76 16.6	T. J. Lee, C. S.	June, 1845	2 32.4 "
Poole's Island,	39 17.1	76 15.6	T. J. Lee, C. S.	June, 1847	2 28.5 "
Rosanne,	39 17.5	76 42.8	T. J. Lee, C. S.	June, 1845	2 12.0 "
Soper,	39 5.1	76 56.7	G. W. Dean, C. S.	July, 1850	2 7.0 "
South Base, Kent Island,	38 53.8	76 21.7	T J. Lee, C. S.	June, 1845	2 26.2 "
Susquehanna Lighthouse, Havre de Grace,	39 32.4	76 4.8	T. J. Lee, C. S.	July, 1847	2 51.1 "
Taylor,	38 59.8	76 27.6	T. J. Lee, C. S.	May, 1847	2 18.4 "
Webb,	39 5.4	76 40.2	G. W. Dean, C. S.	Nov., 1850	2 7.9 "
District of Columbia.					
Causten, Georgetown,	38 55.5	77 4.1	G. W. Dean, C. S.	June, 1851	2 11.3 W.
Washington,	38 53.7	77 2.8	J. M. Gilliss,	June, 1842	1 26.0 "
Virginia.					
Charlottesville,	38 2.0	78 31.0	Prof. Patterson,	1835	0 0.0
Roslyn, Petersburg,	37 14.4	77 23.5	G. W. Dean, C. S.	Aug., 1852	0 26.4 W.
Wheeling,	40 8.0	80 47.0	J. Locke,	April, 1845	2 4.0 E.
North Carolina.					
Bodie's Island,	35 47.5	75 31.6	C. O. Boutelle, C. S.	Dec., 1846	1 13.4 W.
Shellbank,	36 3.3	75 44.1	C. O. Boutelle, C. S.	Mar., 1847	1 44.8 "
Stevenson's Point,	36 6.3	76 11.0	C. O. Boutelle, C. S.	Feb., 1847	1 39.7 "
South Carolina.					
Breach Inlet,	32 46.3	79 48.7	C. O. Boutelle, C. S.	April, 1849	2 16.5 E.
Charleston,	32 41.0	79 53.0	Capt. Barnett,	May, 1841	2 24.0 "
East Base, Edisto,	32 33.3	80 10.0	G. Davidson, C. S.	April, 1850	2 53.6 "
Georgia.					
Athens,	34 0.0	83 20.0	Prof. McCay.	1837	4 31.0 E.
Columbus,	32 28.0	85 10.0	Geol. Survey,	1839	5 30.0 "
Milledgeville,	33 7.0	83 20.0	Geol. Survey,	1838	5 51.0 "
Savannah,	32 5.0	81 5.2	J. E. Hilgard, C. S.	April, 1852	3 45.0 "

Place.	Latitude.	Longitude.	Authority.	Date.	Variation.
Florida.	° ′	° ′			° ′
Cape Florida,	25 39.9	80 9.4	J. E. Hilgard, C. S.	Feb., 1850	4 25.2 E.
Cedar Keys,	29 7.5	83 2.8	J. E. Hilgard, C. S.	Mar., 1852	5 20.5 "
St. Marks Light,	30 4.5	84 12.5	J. E. Hilgard, C. S.	April, 1852	5 29.2 "
Sand Key,	24 27.2	81 52.0	J. E. Hilgard, C. S.	Aug., 1849	5 29.0 "
Alabama.					
Fort Morgan, Mobile Bay,	30 13.8	88 0.4	R.H. Fauntleroy, C.S.	May, 1847	7 3.8 E.
Tuscaloosa,	33 12.0	87 42.0	Prof. Barnard,	1839	7 28.0 "
Mississippi.					
East Pascagoula,	30 20.7	88 31.4	R.H. Fauntleroy, C.S.	June, 1847	7 12.4 E.
Texas.					
Dollar Point, Galveston,	29 26.0	94 53.0	R.H. Fauntleroy, C.S.	April, 1848	8 57.2 E.
Mouth of Sabine,	29 43.9	93 51.5	J. D. Graham,	Feb., 1840	8 40.2 "
Ohio.					
Carrolton,	39 38.0	84 9.0	J Locke,	Sept., 1845	4 45.4 E.
Cincinnati,	39 6.0	84 22.0	J. Locke,	April, 1845	4 4.0 "
Columbus,	39 57.0	83 3.0	J. Locke,	July, 1845	2 29.3 "
Hudson,	41 15.0	81 26.0	E. Loomis,	1840	0 52.0 "
Marietta,	39 26.0	81 29.0	J. Locke,	April, 1845	2 25.0 "
Oxford,	39 30.0	84 38.0	J. Locke,	Aug., 1845	4 50.0 "
St. Mary's,	40 32.0	84 19.[illegible]	J. Locke,	Sept., 1845	3 4.0 "
Tennessee.					
Nashville,	36 10.0	86 4[illegible]	Prof. Hamilton,	1835	7 7.0 E.
Michigan.					
Detroit,	42 24.0	82 58.0	Geol. Report,	1840	2 0.0 E.
Indiana.					
Richmond,	39 19.0	84 47.0	J Locke,	Sept., 1845	4 52.0 E.
South Hanover,	38 45.0	85 23.0	Prof. Dunn,	1837	4 35.0 "
Illinois.					
Alton,	38 52.0	90 12.0	H. Loomis,	1840	7 45.0 E.
Missouri.					
St. Louis,	38 36.0	89 36.0	Col. Nicolls,	1835	8 49.0 E.
Wisconsin.					
Madison,	43 5.0	89 41.0	U. S. Surveyors,	Nov., 1839	7 30.0 E.
Prairie du Chien,	43 1.0	91 8.0	U. S. Surveyors,	Oct., 1839	9 5.0 "
Iowa.					
Brown's Settlement	42 2.0	91 18.0	J. Locke,	Sept., 1839	9 4.0 E.
Davenport,	41 30.0	90 34.0	U. S. Surveyors,	Sept., 1839	7 50.0 "
Farmer's Creek,	42 13.0	90 39.0	J. Locke,	Oct., 1839	9 11.0 "
Wapsipinnicon River,	41 44.0	90 39.0	J. Locke,	Sept., 1839	8 25.0 "
California.					
Point Conception,	34 26.9	120 26.0	G. Davidson, C. S.	Sept., 1850	13 49.5 E.

Place.	Latitude.	Longitude.	Authority.	Date	Variation.
Point Pinos, Monterey,	° ′ 36 38.0	° ′ 121 54.0	G. Davidson, C. S.	Feb., 1851	° ′ 14 58.0 E.
Presidio, San Francisco,	37 47.8	122 27.0	G. Davidson, C. S.	Feb., 1852	15 26.9 "
San Diego,	32 42.0	117 14.0	G. Davidson, C. S.	May, 1851	12 29.0 "
Oregon.					
Cape Disappointment,	46 16.6	124 2.0	G. Davidson, C. S.	July, 1851	20 45.0 E.
Ewing Harbor,	42 44.4	124 21.0	G. Davidson, C. S.	Nov., 1851	18 29.2 "
Washington Territory.					
Scarboro' Harbor,	48 21.8	124 37.2	G. Davidson, C. S.	Aug., 1852	21 30.2 E.
British America.					
Montreal,	45 30.0	73 35.0	Capt. Lefroy,	1842	8 58.0 W.
Quebec,	46 49.0	71 16.0	Capt. Lefroy,	1842	14 12.0 "
St. Johns, C. E.	45 19.0	73 18.0	Capt. Lefroy,	1842	11 22.0 "
Stanstead,	45 0.0	72 13.0	Boundary Survey,	Nov., 1845	11 33.0 "
Toronto,	43 39.6	79 21.5	British Govern.,	Sept., 1844	1 27.2 "
New Grenada.					
Panama,	8 57.2	79 29.4	W. H. Emory,	Mar., 1849	6 54.6 E.
Eastern Hemisphere.					
Greenwich, England,	51 28.0	0 0.0	Prof. Airy,	1841	23 16.0 W.
Makerstoun, Scotland,	55 35.0	2 31.0 W.	J. A. Broun,	1842	25 28.6 "
Paris, France,	48 50.0	2 20.0 E.	Paris Observatory	Nov., 1851	20 25.0 "
Munich, Bavaria,	48 9.0	11 37.0 "		1842	16 48.0 "
St. Petersburg, Russia,	59 56.0	30 19.0 "	Russian Govern.,	1842	6 21.1 "
Catherinenburg Siberia,	56 51.0	60 34.0 "	Russian Govern.,	1842	6 38.9 E.
Nertchinsk, Siberia,	51 56.0	116 31.0 "	Russian Govern.,	1842	3 46.9 W.
St. Helena,	15 56.7 S.	5 40.5 W.	British Govern.,	Dec., 1845	23 36.6 "
Cape of Good Hope,	33 56.0 "	18 28.7 E.	British Govern.,	July, 1846	29 8.0 "
Hobarton, Van Diemen's Ld.,	42 52.5 "	147 27.5 "	British Govern.,	Dec., 1848	10 8.0 E.

TABLE X.

TRIGONOMETRICAL AND MISCELLANEOUS FORMULÆ

LET A (fig. 57) be any acute angle, and let a perpendicular $B\,C$ be drawn from any point in one side to the other side. Then, if the sides

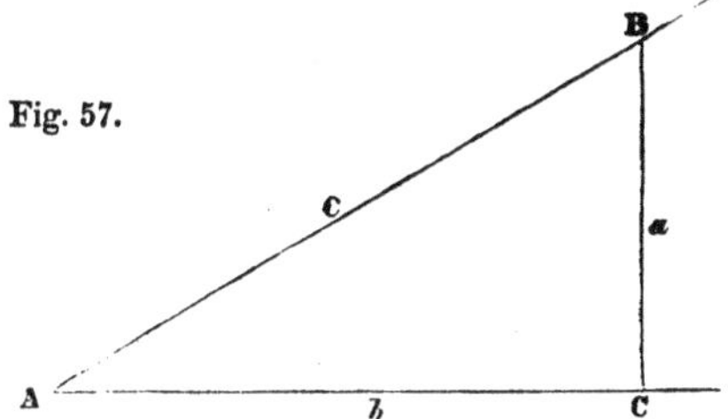

Fig. 57.

of the right triangle thus formed are denoted by letters, as in the figure, we shall have these six formulæ: —

1. $\sin. A = \frac{a}{c}$.
2. $\cos. A = \frac{b}{c}$.
3. $\tan. A = \frac{a}{b}$.
4. $\text{cosec.} A = \frac{c}{a}$.
5. $\sec. A = \frac{c}{b}$.
6. $\cot. A = \frac{b}{a}$.

Solution of Right Triangles (fig. 57).

	Given.	Sought.	Formulæ.
7	a, c	A, B, b	$\sin. A = \frac{a}{c}$, $\cos. B = \frac{a}{c}$, $b = \sqrt{(c+a)\,(c-a)}$
8	a, b	A, B, c	$\tan. A = \frac{a}{b}$, $\cot. B = \frac{a}{b}$, $c = \sqrt{a^2 + b^2}$.
9	A, a	B, b, c	$B = 90° - A$, $b = a \cot. A$, $c = \frac{a}{\sin. A}$.
10	A, b	B, a, c	$B = 90° - A$, $a = b \tan. A$, $c = \frac{b}{\cos. A}$.
11	A, c	B, a, b	$B = 90° - A$, $a = c \sin. A$, $b = c \cos. A$.

Solution of Oblique Triangles (fig. 58).

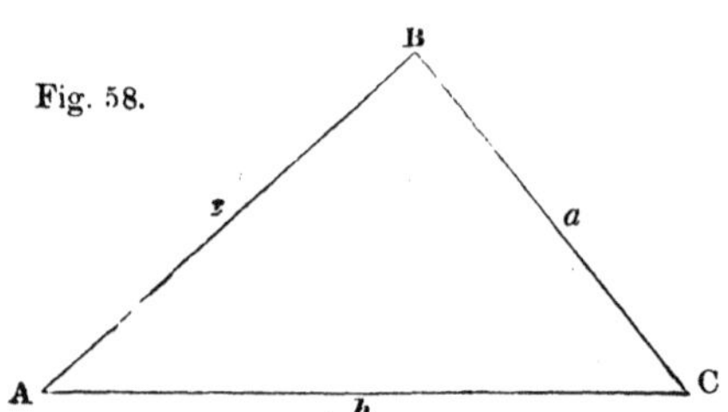

Fig. 58.

	Given.	Sought.	Formulæ.
12	A, B, a	b	$b = \frac{a \sin. B}{\sin. A}$.
13	A, a, b	B	$\sin. B = \frac{b \sin. A}{a}$.
14	a, b, C	$A - B$	$\tan. \frac{1}{2}(A - B) = \frac{(a - b) \tan. \frac{1}{2}(A + B)}{a + b}$.
15	a, b, c	A	If $s = \frac{1}{2}(a + b + c)$, $\sin. \frac{1}{2} A = \sqrt{\frac{(s - b)(s - c)}{bc}}$ $\cos. \frac{1}{2} A = \sqrt{\frac{s(s - a)}{bc}}$, $\tan. \frac{1}{2} A = \sqrt{\frac{(s - b)(s - c)}{s(s - a)}}$, $\sin. A = \frac{2\sqrt{s(s - a)(s - b)(s - c)}}{bc}$.
16	A, B, C, a	area	area $= \frac{a^2 \sin. B \sin C}{2 \sin. A}$.
17	A, b, c	area	area $= \frac{1}{2} b c \sin. A$.
18	a, b, c	area	$s = \frac{1}{2}(a + b + c)$, area $= \sqrt{s(s - a)(s - b)(s - c)}$.

General Trigonometrical Formulæ.

19 $\sin.^2 A + \cos.^2 A = 1$.

20 $\sin. (A \pm B) = \sin. A \cos. B \pm \sin. B \cos. A$.

21 $\cos. (A \pm B) = \cos. A \cos. B \mp \sin. A \sin. B$.

22 $\sin. 2A = 2 \sin. A \cos. A$.

23 $\cos. 2A = \cos.^2 A - \sin.^2 A = 1 - 2 \sin.^2 A = 2 \cos.^2 A - 1$.

24 $\sin.^2 A = \frac{1}{2} - \frac{1}{2} \cos. 2A$.

25 $\cos.^2 A = \frac{1}{2} + \frac{1}{2} \cos. 2A$.

26 $\sin. A + \sin. B = 2 \sin. \frac{1}{2}(A + B) \cos. \frac{1}{2}(A - B)$.

27 $\sin. A - \sin. B = 2 \cos. \frac{1}{2}(A + B) \sin. \frac{1}{2}(A - B)$.

28 $\cos. A + \cos. B = 2 \cos. \frac{1}{2}(A + B) \cos. \frac{1}{2}(A - B)$.

29 $\cos. B - \cos. A = 2 \sin. \frac{1}{2}(A + B) \sin. \frac{1}{2}(A - B)$.

30 $\sin.^2 A - \sin.^2 B = \cos.^2 B - \cos.^2 A = \sin. (A + B) \sin. (A - B)$.

31 $\cos.^2 A - \sin.^2 B = \cos. (A + B) \cos. (A - B)$.

32	$\tan. A = \frac{\sin. A}{\cos. A}.$
33	$\cot. A = \frac{\cos. A}{\sin. A}.$
34	$\tan. (A \pm B) = \frac{\tan A \pm \tan B}{1 \mp \tan. A \tan. B}.$
35	$\tan. A \pm \tan. B = \frac{\sin (A \pm B)}{\cos. A \cos. B}.$
36	$\cot. A \pm \cot. B = \pm \frac{\sin (A \pm B)}{\sin. A \sin. B}.$
37	$\frac{\sin A + \sin B}{\sin. A - \sin. B} = \frac{\tan \frac{1}{2} (A + B)}{\tan. \frac{1}{2} (A - B)}.$
38	$\frac{\sin A + \sin. B}{\cos. A + \cos. B} = \tan. \frac{1}{2} (A + B).$
39	$\frac{\sin A + \sin B}{\cos. B - \cos. A} = \cot. \frac{1}{2} (A - B).$
40	$\frac{\sin. A - \sin. B}{\cos. A + \cos. B} = \tan. \frac{1}{2} (A - B).$
41	$\frac{\sin. A - \sin. B}{\cos. B - \cos. A} = \cot. \frac{1}{2} (A + B).$
42	$\tan. \frac{1}{2} A = \frac{\sin A}{1 + \cos. A}.$
43	$\cot. \frac{1}{2} A = \frac{\sin. A}{1 - \cos A}.$

Miscellaneous Formulæ.

	Sought.	Given.	Formulæ.
	Area of		
44	Circle	Radius $= r$	πr^2.
45	Ellipse	Semi-axes $= a$ and b	$\pi a b$.
46	Parabola	Chord $= c$, height $= h$	$\frac{2}{3} c h$.*
47	Regular Polygon	Side $= a$, number of sides $= n$	$\frac{1}{4} a^2 n \cot. \frac{180°}{n}$.
	Surface of		
48	Sphere	Radius $= r$	$4 \pi r^2$.
49	Zone	Radius $= r$, height $= h$	$2 \pi r h$.
50	Spherical Polygon	Radius of sphere $= r$, sum of angles $= S$, number of sides $= n$	$\pi r^2 \times \frac{S - (n - 2) 180°}{180°}$
	Solidity of		
51	Prism or Cylinder	Base $= b$, height $= h$	$b h$.
52	Pyramid or Cone	Base $= b$, height $= h$	$\frac{1}{3} b h$.
53	Frustum of Pyramid or Cone	Bases $= b$ and b_1, height $= h$	$\frac{1}{3} h (b + b_1 + \sqrt{b b_1})$

* The area of a circular segment on railroad curves, where the chord is very long in proportion to the height, may be found with great accuracy by the above formula.

	Sought.	Given.	Formulæ.
	Solidity of		
54	Sphere	Radius $= r$	$\frac{4}{3}\pi r^3$.
55	Spherical Segment	Radii of bases $= r$ and r_1, height $= h$	$\frac{1}{2}\pi h(r^2 + r_1^2 + \frac{1}{3}h^2)$
56	Prolate Spheroid	Semi-transverse axis of ellipse $= a$	$\frac{4}{3}\pi a b^2$.
57	Oblate Spheroid	Semi-conjugate axis of ellipse $= b$	$\frac{4}{3}\pi a^2 b$.
58	Paraboloid	Radius of base $= r$, height $= h$	$\frac{1}{2}\pi r^2 h$.

π = 3.14159 26535 89793 23846 26433 83280.

Log. π = 0.49714 98726 94133 85435 12682 88291

United States Standard Gallon	= 231 cub. in.	=	0.133681 cub. ft
" " " Bushel	= 2150.42 "	=	1.244456 "
British Imperial Gallon	= 277.27384 "	=	0.160459 "

	According to Hassler.		As usually given.
French Metre,	= 3.2817431 ft.,	=	3.280899 ft.
" Litre,	= 61.0741569 cub. in.,	=	61.02705 cub. in.
" Kilogram,	= 2.204737 lb. avoir.,	=	2.204597 lb. avoir

Weight of Cubic Foot of Water,

Barom. 30 inches, Therm. Fahr. 39.83°,	=	62.379 lb. avoir.
" " " 62°,	=	62.321 "

Length of Seconds Pendulum at New York	=	39.10120 inches.
" " " " " London	=	39.13908 "
" " " " " Paris	=	39.12843 "

Equatorial Radius of Earth according to Bessel	=	20,923,597.017 feet
Polar " " " "	=	20,853,654.177 "

TABLE XI.

SQUARES, CUBES, SQUARE ROOTS, CUBE ROOTS,

AND

RECIPROCALS OF NUMBERS

FROM 1 TO 1054.

No.	Squares.	Cubes.	Square Roots.	Cube Roots.	Reciprocals.
1	1	1	1.0000000	1.0000000	1.000000000
2	4	8	1.4142136	1.2599210	.500000000
3	9	27	1.7320508	1.4422496	.333333333
4	16	64	2.0000000	1.5874011	.250000000
5	25	125	2.2360680	1.7099759	.200000000
6	36	216	2.4494897	1.8171206	.166666667
7	49	343	2.6457513	1.9129312	.142857143
8	64	512	2.8284271	2.0000000	.125000000
9	81	729	3.0000000	2.0800837	.111111111
10	100	1000	3 1622777	2.1544347	.100000000
11	121	1331	3.3166248	2.2239801	.090909091
12	144	1728	3.4641016	2.2894286	.083333333
13	169	2197	3.6055513	2.3513347	.076923077
14	196	2744	3.7416574	2.4101422	.071428571
15	225	3375	3.8729833	2.4662121	.066666667
16	256	4096	4.0000000	2.5198421	.062500000
17	289	4913	4.1231056	2.5712816	.058823529
18	324	5832	4.2426407	2.6207414	.055555556
19	361	6859	4.3588989	2.6684016	.052631579
20	400	8000	4.4721360	2.7144177	.050000000
21	441	9261	4.5825757	2.7589243	.047619048
22	484	10648	4.6904158	2.8020393	.045454545
23	529	12167	4.7958315	2.8438670	.04347826
24	576	13824	4.8989795	2.8844991	.041666667
25	625	15625	5.0000000	2.9240177	.040000000
26	676	17576	5.0990195	2.9624960	.038461538
27	729	19683	5.1961524	3.0000000	.037037037
28	784	21952	5.2915026	3.0365889	.035714286
29	841	24389	5.3851648	3.0723168	.034482759
30	900	27000	5.4772256	3.1072325	.033333333
31	961	29791	5.5677644	3.1413806	.032258065
32	1024	32768	5.6568542	3.1748021	.031250000
33	1089	35937	5.7445626	3.2075343	.030303030
34	1156	39304	5.8309519	3.2396118	.029411765
35	1225	42875	5.9160798	3.2710663	.028571429
36	1296	46656	6.0000000	3.3019272	.027777778
37	1369	50653	6.0827625	3.3322218	.027027027
38	1444	54872	6.1644140	3.3619754	.026315789
39	1521	59319	6.2449980	3.3912114	.025641026
40	1600	64000	6.3245553	3.4199519	.025000000
41	1681	68921	6.4031242	3.4482172	.024390244
42	1764	74088	6.4807407	3.4760266	.023809524
43	1849	79507	6.5574385	3.5033981	.023255814
44	1936	85184	6.6332496	3.5303483	.022727273
45	2025	91125	6.7082039	3.5568933	.022222222
46	2116	97336	6.7823300	3.5830479	.021739130
47	2209	103823	6.8556546	3.6088261	.021276600
48	2304	110592	6.9282032	3.6342411	.020833333
49	2401	117649	7.0000000	3.6593057	.020408163
50	2500	125000	7.0710678	3.6840314	.020000000
51	2601	132651	7.1414284	3.7084298	.019607843
52	2704	140608	7.2111026	3.7325111	.019230769
53	2809	148877	7.2801099	3.7562858	.018867925
54	2916	157464	7.3484692	3.7797631	.018518519
55	3025	166375	7.4161985	3.8029525	.018181818
56	3136	175616	7.4833148	3.8258624	.017857143
57	3249	185193	7.5498344	3.8485011	.017543860
58	3364	195112	7.6157731	3.8708766	.017241379
59	3481	205379	7.6811457	3.8929965	.016949153
60	3600	216000	7.7459667	3.9148676	.016666667
61	3721	226981	7.8102497	3.9364972	.016393443
62	3844	238328	7.8740079	3.9578915	.016129032

No.	Squares.	Cubes.	Square Roots.	Cube Roots.	Reciprocals.
63	3969	250047	7.9372539	3.9790571	.015873016
64	4096	262144	8.0000000	4.0000000	.015625000
65	4225	274625	8.0622577	4.0207256	.015384615
66	4356	287496	8.1240384	4.0412401	.015151515
67	4489	300763	8.1853528	4.0615480	.014925373
68	4624	314432	8.2462113	4.0816551	.014705882
69	4761	328509	8.3066239	4.1015661	.014492754
70	4900	343000	8.3666003	4.1212853	.014285714
71	5041	357911	8.4261498	4.1408178	.014084507
72	5184	373248	8.4852814	4.1601676	.013888889
73	5329	389017	8.5440037	4.1793390	.013698630
74	5476	405224	8.6023253	4.1983364	.013513514
75	5625	421875	8.6602540	4.2171633	.013333333
76	5776	438976	8.7177979	4.2358236	.013157895
77	5929	456533	8.7749644	4.2543210	.012987013
78	6084	474552	8.8317609	4.2726586	.012820513
79	6241	493039	8.8881944	4.2908404	.012658228
80	6400	512000	8.9442719	4.3088695	.012500000
81	6561	531441	9.0000000	4.3267487	.012345679
82	6724	551368	9.0553851	4.3444815	.012195122
83	6889	571787	9.1104336	4.3620707	.012048193
84	7056	592704	9.1651514	4.3795191	.011904762
85	7225	614125	9.2195445	4.3968296	.011764706
86	7396	636056	9.2736185	4.4140049	.011627907
87	7569	658503	9.3273791	4.4310476	.011494253
88	7744	681472	9.3808315	4.4479602	.011363636
89	7921	704969	9.4339811	4.4647451	.011235955
90	8100	729000	9.4868330	4.4814047	.011111111
91	8281	753571	9.5393920	4.4979414	.010989011
92	8464	778688	9.5916630	4.5143574	.010869565
93	8649	804357	9.6436508	4.5306549	.010752688
94	8836	830584	9.6953597	4.5468359	.010638298
95	9025	857375	9.7467943	4.5629026	.010526316
96	9216	884736	9.7979590	4.5788570	.010416667
97	9409	912673	9.8488578	4.5947009	.010309278
98	9604	941192	9.8994949	4.6104363	.010204082
99	9801	970299	9.9498744	4.6260650	.010101010
100	10000	1000000	10.0000000	4.6415888	.010000000
101	10201	1030301	10.0498756	4.6570095	.009900990
102	10404	1061208	10.0995049	4.6723287	.009803922
103	10609	1092727	10.1488916	4.6875482	.009708738
104	10816	1124864	10.1980390	4.7026694	.009615385
105	11025	1157625	10.2469508	4.7176940	.009523810
106	11236	1191016	10.2956301	4.7326235	.009433962
107	11449	1225043	10.3440804	4.7474594	.009345794
108	11664	1259712	10.3923048	4.7622032	.009259259
109	11881	1295029	10.4403065	4.7768562	.009174312
110	12100	1331000	10.4880085	4.7914199	.009090909
111	12321	1367631	10.5356538	4.8058955	.009009009
112	12544	1404928	10.5830052	4.8202845	.008928571
113	12769	1442897	10.6301458	4.8345881	.008849558
114	12996	1481544	10.6770783	4.8488076	.008771930
115	13225	1520875	10.7238053	4.8629442	.008695652
116	13456	1560896	10.7703296	4.8769990	.008620690
117	13689	1601613	10.8166538	4.8909732	.008547009
118	13924	1643032	10.8627805	4.9048681	.008474576
119	14161	1685159	10.9087121	4.9186847	.008403361
120	14400	1728000	10.9544512	4.9324242	.008333333
121	14641	1771561	11.0000000	4.9460874	.008264463
122	14884	1815848	11.0453610	4.9596757	.008196721
123	15129	1860867	11.0905365	4.9731898	.008130081
124	15376	1906624	11.1355287	4.9866310	.008064516

No.	Squares.	Cubes.	Square Roots.	Cube Roots.	Reciprocals.
125	15625	1953125	11.1803399	5.0000000	.008000000
126	15876	2000376	11.2249722	5.0132979	.007936508
127	16129	2048383	11.2694277	5.0265257	.007874016
128	16384	2097152	11.3137085	5.0396842	.007812500
129	16641	2146689	11.3578167	5.0527743	.007751938
130	16900	2197000	11.4017543	5.0657970	.007692308
131	17161	2248091	11.4455231	5.0787531	.007633588
132	17424	2299968	11.4891253	5.0916434	.007575758
133	17689	2352637	11.5325626	5.1044687	.007518797
134	17956	2406104	11.5758369	5.1172299	.007462687
135	18225	2460375	11.6189500	5.1299278	.007407407
136	18496	2515456	11.6619038	5.1425632	.007352941
137	18769	2571353	11.7046999	5.1551367	.007299270
138	19044	2628072	11.7473444	5.1676493	.007246377
139	19321	2685619	11.7898261	5.1801015	.007194245
140	19600	2744000	11.8321596	5.1924941	.007142857
141	19881	2803221	11.8743421	5.2048279	.007092199
142	20164	2863288	11.9163753	5.2171034	.007042254
143	20449	2924207	11.9582607	5.2293215	.006993007
144	20736	2985984	12.0000000	5.2414828	.006944444
145	21025	3048625	12.0415946	5.2535879	.006896552
146	21316	3112136	12.0830460	5.2656374	.006849315
147	21609	3176523	12.1243557	5.2776321	.006802721
148	21904	3241792	12.1655251	5.2895725	.006756757
149	22201	3307949	12.2065556	5.3014592	.006711409
150	22500	3375000	12.2474487	5.3132928	.006666667
151	22801	3442951	12.2882057	5.3250740	.006622517
152	23104	3511808	12.3288280	5.3368033	.006578947
153	23409	3581577	12.3693169	5.3484812	.006535948
154	23716	3652264	12.4096736	5.3601084	.006493506
155	24025	3723875	12.4498996	5.3716854	.006451613
156	24336	3796416	12.4899960	5.3832126	.006410256
157	24649	3869893	12.5299641	5.3946907	.006369427
158	24964	3944312	12.5698051	5.4061202	.006329114
159	25281	4019679	12.6095202	5.4175015	.006289308
160	25600	4096000	12.6491106	5.4288352	.006250000
161	25921	4173281	12.6885775	5.4401218	.006211180
162	26244	4251528	12.7279221	5.4513618	.006172840
163	26569	4330747	12.7671453	5.4625556	.006134969
164	26896	4410944	12.8062485	5.4737037	.006097561
165	27225	4492125	12.8452326	5.4848066	.006060606
166	27556	4574296	12.8840987	5.4958647	.006024096
167	27889	4657463	12.9228480	5.5068784	.005988024
168	28224	4741632	12.9614814	5.5178484	.005952381
169	28561	4826809	13.0000000	5.5287748	.005917160
170	28900	4913000	13.0384048	5.5396583	.005882353
171	29241	5000211	13.0766968	5.5504991	.005847953
172	29584	5088448	13.1148770	5.5612978	.005813953
173	29929	5177717	13.1529464	5.5720546	.005780347
174	30276	5268024	13.1909060	5.5827702	.005747126
175	30625	5359375	13.2287566	5.5934447	.005714286
176	30976	5451776	13.2664992	5.6040787	.005681818
177	31329	5545233	13.3041347	5.6146724	.005649718
178	31684	5639752	13.3416641	5.6252263	.005617978
179	32041	5735339	13.3790882	5.6357408	.005586592
180	32400	5832000	13.4164079	5.6462162	.005555556
181	32761	5929741	13.4536240	5.6566528	.005524862
182	33124	6028568	13.4907376	5.6670511	.005494505
183	33489	6128487	13.5277493	5.6774114	.005464481
184	33856	6229504	13.5646600	5.6877340	.005434783
185	34225	6331625	13.6014705	5.6980192	.005405405
186	34596	6434856	13.6381817	5.7082675	.005376344

No.	Squares.	Cubes.	Square Roots.	Cube Roots.	Reciprocals.
187	34969	6539203	13.6747943	5.7184791	.005347594
188	35344	6644672	13.7113092	5.7286543	.005319149
189	35721	6751269	13.7477271	5.7387936	.005291005
190	36100	6859000	13.7840488	5.7488971	.005263158
191	36481	6967871	13.8202750	5.7589652	.005235602
192	36864	7077888	13.8564065	5.7689982	.005208333
193	37249	7189057	13.8924440	5.7789966	.005181347
194	37636	7301384	13.9283883	5.7889604	.005154639
195	38025	7414875	13.9642400	5.7988900	.005128205
196	38416	7529536	14.0000000	5.8087857	.005102041
197	38809	7645373	14.0356688	5.8186479	.005076142
198	39204	7762392	14.0712473	5.8284767	.005050505
199	39601	7880599	14.1067360	5.8382725	.005025126
200	40000	8000000	14.1421356	5.8480355	.005000000
201	40401	8120601	14.1774469	5.8577660	.004975124
202	40804	8242408	14.2126704	5.8674643	.004950495
203	41209	8365427	14.2478068	5.8771307	.004926108
204	41616	8489664	14.2828569	5.8867653	.004901961
205	42025	8615125	14.3178211	5.8963685	.004878049
206	42436	8741816	14.3527001	5.9059406	.004854369
207	42849	8869743	14.3874946	5.9154817	.004830918
208	43264	8998912	14.4222051	5.9249921	.004807692
209	43681	9129329	14.4568323	5.9344721	.004784689
210	44100	9261000	14.4913767	5.9439220	.004761905
211	44521	9393931	14.5258390	5.9533418	.004739336
212	44944	9528128	14.5602198	5.9627320	.004716981
213	45369	9663597	14.5945195	5.9720926	.004694836
214	45796	9800344	14.6287388	5.9814240	.004672897
215	46225	9938375	14.6628783	5.9907264	.004651163
216	46656	10077696	14.6969385	6.0000000	.004629630
217	47089	10218313	14.7309199	6.0092450	.004608295
218	47524	10360232	14.7648231	6.0184617	.004587156
219	47961	10503459	14.7986486	6.0276502	.004566210
220	48400	10648000	14.8323970	6.0368107	.004545455
221	48841	10793861	14.8660687	6.0459435	.004524887
222	49284	10941048	14.8996644	6.0550489	.004504505
223	49729	11089567	14.9331845	6.0641270	.004484305
224	50176	11239424	14.9666295	6.0731779	.004464286
225	50625	11390625	15.0000000	6.0822020	.004444444
226	51076	11543176	15.0332964	6.0911994	.004424779
227	51529	11697083	15.0665192	6.1001702	.004405286
228	51984	11852352	15.0996689	6.1091147	.004385965
229	52441	12008989	15.1327460	6.1180332	.004366812
230	52900	12167000	15.1657509	6.1269257	.004347826
231	53361	12326391	15.1986842	6.1357924	.004329004
232	53824	12487168	15.2315462	6.1446337	.004310345
233	54289	12649337	15.2643375	6.1534495	.004291845
234	54756	12812904	15.2970585	6.1622401	.004273504
235	55225	12977875	15.3297097	6.1710058	.004255319
236	55696	13144256	15.3622915	6.1797466	.004237288
237	56169	13312053	15.3948043	6.1884628	.004219409
238	56644	13481272	15.4272486	6.1971544	.004201681
239	57121	13651919	15.4596248	6.2058218	.004184100
240	57600	13824000	15.4919334	6.2144650	.004166667
241	58081	13997521	15.5241747	6.2230843	.004149378
242	58564	14172488	15.5563492	6.2316797	.004132231
243	59049	14348907	15.5884573	6.2402515	.004115226
244	59536	14526784	15.6204994	6.2487998	.004098361
245	60025	14706125	15.6524758	6.2573248	.004081633
246	60516	14886936	15.6843871	6.2658266	.004065041
247	61009	15069223	15.7162336	6.2743054	.004048583
248	61504	15252992	15.7480157	6.2827613	.004032258

No.	Squares.	Cubes.	Square Roots.	Cube Roots.	Reciprocals.
249	62001	15438249	15.7797338	6.2911946	.004016064
250	62500	15625000	15.8113883	6.2996053	.004000000
251	63001	15813251	15.8429795	6.3079935	.003984064
252	63504	16003008	15.8745079	6.3163596	.003968254
253	64009	16194277	15.9059737	6.3247035	.003952569
254	64516	16387064	15.9373775	6.3330256	.003937008
255	65025	16581375	15.9687194	6.3413257	.003921569
256	65536	16777216	16.0000000	6.3496042	.003906250
257	66049	16974593	16.0312195	6.3578611	.003891051
258	66564	17173512	16.0623784	6.3660968	.003875969
259	67081	17373979	16.0934769	6.3743111	.003861004
260	67600	17576000	16.1245155	6.3825043	.003846154
261	68121	17779581	16.1554944	6.3906765	.003831418
262	68644	17984728	16.1864141	6.3988279	.003816794
263	69169	18191447	16.2172747	6.4069585	.003802281
264	69696	18399744	16.2480768	6.4150687	.003787879
265	70225	18609625	16.2788206	6.4231583	.003773585
266	70756	18821096	16.3095064	6.4312276	.003759398
267	71289	19034163	16.3401346	6.4392767	.003745318
268	71824	19248832	16.3707055	6.4473057	.003731343
269	72361	19465109	16.4012195	6.4553148	.003717472
270	72900	19683000	16.4316767	6.4633041	.003703704
271	73441	19902511	16.4620776	6.4712736	.003690037
272	73984	20123648	16.4924225	6.4792236	.003676471
273	74529	20346417	16.5227116	6.4871541	.003663004
274	75076	20570824	16.5529454	6.4950653	.003649635
275	75625	20796875	16.5831240	6.5029572	.003636364
276	76176	21024576	16.6132477	6.5108300	.003623188
277	76729	21253933	16.6433170	6.5186839	.003610108
278	77284	21484952	16.6733320	6.5265189	.003597122
279	77841	21717639	16.7032931	6.5343351	.003584229
280	78400	21952000	16.7332005	6.5421326	.003571429
281	78961	22188041	16.7630546	6.5499116	.003558719
282	79524	22425768	16.7928556	6.5576722	.003546099
283	80089	22665187	16.8226038	6.5654144	.003533569
284	80656	22906304	16.8522995	6.5731385	.003521127
285	81225	23149125	16.8819430	6.5808443	.003508772
286	81796	23393656	16.9115345	6.5885323	.003496503
287	82369	23639903	16.9410743	6.5962023	003484321
288	82944	23887872	16.9705627	6.6038545	.003472222
289	83521	24137569	17.0000000	6.6114890	.003460208
290	84100	24389000	17.0293864	6.6191060	.003448276
291	84681	24642171	17.0587221	6.6267054	.003436426
292	85264	24897088	17.0880075	6.6342874	.003424658
293	85849	25153757	17.1172428	6.6418522	.003412969
294	86436	25412184	17.1464282	6.6493998	.003401361
295	87025	25672375	17.1755640	6.6569302	.003389831
296	87616	25934336	17.2046505	6.6644437	.003378378
297	88209	26198073	17.2336879	6 6719403	.003367003
298	88804	26463592	17.2626765	6.6794200	.003355705
299	89401	26730899	17.2916165	6.6868831	.003344482
300	90000	27000000	17.3205081	6.6943295	.003333333
301	90601	27270901	17.3493516	6.7017593	.003322259
302	91204	27543608	17.3781472	6.7091729	.003311258
303	91809	27818127	17.4068952	6.7165700	.003300330
304	92416	28094464	17.4355958	6.7239508	.003289474
305	93025	28372625	17.4642492	6.7313155	.003278689
306	93636	28652616	17.4928557	6.7386641	.003267974
307	94249	28934443	17.5214155	6.7459967	.003257329
308	94864	29218112	17.5499288	6.7533134	.003246753
309	95481	29503629	17.5783958	6.7606143	.003236246
310	96100	29791000	17.6068169	6.7678995	.003225806

No.	Squares.	Cubes.	Square Roots.	Cube Roots.	Reciprocals.
311	96721	30080231	17.6351921	6.7751690	.003215434
312	97344	30371328	17.6635217	6.7824229	.003205128
313	97969	30664297	17.6918060	6.7896613	.003194888
314	98596	30959144	17.7200451	6.7968844	.003184713
315	99225	31255875	17.7482393	6.8040921	.003174603
316	99856	31554496	17.7763888	6.8112847	.003164557
317	100489	31855013	17.8044938	6.8184620	.003154574
318	101124	32157432	17.8325545	6.8256242	.003144654
319	101761	32461759	17.8605711	6.8327714	.003134796
320	102400	32768000	17.8885438	6.8399037	.003125000
321	103041	33076161	17.9164729	6.8470213	.003115265
322	103684	33386248	17.9443584	6.8541240	.003105590
323	104329	33698267	17.9722008	6.8612120	.003095975
324	104976	34012224	18.0000000	6.8682855	.003086420
325	105625	34328125	18.0277564	6.8753443	.003076923
326	106276	34645976	18.0554701	6.8823888	.003067485
327	106929	34965783	18.0831413	6.8894188	.003058104
328	107584	35287552	18.1107703	6.8964345	.003048780
329	108241	35611289	18.1383571	6.9034359	.003039514
330	108900	35937000	18.1659021	6.9104232	.003030303
331	109561	36264691	18.1934054	6.9173964	.003021148
332	110224	36594368	18.2208672	6.9243556	.003012048
333	110889	36926037	18.2482876	6.9313008	.003003003
334	111556	37259704	18.2756669	6.9382321	.002994012
335	112225	37595375	18.3030052	6.9451496	.002985075
336	112896	37933056	18.3303028	6.9520533	.002976190
337	113569	38272753	18.3575598	6.9589434	.002967359
338	114244	38614472	18.3847763	6.9658198	.002958580
339	114921	38958219	18.4119526	6.9726826	.002949853
340	115600	39304000	18.4390889	6.9795321	.002941176
341	116281	39651821	18.4661853	6.9863681	.002932551
342	116964	40001688	18.4932420	6.9931906	.002923977
343	117649	40353607	18.5202592	7.0000000	.002915452
344	118336	40707584	18.5472370	7.0067962	.002906977
345	119025	41063625	18.5741756	7.0135791	.002898551
346	119716	41421736	18.6010752	7.0203490	.002890173
347	120409	41781923	18.6279360	7.0271058	.002881844
348	121104	42144192	18.6547581	7.0338497	.002873563
349	121801	42508549	18.6815417	7.0405806	.002865330
350	122500	42875000	18.7082869	7.0472987	.002857143
351	123201	43243551	18.7349940	7.0540041	.002849003
352	123904	43614208	18.7616630	7.0606967	.002840909
353	124609	43986977	18.7882942	7.0673767	.002832861
354	125316	44361864	18.8148877	7.0740440	.002824859
355	126025	44738875	18.8414437	7.0806988	.002816901
356	126736	45118016	18.8679623	7.0873411	.002808989
357	127449	45499293	18.8944436	7.0939709	.002801120
358	128164	45882712	18.9208879	7.1005885	.002793296
359	128881	46268279	18.9472953	7.1071937	.002785515
360	129600	46656000	18.9736660	7.1137866	.002777778
361	130321	47045881	19.0000000	7.1203674	.002770083
362	131044	47437928	19.0262976	7.1269360	.002762431
363	131769	47832147	19.0525589	7.1334925	.002754821
364	132496	48228544	19.0787840	7.1400370	.002747253
365	133225	48627125	19.1049732	7.1465695	.002739726
366	133956	49027896	19.1311265	7.1530901	.002732240
367	134689	49430863	19.1572441	7.1595988	.002724796
368	135424	49836032	19.1833261	7.1660957	.002717391
369	136161	50243409	19.2093727	7.1725809	.002710027
370	136900	50653000	19.2353841	7.1790544	.002702703
371	137641	51064811	19.2613603	7.1855162	.002695418
372	138384	51478848	19.2873015	7.1919663	.002688172

No.	Squares.	Cubes.	Square Roots.	Cube Roots.	Reciprocals.
373	139129	51895117	19.3132079	7.1984050	.002680965
374	139876	52313624	19.3390796	7.2048322	.002673797
375	140625	52734375	19.3649167	7.2112479	.002666667
376	141376	53157376	19.3907194	7.2176522	.002659574
377	142129	53582633	19.4164878	7.2240450	.002652520
378	142884	54010152	19.4422221	7.2304263	.002645503
379	143641	54439939	19.4679223	7.2367972	.002638522
380	144400	54872000	19.4935887	7.2431565	.002631579
381	145161	55306341	19.5192213	7.2495045	.002624672
382	145924	55742968	19.5448203	7.2558415	.002617801
383	146689	56181887	19.5703858	7.2621675	.002610966
384	147456	56623104	19.5959179	7.2684824	.002604167
385	148225	57066625	19.6214169	7.2747864	.002597403
386	148996	57512456	19.6468827	7.2810794	.002590674
387	149769	57960603	19.6723156	7.2873617	.002583979
388	150544	58411072	19.6977156	7.2936330	.002577320
389	151321	58863869	19.7230829	7.2998936	.002570694
390	152100	59319000	19.7484177	7.3061436	.002564103
391	152881	59776471	19.7737199	7.3123828	.002557545
392	153664	60236288	19.7989899	7.3186114	.002551020
393	154449	60698457	19.8242276	7.3248295	.002544529
394	155236	61162984	19.8494332	7.3310369	.002538071
395	156025	61629875	19.8746069	7.3372339	.002531646
396	156816	62099136	19.8997487	7.3434205	.002525253
397	157609	62570773	19.9248588	7.3495966	.002518892
398	158404	63044792	19.9499373	7.3557624	.002512563
399	159201	63521199	19.9749844	7.3619178	.002506266
400	160000	64000000	20.0000000	7.3680630	.002500000
401	160801	64481201	20.0249844	7.3741979	.002493766
402	161604	64964808	20.0499377	7.3803227	.002487562
403	162409	65450827	20.0748599	7.3864373	.002481390
404	163216	65939264	20.0997512	7.3925418	.002475248
405	164025	66430125	20.1246118	7.3986363	.002469136
406	164836	66923416	20.1494417	7.4047206	.002463054
407	165649	67419143	20.1742410	7.4107950	.002457002
408	166464	67917312	20.1990099	7.4168595	.002450980
409	167281	68417929	20.2237484	7.4229142	.002444988
410	168100	68921000	20.2484567	7.4289589	.002439024
411	168921	69426531	20.2731349	7.4349938	.002433090
412	169744	69934528	20.2977831	7.4410189	.002427184
413	170569	70444997	20.3224014	7.4470342	.002421308
414	171396	70957944	20.3469899	7.4530399	.002415459
415	172225	71473375	20.3715488	7.4590359	.002409639
416	173056	71991296	20.3960781	7.4650223	.002403846
417	173889	72511713	20.4205779	7.4709991	.002398082
418	174724	73034632	20.4450483	7.4769664	.002392344
419	175561	73560059	20.4694895	7.4829242	.002386635
420	176400	74088000	20.4939015	7.4888724	.002380952
421	177241	74618461	20.5182845	7.4948113	.002375297
422	178084	75151448	20.5426386	7.5007406	.002369668
423	178929	75686967	20.5669638	7.5066607	.002364066
424	179776	76225024	20.5912603	7.5125715	.002358491
425	180625	76765625	20.6155281	7.5184730	.002352941
426	181476	77308776	20.6397674	7.5243652	.002347418
427	182329	77854483	20.6639783	7.5302482	.002341920
428	183184	78402752	20.6881609	7.5361221	.002336449
429	184041	78953589	20.7123152	7.5419867	.002331002
430	184900	79507000	20.7364414	7.5478423	.002325581
431	185761	80062991	20.7605395	7.5536888	.002320186
432	186624	80621568	20.7846097	7.5595263	.002314815
433	187489	81182737	20.8086520	7.5653548	.002309469
434	188356	81746504	20.8326667	7.5711743	.002304147

No.	Squares.	Cubes.	Square Roots.	Cube Roots.	Reciprocals.
435	189225	82312875	20.8566536	7.5769349	.002298851
436	190096	82881856	20.8806130	7.5827865	.002293578
437	190969	83453453	20.9045450	7.5885793	.002288330
438	191844	84027672	20.9284495	7.5943633	.002283105
439	192721	84604519	20.9523268	7.6001385	.002277904
440	193600	85184000	20.9761770	7.6059049	.002272727
441	194481	85766121	21.0000000	7.6116626	.002267574
442	195364	86350888	21.0237960	7.6174116	.002262443
443	196249	86938307	21.0475652	7.6231519	.002257336
444	197136	87528384	21.0713075	7.6288837	.002252252
445	198025	88121125	21.0950231	7.6346067	.002247191
446	198916	88716536	21.1187121	7.6403213	.002242152
447	199809	89314623	21.1423745	7.6460272	.002237136
448	200704	89915392	21.1660105	7.6517247	.002232143
449	201601	90518849	21.1896201	7.6574138	.002227171
450	202500	91125000	21.2132034	7.6630943	.002222222
451	203401	91733851	21.2367606	7.6687665	.002217295
452	204304	92345408	21.2602916	7.6744303	.002212389
453	205209	92959677	21.2837967	7.6800857	.002207506
454	206116	93576664	21.3072758	7.6857328	.002202643
455	207025	94196375	21.3307290	7.6913717	.002197802
456	207936	94818816	21.3541565	7.6970023	.002192982
457	208849	95443993	21.3775583	7.7026246	.002188184
458	209764	96071912	21.4009346	7.7082388	.002183406
459	210681	96702579	21.4242853	7.7138448	.002178649
460	211600	97336000	21.4476106	7.7194426	.002173913
461	212521	97972181	21.4709106	7.7250325	.002169197
462	213444	98611128	21.4941853	7.7306141	.002164502
463	214369	99252847	21.5174348	7.7361877	.002159827
464	215296	99897344	21.5406592	7.7417532	.002155172
465	216225	100544625	21.5638587	7.7473109	.002150538
466	217156	101194696	21.5870331	7.7528606	.002145923
467	218089	101847563	21.6101828	7.7584023	.002141328
468	219024	102503232	21.6333077	7.7639361	.002136752
469	219961	103161709	21.6564078	7.7694620	.002132196
470	220900	103823000	21.6794834	7.7749801	.002127660
471	221841	104487111	21.7025344	7.7804904	.002123142
472	222784	105154048	21.7255610	7.7859928	.002118644
473	223729	105823817	21.7485632	7.7914875	.002114165
474	224676	106496424	21.7715411	7.7969745	.002109705
475	225625	107171875	21.7944947	7.8024538	.002105263
476	226576	107850176	21.8174242	7.8079254	.002100840
477	227529	108531333	21.8403297	7.8133892	.002096436
478	228484	109215352	21.8632111	7.8188456	.002092050
479	229441	109902239	21.8860686	7.8242942	.002087683
480	230400	110592000	21.9089023	7.8297353	.002083333
481	231361	111284641	21.9317122	7.8351688	.002079002
482	232324	111980168	21.9544984	7.8405949	.002074689
483	233289	112678587	21.9772610	7.8460134	.002070393
484	234256	113379904	22.0000000	7.8514244	.002066116
485	235225	114084125	22.0227155	7.8568281	.002061856
486	236196	114791256	22.0454077	7.8622242	.002057613
487	237169	115501303	22.0680765	7.8676130	.002053388
488	238144	116214272	22.0907220	7.8729944	.002049180
489	239121	116930169	22.1133444	7.8783684	.002044990
490	240100	117649000	22.1359436	7.8837352	.002040816
491	241081	118370771	22.1585198	7.8890946	.002036660
492	242064	119095488	22.1810730	7.8944468	.002032520
493	243049	119823157	22.2036033	7.8997917	.002028398
494	244036	120553784	22.2261108	7.9051294	.002024291
495	245025	121287375	22.2485955	7.9104599	.002020202
496	246016	122023936	22.2710575	7.9157832	.002016129

No.	Squares.	Cubes.	Square Roots.	Cube Roots.	Reciprocals.
497	247009	122763473	22.2934968	7.9210994	.002012072
498	248004	123505992	22.3159136	7.9264085	.002008032
499	249001	124251499	22.3383079	7.9317104	.002004008
500	250000	125000000	22.3606798	7.9370053	.002000000
501	251001	125751501	22.3830293	7.9422931	.001996008
502	252004	126506008	22.4053565	7.9475739	.001992032
503	253009	127263527	22.4276615	7.9528477	.001988072
504	254016	128024064	22.4499443	7.9581144	.001984127
505	255025	128787625	22.4722051	7.9633743	.001980198
506	256036	129554216	22.4944438	7.9686271	.001976285
507	257049	130323843	22.5166605	7.9738731	.001972387
508	258064	131096512	22.5388553	7.9791122	.001968504
509	259081	131872229	22.5610283	7.9843444	.001964637
510	260100	132651000	22.5831796	7.9895697	.001960784
511	261121	133432831	22.6053091	7.9947883	.001956947
512	262144	134217728	22.6274170	8.0000000	.001953125
513	263169	135005697	22.6495033	8.0052049	.001949318
514	264196	135796744	22.6715681	8.0104032	.001945525
515	265225	136590875	22.6936114	8.0155946	.001941748
516	266256	137388096	22.7156334	8.0207794	.001937984
517	267289	138188413	22.7376340	8.0259574	.001934236
518	268324	138991832	22.7596134	8.0311287	.001930502
519	269361	139798359	22.7815715	8.0362935	.001926782
520	270400	140608000	22.8035085	8.0414515	.001923077
521	271441	141420761	22.8254244	8.0466030	.001919386
522	272484	142236648	22.8473193	8.0517479	.001915709
523	273529	143055667	22.8691933	8.0568862	.001912046
524	274576	143877824	22.8910463	8.0620180	.001908397
525	275625	144703125	22.9128785	8.0671432	.001904762
526	276676	145531576	22.9346899	8.0722620	.001901141
527	277729	146363183	22.9564806	8.0773743	.001897533
528	278784	147197952	22.9782506	8.0824800	.001893939
529	279841	148035889	23.0000000	8.0875794	.001890359
530	280900	148877000	23.0217289	8.0926723	.001886792
531	281961	149721291	23.0434372	8.0977589	.001883239
532	283024	150568768	23.0651252	8.1028390	.001879699
533	284089	151419437	23.0867928	8.1079128	.001876173
534	285156	152273304	23.1084400	8.1129803	.001872659
535	286225	153130375	23.1300670	8.1180414	.001869159
536	287296	153990656	23.1516738	8.1230962	.001865672
537	288369	154854153	23.1732605	8.1281447	.001862197
538	289444	155720872	23.1948270	8.1331870	.001858736
539	290521	156590819	23.2163735	8.1382230	.001855288
540	291600	157464000	23.2379001	8.1432529	.001851852
541	292681	158340421	23.2594067	8.1482765	.001848429
542	293764	159220088	23.2808935	8.1532939	.001845018
543	294849	160103007	23.3023604	8.1583051	.001841621
544	295936	160989184	23.3238076	8.1633102	.001838235
545	297025	161878625	23.3452351	8.1683092	.001834862
546	298116	162771336	23.3666429	8.1733020	.001831502
547	299209	163667323	23.3880311	8.1782888	.001828154
548	300304	164566592	23.4093998	8.1832695	.001824818
549	301401	165469149	23.4307490	8.1882441	.001821494
550	302500	166375000	23.4520788	8.1932127	.001818182
551	303601	167284151	23.4733892	8.1981753	.001814882
552	304704	168196608	23.4946802	8.2031319	.001811594
553	305809	169112377	23.5159520	8.2080825	.001808318
554	306916	170031464	23.5372046	8.2130271	.001805054
555	308025	170953875	23.5584380	8.2179657	.001801802
556	309136	171879616	23.5796522	8.2228985	.001798561
557	310249	172808693	23.6008474	8.2278254	.001795332
558	311364	173741112	23.6220236	8.2327463	.001792115

No.	Squares.	Cubes.	Square Roots.	Cube Roots.	Reciprocals.
559	312481	174676879	23.6431808	8.2376614	.001788909
560	313600	175616000	23.6643191	8.2425706	.001785714
561	314721	176558481	23.6854386	8.2474740	.001782531
562	315844	177504328	23.7065392	8.2523715	.001779359
563	316969	178453547	23.7276210	8.2572633	.001776199
564	318096	179406144	23.7486842	8.2621492	.001773050
565	319225	180362125	23.7697286	8.2670294	.001769912
566	320356	181321496	23.7907545	8.2719039	.001766784
567	321489	182284263	23.8117618	8.2767726	.001763668
568	322624	183250432	23.8327506	8.2816355	.001760563
569	323761	184220009	23.8537209	8.2864928	.001757469
570	324900	185193000	23.8746728	8.2913444	.001754386
571	326041	186169411	23.8956063	8.2961903	.001751313
572	327184	187149248	23.9165215	8.3010304	.001748252
573	328329	188132517	23.9374184	8.3058651	.001745201
574	329476	189119224	23.9582971	8.3106941	.001742160
575	330625	190109375	23.9791576	8.3155175	.001739130
576	331776	191102976	24.0000000	8.3203353	.001736111
577	332929	192100033	24.0208243	8.3251475	.001733102
578	334084	193100552	24.0416306	8.3299542	.001730104
579	335241	194104539	24.0624188	8.3347553	.001727116
580	336400	195112000	24.0831891	8.3395509	.001724138
581	337561	196122941	24.1039416	8.3443410	.001721170
582	338724	197137368	24.1246762	8.3491256	.001718213
583	339889	198155287	24.1453929	8.3539047	.001715266
584	341056	199176704	24.1660919	8.3586784	.001712329
585	342225	200201625	24.1867732	8.3634466	.001709402
586	343396	201230056	24.2074369	8.3682095	.001706485
587	344569	202262003	24.2280829	8.3729668	.001703578
588	345744	203297472	24.2487113	8.3777188	.001700680
589	346921	204336469	24.2693222	8.3824653	.001697793
590	348100	205379000	24.2899156	8.3872065	.001694915
591	349281	206425071	24.3104916	8.3919423	.001692047
592	350464	207474688	24.3310501	8.3966729	.001689189
593	351649	208527857	24.3515913	8.4013981	.001686341
594	352836	209584584	24.3721152	8.4061180	.001683502
595	354025	210644875	24.3926218	8.4108326	.001680672
596	355216	211708736	24.4131112	8.4155419	.001677852
597	356409	212776173	24.4335834	8.4202460	.001675042
598	357604	213847192	24.4540385	8.4249448	.001672241
599	358801	214921799	24.4744765	8.4296383	.001669449
600	360000	216000000	24.4948974	8.4343267	.001666667
601	361201	217081801	24.5153013	8.4390098	.001663894
602	362404	218167208	24.5356883	8.4436877	.001661130
603	363609	219256227	24.5560583	8.4483605	.001658375
604	364816	220348864	24.5764115	8.4530281	.001655629
605	366025	221445125	24.5967478	8.4576906	.001652893
606	367236	222545016	24.6170673	8.4623479	.001650165
607	368449	223648543	24.6373700	8.4670001	.001647446
608	369664	224755712	24.6576560	8.4716471	.001644737
609	370881	225866529	24.6779254	8.4762892	.001642036
610	372100	226981000	24.6981781	8.4809261	.001639344
611	373321	228099131	24.7184142	8.4855579	.001636661
612	374544	229220928	24.7386338	8.4901848	.001633987
613	375769	230346397	24.7588368	8.4948065	.001631321
614	376996	231475544	24.7790234	8.4994233	.001628664
615	378225	232608375	24.7991935	8.5040350	.001626016
616	379456	233744896	24.8193473	8.5086417	.001623377
617	380689	234885113	24.8394847	8.5132435	.001620746
618	381924	236029032	24.8596058	8.5178403	.001618123
619	383161	237176659	24.8797106	8.5224321	.001615509
620	384400	238328000	24.8997992	8.5270189	.001612903

No.	Squares.	Cubes.	Square Roots.	Cube Roots.	Reciprocals.
621	385641	239483061	24.9198716	8.5316009	.001610306
622	386884	240641848	24.9399278	8.5361780	.001607717
623	388129	241804367	24.9599679	8.5407501	.001605136
624	389376	242970624	24.9799920	8.5453173	.001602564
625	390625	244140625	25.0000000	8.5498797	.001600000
626	391876	245314376	25.0199920	8.5544372	.001597444
627	393129	246491883	25.0399681	8.5589899	.001594896
628	394384	247673152	25.0599282	8.5635377	.001592357
629	395641	248858189	25.0798724	8.5680807	.001589825
630	396900	250047000	25.0998008	8.5726189	.001587302
631	398161	251239591	25.1197134	8.5771523	.001584786
632	399424	252435968	25.1396102	8.5816809	.001582278
633	400689	253636137	25.1594913	8.5862047	.001579779
634	401956	254840104	25.1793566	8.5907238	.001577287
635	403225	256047875	25.1992063	8.5952380	.001574803
636	404496	257259456	25.2190404	8.5997476	.001572327
637	405769	258474853	25.2388589	8.6042525	.001569859
638	407044	259694072	25.2586619	8.6087526	.001567398
639	408321	260917119	25.2784493	8.6132480	.001564945
640	409600	262144000	25.2982213	8.6177388	.001562500
641	410881	263374721	25.3179778	8.6222248	.001560062
642	412164	264609288	25.3377189	8.6267063	.001557632
643	413449	265847707	25.3574447	8.6311830	.001555210
644	414736	267089984	25.3771551	8.6356551	.001552795
645	416025	268336125	25.3968502	8.6401226	.001550388
646	417316	269586136	25.4165301	8.6445855	.001547988
647	418609	270840023	25.4361947	8.6490437	.001545595
648	419904	272097792	25.4558441	8.6534974	.001543210
649	421201	273359449	25.4754784	8.6579465	.001540832
650	422500	274625000	25.4950976	8.6623911	.001538462
651	423801	275894451	25.5147016	8.6668310	.001536098
652	425104	277167808	25.5342907	8.6712665	.001533742
653	426409	278445077	25.5538647	8.6756974	.001531394
654	427716	279726264	25.5734237	8.6801237	.001529052
655	429025	281011375	25.5929678	8.6845456	.001526718
656	430336	282300416	25.6124969	8.6889630	.001524390
657	431649	283593393	25.6320112	8.6933759	.001522070
658	432964	284890312	25.6515107	8.6977843	.001519757
659	434281	286191179	25.6709953	8.7021882	.001517451
660	435600	287496000	25.6904652	8.7065877	.001515152
661	436921	288804781	25.7099203	8.7109827	.001512859
662	438244	290117528	25.7293607	8.7153734	.001510574
663	439569	291434247	25.7487864	8.7197596	.001508296
664	440896	292754944	25.7681975	8.7241414	.001506024
665	442225	294079625	25.7875939	8.7285187	.001503759
666	443556	295408296	25.8069758	8.7328918	.001501502
667	444889	296740963	25.8263431	8.7372604	.001499250
668	446224	298077632	25.8456960	8.7416246	.001497006
669	447561	299418309	25.8650343	8.7459846	.001494768
670	448900	300763000	25.8843582	8.7503401	.001492537
671	450241	302111711	25.9036677	8.7546913	.001490313
672	451584	303464448	25.9229628	8.7590383	.001488095
673	452929	304821217	25.9422435	8.7633809	.001485884
674	454276	306182024	25.9615100	8.7677192	.001483680
675	455625	307546875	25.9807621	8.7720532	.001481481
676	456976	308915776	26.0000000	8.7763830	.001479290
677	458329	310288733	26.0192237	8.7807084	.001477105
678	459684	311665752	26.0384331	8.7850296	.001474926
679	461041	313046839	26.0576284	8.7893466	.001472754
680	462400	314432000	26.0768096	8.7936593	.001470588
681	463761	315821241	26.0959767	8.7979679	.001468429
682	465124	317214568	26.1151297	8.8022721	.001466276

No.	Squares.	Cubes.	Square Roots.	Cube Roots.	Reciprocals.
683	466489	318611987	26.1342687	8.8065722	.001464129
684	467856	320013504	26.1533937	8.8108681	.001461988
685	469225	321419125	26.1725047	8.8151598	.001459854
686	470596	322828856	26.1916017	8.8194474	.001457726
687	471969	324242703	26.2106848	8.8237307	.001455604
688	473344	325660672	26.2297541	8.8280099	.001453488
689	474721	327082769	26.2488095	8.8322850	.001451379
690	476100	328509000	26.2678511	8.8365559	.001449275
691	477481	329939371	26.2868789	8.8408227	.001447178
692	478864	331373888	26.3058929	8.8450854	.001445087
693	480249	332812557	26.3248932	8.8493440	.001443001
694	481636	334255384	26.3438797	8.8535985	.001440922
695	483025	335702375	26.3628527	8.8578489	.001438849
696	484416	337153536	26.3818119	8.8620952	.001436782
697	485809	338608873	26.4007576	8.8663375	.001434720
698	487204	340068392	26.4196896	8.8705757	.001432665
699	488601	341532099	26.4386081	8.8748099	.001430615
700	490000	343000000	26.4575131	8.8790400	.001428571
701	491401	344472101	26.4764046	8.8832661	.001426534
702	492804	345948408	26.4952826	8.8874882	.001424501
703	494209	347428927	26.5141472	8.8917063	.001422475
704	495616	348913664	26.5329983	8.8959204	.001420455
705	497025	350402625	26.5518361	8.9001304	.001418440
706	498436	351895816	26.5706605	8.9043366	.001416431
707	499849	353393243	26.5894716	8.9085387	.001414427
708	501264	354894912	26.6082694	8.9127369	.001412429
709	502681	356400829	26.6270539	8.9169311	.001410437
710	504100	357911000	26.6458252	8.9211214	.001408451
711	505521	359425431	26.6645833	8.9253078	.001406470
712	506944	360944128	26.6833281	8.9294902	.001404494
713	508369	362467097	26.7020598	8.9336687	.001402525
714	509796	363994344	26.7207784	8.9378433	.001400560
715	511225	365525875	26.7394839	8.9420140	.001398601
716	512656	367061696	26.7581763	8.9461809	.001396648
717	514089	368601813	26.7768557	8.9503438	.001394700
718	515524	370146232	26.7955220	8.9545029	001392758
719	516961	371694959	26.8141754	8.9586581	.001390821
720	518400	373248000	26.8328157	8.9628095	.001388889
721	519841	374805361	26.8514432	8.9669570	.001386963
722	521284	376367048	26.8700577	8.9711007	.001385042
723	522729	377933067	26.8886593	8.9752406	.001383126
724	524176	379503424	26.9072481	8.9793766	.001381215
725	525625	381078125	26.9258240	8.9835089	.001379310
726	527076	382657176	26.9443872	8.9876373	.001377410
727	528529	384240583	26.9629375	8.9917620	.001375516
728	529984	385828352	26.9814751	8.9958829	.001373626
729	531441	387420489	27.0000000	9.0000000	.001371742
730	532900	389017000	27.0185122	9.0041134	.001369863
731	534361	390617891	27.0370117	9.0082229	.001367989
732	535824	392223168	27.0554985	9.0123288	.001366120
733	537289	393832837	27.0739727	9.0164309	.001364256
734	538756	395446904	27.0924344	9.0205293	.001362398
735	540225	397065375	27.1108834	9.0246239	.001360544
736	541696	398688256	27.1293199	9.0287149	.001358696
737	543169	400315553	27.1477439	9.0328021	.001356852
738	544644	401947272	27.1661554	9.0368857	.001355014
739	546121	403583419	27.1845544	9.0409655	.001353180
740	547600	405224000	27.2029410	9.0450419	.001351351
741	549081	406869021	27.2213152	9.0491142	.001349528
742	550564	408518488	27.2396769	9.0531831	.001347709
743	552049	410172407	27.2580263	9.0572482	.001345895
744	553536	411830784	27.2763634	9.0613098	.001344086

No.	Squares.	Cubes.	Square Roots.	Cube Roots.	Reciprocals.
745	555025	413493625	27.2946881	9.0653677	.001342282
746	556516	415160936	27.3130006	9.0694220	.001340483
747	558009	416832723	27.3313007	9.0734726	.001338688
748	559504	418508992	27.3495887	9.0775197	.001336898
749	561001	420189749	27.3678644	9.0815631	.001335113
750	562500	421875000	27.3861279	9.0856030	.001333333
751	564001	423564751	27.4043792	9.0896392	.001331558
752	565504	425259008	27.4226184	9.0936719	.001329787
753	567009	426957777	27.4408455	9.0977010	.001328021
754	568516	428661064	27.4590604	9.1017265	.001326260
755	570025	430368875	27.4772633	9.1057485	.001324503
756	571536	432081216	27.4954542	9.1097669	.001322751
757	573049	433798093	27.5136330	9.1137818	.001321004
758	574564	435519512	27.5317998	9.1177931	.001319261
759	576081	437245479	27.5499546	9.1218010	.001317523
760	577600	438976000	27.5680975	9.1258053	.001315789
761	579121	440711081	27.5862284	9.1298061	.001314060
762	580644	442450728	27.6043475	9.1338034	.001312336
763	582169	444194947	27.6224546	9.1377971	.001310616
764	583696	445943744	27.6405499	9.1417874	.001308901
765	585225	447697125	27.6586334	9.1457742	.001307190
766	586756	449455096	27.6767050	9.1497576	.001305483
767	588289	451217663	27.6947648	9.1537375	.001303781
768	589824	452984832	27.7128129	9.1577139	.001302083
769	591361	454756609	27.7308492	9.1616869	.001300390
770	592900	456533000	27.7488739	9.1656565	.001298701
771	594441	458314011	27.7668868	9.1696225	.001297017
772	595984	460099648	27.7848880	9.1735852	.001295337
773	597529	461889917	27.8028775	9.1775445	.001293661
774	599076	463684824	27.8208555	9.1815003	.001291990
775	600625	465484375	27.8388218	9.1854527	.001290323
776	602176	467288576	27.8567766	9.1894018	.001288660
777	603729	469097433	27.8747197	9.1933474	.001287001
778	605284	470910952	27.8926514	9.1972897	.001285347
779	606841	472729139	27.9105715	9.2012286	.001283697
780	608400	474552000	27.9284801	9.2051641	.001282051
781	609961	476379541	27.9463772	9.2090962	.001280410
782	611524	478211768	27.9642629	9.2130250	.001278772
783	613089	480048687	27.9821372	9.2169505	.001277139
784	614656	481890304	28.0000000	9.2208726	.001275510
785	616225	483736625	28.0178515	9.2247914	.001273885
786	617796	485587656	28.0356915	9.2287068	.001272265
787	619369	487443403	28.0535203	9.2326189	.001270648
788	620944	489303872	28.0713377	9.2365277	.001269036
789	622521	491169069	28.0891438	9.2404333	.001267427
790	624100	493039000	28.1069386	9.2443355	.001265823
791	625681	494913671	28.1247222	9.2482344	.001264223
792	627264	496793088	28.1424946	9.2521300	.001262626
793	628849	498677257	28.1602557	9.2560224	.001261034
794	630436	500566184	28.1780056	9.2599114	.001259446
795	632025	502459875	28.1957444	9.2637973	.001257862
796	633616	504358336	28.2134720	9.2676798	.001256281
797	635209	506261573	28.2311884	9.2715592	.001254705
798	636804	508169592	28.2488938	9.2754352	.001253133
799	638401	510082399	28.2665881	9.2793081	.001251564
800	640000	512000000	28.2842712	9.2831777	.001250000
801	641601	513922401	28.3019434	9.2870440	.001248439
802	643204	515849608	28.3196045	9.2909072	.001246883
803	644809	517781627	28.3372546	9.2947671	.001245330
804	646416	519718464	28.3548938	9.2986239	.001243781
805	648025	521660125	28.3725219	9.3024775	.001242236
806	649636	523606616	28.3901391	9.3063278	.001240695

No.	Squares.	Cubes.	Square Roots.	Cube Roots.	Reciprocals.
807	651249	525557943	28.4077454	9.3101750	.001239157
808	652864	527514112	28.4253408	9.3140190	.001237624
809	654481	529475129	28.4429253	9.3178599	.001236094
810	656100	531441000	28.4604989	9.3216975	.001234568
811	657721	533411731	28.4780617	9.3255320	.001233046
812	659344	535387328	28.4956137	9.3293634	.001231527
813	660969	537367797	28.5131549	9.3331916	.001230012
814	662596	539353144	28.5306852	9.3370167	.001228501
815	664225	541343375	28.5482048	9.3408386	.001226994
816	665856	543338196	28.5657137	9.3446575	.001225490
817	667489	545338513	28.5832119	9.3484731	.001223990
818	669124	547343432	28.6006993	9.3522857	.001222494
819	670761	549353259	28.6181760	9.3560952	.001221001
820	672400	551368000	28.6356421	9.3599016	.001219512
821	674041	553387661	28.6530976	9.3637049	.001218027
822	675684	555412248	28.6705424	9.3675051	.001216545
823	677329	557441767	28.6879766	9.3713022	.001215067
824	678976	559476224	28.7054002	9.3750963	.001213592
825	680625	561515625	28.7228132	9.3788873	.001212121
826	682276	563559976	28.7402157	9.3826752	.001210654
827	683929	565609283	28.7576077	9.3864600	.001209190
828	685584	567663552	28.7749891	9.3902419	.001207729
829	687241	569722789	28.7923601	9.3940206	.001206273
830	688900	571787000	28.8097206	9.3977964	.001204819
831	690561	573856191	28.8270706	9.4015691	.001203369
832	692224	575930368	28.8444102	9.4053387	.001201923
833	693889	578009537	28.8617394	9.4091054	.001200480
834	695556	580093704	28.8790582	9.4128690	.001199041
835	697225	582182875	28.8963666	9.4166297	.001197605
836	698896	584277056	28.9136646	9.4203873	.001196172
837	700569	586376253	28.9309523	9.4241420	.001194743
838	702244	588480472	28.9482297	9.4278936	.001193317
839	703921	590589719	28.9654967	9.4316423	.001191895
840	705600	592704000	28.9827535	9.4353880	.001190476
841	707281	594823321	29.0000000	9.4391307	.001189061
842	708964	596947688	29.0172363	9.4428704	.001187648
843	710649	599077107	29.0344623	9.4466072	.001186240
844	712336	601211584	29.0516781	9.4503410	.001184834
845	714025	603351125	29.0688837	9.4540719	.001183432
846	715716	605495736	29.0860791	9.4577999	.001182033
847	717409	607645423	29.1032644	9.4615249	.001180638
848	719104	609800192	29.1204396	9.4652470	.001179245
849	720801	611960049	29.1376046	9.4689661	.001177856
850	722500	614125000	29.1547595	9.4726824	.001176471
851	724201	616295051	29.1719043	9.4763957	.001175088
852	725904	618470208	29.1890390	9.4801061	.001173709
853	727609	620650477	29.2061637	9.4838136	.001172333
854	729316	622835864	29.2232784	9.4875182	.001170960
855	731025	625026375	29.2403830	9.4912200	.001169591
856	732736	627222016	29.2574777	9.4949188	.001168224
857	734449	629422793	29.2745623	9.4986147	.001166861
858	736164	631628712	29.2916370	9.5023078	.001165501
859	737881	633839779	29.3087018	9.5059980	.001164144
860	739600	636056000	29.3257566	9.5096854	.001162791
861	741321	638277381	29.3428015	9.5133699	.001161440
862	743044	640503928	29.3598365	9.5170515	.001160093
863	744769	642735647	29.3768616	9.5207303	.001158749
864	746496	644972544	29.3938769	9.5244063	.001157407
865	748225	647214625	29.4108823	9.5280794	.001156069
866	749956	649461896	29.4278779	9.5317497	.001154734
867	751689	651714363	29.4448637	9.5354172	.001153403
868	753424	653972032	29.4618397	9.5390818	.001152074

No.	Squares.	Cubes.	Square Roots.	Cube Roots.	Reciprocals.
869	755161	656234909	29.4788059	9.5427437	.001150748
870	756900	658503000	29.4957624	9.5464027	.001149425
871	758641	660776311	29.5127091	9.5500589	.001148106
872	760384	663054848	29.5296461	9.5537123	.001146789
873	762129	665338617	29.5465734	9.5573630	.001145475
874	763876	667627624	29.5634910	9.5610108	.001144165
875	765625	669921875	29.5803989	9.5646559	.001142857
876	767376	672221376	29.5972972	9.5682982	.001141553
877	769129	674526133	29.6141858	9.5719377	.001140251
878	770884	676836152	29.6310648	9.5755745	.001138952
879	772641	679151439	29.6479342	9.5792085	.001137656
880	774400	681472000	29.6647939	9.5828397	.001136364
881	776161	683797841	29.6816442	9.5864682	.001135074
882	777924	686128968	29.6984848	9.5900939	.001133787
883	779689	688465387	29.7153159	9.5937169	.001132503
884	781456	690807104	29.7321375	9.5973373	.001131222
885	783225	693154125	29.7489496	9.6009548	.001129944
886	784996	695506456	29.7657521	9.6045696	.001128668
887	786769	697864103	29.7825452	9.6081817	.001127396
888	788544	700227072	29.7993289	9.6117911	.001126126
889	790321	702595369	29.8161030	9.6153977	.001124859
890	792100	704969000	29.8328678	9.6190017	.001123596
891	793881	707347971	29.8496231	9.6226030	.001122334
892	795664	709732288	29.8663690	9.6262016	.001121076
893	797449	712121957	29.8831056	9.6297975	.001119821
894	799236	714516984	29.8998328	9.6333907	.001118568
895	801025	716917375	29.9165506	9.6369812	.001117318
896	802816	719323136	29.9332591	9.6405690	.001116071
897	804609	721734273	29.9499583	9.6441542	.001114827
898	806404	724150792	29.9666481	9.6477367	.001113586
899	808201	726572699	29.9833287	9.6513166	.001112347
900	810000	729000000	30.0000000	9.6548938	.001111111
901	811801	731432701	30.0166620	9.6584684	.001109878
902	813604	733870808	30.0333148	9.6620403	.001108647
903	815409	736314327	30.0499584	9.6656096	.001107420
904	817216	738763264	30.0665928	9.6691762	.001106195
905	819025	741217625	30.0832179	9.6727403	.001104972
906	820836	743677416	30.0998339	9.6763017	.001103753
907	822649	746142643	30.1164407	9.6798604	.001102536
908	824464	748613312	30.1330383	9.6834166	.001101322
909	826281	751089429	30.1496269	9.6869701	.001100110
910	828100	753571000	30.1662063	9.6905211	.001098901
911	829921	756058031	30.1827765	9.6940694	.001097695
912	831744	758550528	30.1993377	9.6976151	.001096491
913	833569	761048497	30.2158899	9.7011583	.001095290
914	835396	763551944	30.2324329	9.7046989	.001094092
915	837225	766060875	30.2489669	9.7082369	.001092896
916	839056	768575296	30.2654919	9.7117723	.001091703
917	840889	771095213	30.2820079	9.7153051	.001090513
918	842724	773620632	30.2985148	9.7188354	.001089325
919	844561	776151559	30.3150128	9.7223631	.001088139
920	846400	778688000	30.3315018	9.7258883	.001086957
921	848241	781229961	30.3479818	9.7294109	.001085776
922	850084	783777448	30.3644529	9.7329309	.001084599
923	851929	786330467	30.3809151	9.7364484	.001083423
924	853776	788889024	30.3973683	9.7399634	.001082251
925	855625	791453125	30.4138127	9.7434758	.001081081
926	857476	794022776	30.4302481	9.7469857	.001079914
927	859329	796597983	30.4466747	9.7504980	.001078749
928	861184	799178752	30.4630924	9.7539979	.001077586
929	863041	801765089	30.4795013	9.7575002	.001076426
930	864900	804357000	30.4959014	9.7610001	.001075269

No.	Squares.	Cubes.	Square Roots.	Cube Roots.	Reciprocals.
931	866761	806954491	30.5122926	9.7644974	.001074114
932	868624	809557568	30.5286750	9.7679922	.001072961
933	870489	812166237	30.5450487	9.7714845	.001071811
934	872356	814780504	30.5614136	9.7749743	.001070664
935	874225	817400375	30.5777697	9.7784616	.001069519
936	876096	820025856	30.5941171	9.7829466	.001068376
937	877969	822656953	30.6104557	9.7854288	.001067236
938	879844	825293672	30.6267857	9.7889087	.001066098
939	881721	827936019	30.6431069	9.7923861	.001064963
940	883600	830584000	30.6594194	9.7958611	.001063830
941	885481	833237621	30.6757233	9.7993336	.001062699
942	887364	835896888	30.6920185	9.8028036	.001061571
943	889249	838561807	30.7083051	9.8062711	.001060445
944	891136	841232384	30.7245830	9.8097362	.001059322
945	893025	843908625	30.7408523	9.8131989	.001058201
946	894916	846590536	30.7571130	9.8166591	.001057082
947	896809	849278123	30.7733651	9.8201169	.001055966
948	898704	851971392	30.7896086	9.8235723	.001054852
949	900601	854670349	30.8058436	9.8270252	.001053741
950	902500	857375000	30.8220700	9.8304757	.001052632
951	904401	860085351	30.8382879	9.8339238	.001051525
952	906304	862801408	30.8544972	9.8373695	.001050420
953	908209	865523177	30.8706981	9.8408127	.001049318
954	910116	868250664	30.8868904	9.8442536	.001048218
955	912025	870983875	30.9030743	9.8476920	.001047120
956	913936	873722816	30.9192497	9.8511280	.001046025
957	915849	876467493	30.9354166	9.8545617	.001044932
958	917764	879217912	30.9515751	9.8579929	.001043841
959	919681	881974079	30.9677251	9.8614218	.001042753
960	921600	884736000	30.9838668	9.8648483	.001041667
961	923521	887503681	31.0000000	9.8682724	.001040583
962	925444	890277128	31.0161248	9.8716941	.001039501
963	927369	893056347	31.0322413	9.8751135	.001038422
964	929296	895841344	31.0483494	9.8785305	.001037344
965	931225	898632125	31.0644491	9.8819451	.001036269
966	933156	901428696	31.0805405	9.8853574	.001035197
967	935089	904231063	31.0966236	9.8887673	.001034126
968	937024	907039232	31.1126984	9.8921749	.001033058
969	938961	909853209	31.1287648	9.8955801	.001031992
970	940900	912673000	31.1448230	9.8989830	.001030928
971	942841	915498611	31.1608729	9.9023835	.001029866
972	944784	918330048	31.1769145	9.9057817	.001028807
973	946729	921167317	31.1929479	9.9091776	.001027749
974	948676	924010424	31.2089731	9.9125712	.001026694
975	950625	926859375	31.2249900	9.9159624	.001025641
976	952576	929714176	31.2409987	9.9193513	.001024590
977	954529	932574833	31.2569992	9.9227379	.001023541
978	956484	935441352	31.2729915	9.9261222	.001022495
979	958441	938313739	31.2889757	9.9295042	.001021450
980	960400	941192000	31.3049517	9.9328839	.001020408
981	962361	944076141	31.3209195	9.9362613	.001019368
982	964324	946966168	31.3368792	9.9396363	.001018330
983	966289	949862087	31.3528308	9.9430092	.001017294
984	968256	952763904	31.3687743	9.9463797	.001016260
985	970225	955671625	31.3847097	9.9497479	.001015228
986	972196	958585256	31.4006369	9.9531138	.001014199
987	974169	961504803	31.4165561	9.9564775	.001013171
988	976144	964430272	31.4324673	9.9598389	.001012146
989	978121	967361669	31.4483704	9.9631981	.001011122
990	980100	970299000	31.4642654	9.9665549	.001010101
991	982081	973242271	31.4801525	9.9699095	.001009082
992	984064	976191488	31.4960315	9.9732619	.001008065

No.	Squares.	Cubes.	Square Roots.	Cube Roots.	Reciprocals.
993	986049	979146657	31.5119025	9.9766120	.001007049
994	988036	982107784	31.5277655	9.9799599	.001006036
995	990025	985074875	31.5436206	9.9833055	.001005025
996	992016	988047936	31.5594677	9.9866488	.001004016
997	994009	991026973	31.5753068	9.9899900	.001003009
998	996004	994011992	31.5911380	9.9933289	.001002004
999	998001	997002999	31.6069613	9.9966656	.001001001
1000	1000000	1000000000	31.6227766	10.0000000	.001000000
1001	1002001	1003003001	31.6385840	10.0033322	.0009990010
1002	1004004	1006012008	31.6543836	10.0066622	.0009980040
1003	1006009	1009027027	31.6701752	10.0099899	.0009970090
1004	1008016	1012048064	31.6859590	10.0133155	.0009960159
1005	1010025	1015075125	31.7017349	10.0166389	.0009950249
1006	1012036	1018108216	31.7175030	10.0199601	.0009940358
1007	1014049	1021147343	31.7332633	10.0232791	.0009930487
1008	1016064	1024192512	31.7490157	10.0265958	.0009920635
1009	1018081	1027243729	31.7647603	10.0299104	.0009910803
1010	1020100	1030301000	31.7804972	10.0332228	.0009900990
1011	1022121	1033364331	31.7962262	10.0365330	.0009891197
1012	1024144	1036433728	31.8119474	10.0398410	.0009881423
1013	1026169	1039509197	31.8276609	10.0431469	.0009871668
1014	1028196	1042590744	31.8433666	10.0464506	.0009861933
1015	1030225	1045678375	31.8590646	10.0497521	.0009852217
1016	1032256	1048772096	31.8747549	10.0530514	.0009842520
1017	1034289	1051871913	31.8904374	10.0563485	.0009832842
1018	1036324	1054977832	31.9061123	10.0596435	.0009823183
1019	1038361	1058089859	31.9217794	10.0629364	.0009813543
1020	1040400	1061208000	31.9374388	10.0662271	.0009803922
1021	1042441	1064332261	31.9530906	10.0695156	.0009794319
1022	1044484	1067462648	31.9687347	10.0728020	.0009784736
1023	1046529	1070599167	31.9843712	10.0760863	.0009775171
1024	1048576	1073741824	32.0000000	10.0793684	.0009765625
1025	1050625	1076890625	32.0156212	10.0826484	.0009756098
1026	1052676	1080045576	32.0312348	10.0859262	.0009746589
1027	1054729	1083206683	32.0468407	10.0892019	.0009737098
1028	1056784	1086373952	32.0624391	10.0924755	.0009727626
1029	1058841	1089547389	32.0780298	10.0957469	.0009718173
1030	1060900	1092727000	32.0936131	10.0990163	.0009708738
1031	1062961	1095912791	32.1091887	10.1022835	.0009699321
1032	1065024	1099104768	32.1247568	10.1055487	.0009689922
1033	1067089	1102302937	32.1403173	10.1088117	.0009680542
1034	1069156	1105507304	32.1558704	10.1120726	.0009671180
1035	1071225	1108717875	32.1714159	10.1153314	.0009661836
1036	1073296	1111934656	32.1869539	10.1185882	.0009652510
1037	1075369	1115157653	32.2024844	10.1218428	.0009643202
1038	1077444	1118386872	32.2180074	10.1250953	.0009633911
1039	1079521	1121622319	32.2335229	10.1283457	.0009624639
1040	1081600	1124864000	32.2490310	10.1315941	.0009615385
1041	1083681	1128111921	32.2645316	10.1348403	.0009606148
1042	1085764	1131366088	32.2800248	10.1380845	.0009596929
1043	1087849	1134626507	32.2955105	10.1413266	.0009587738
1044	1089936	1137893184	32.3109888	10.1445667	.0009578544
1045	1092025	1141166125	32.3264598	10.1478047	.0009569378
1046	1094116	1144445336	32.3419233	10.1510406	.0009560229
1047	1096209	1147730823	32.3573794	10.1542744	.0009551098
1048	1098304	1151022592	32.3728281	10.1575062	.0009541985
1049	1100401	1154320649	32.3882695	10.1607359	.0009532888
1050	1102500	1157625000	32.4037035	10.1639636	.0009523810
1051	1104601	1160935651	32.4191301	10.1671893	.0009514748
1052	1106704	1164252608	32.4345495	10.1704129	.0009505703
1053	1108809	1167575877	32.4499615	10.1736344	.0009496676
1054	1110916	1170905464	32.4653662	10.1768539	.0009487666

TABLE XII.

LOGARITHMS OF NUMBERS

FROM 1 TO 10,000

No.	0	1	2	3	4	5	6	7	8	9	Diff.
100	000000	000434	000868	001301	001734	002166	002598	003029	003461	003891	432
1	4321	4751	5181	5609	6038	6466	6894	7321	7748	8174	428
2	8600	9026	9451	9876	010300	010724	011147	011570	011993	012415	424
3	012837	013259	013680	014100	4521	4940	5360	5779	6197	6616	420
4	7033	7451	7868	8284	8700	9116	9532	9947	020361	020775	416
5	021189	021603	022016	022428	022841	023252	023664	024075	4486	4896	412
6	5306	5715	6125	6533	6942	7350	7757	8164	8571	8978	408
7	9384	9789	030195	030600	031004	031408	031812	032216	032619	033021	404
8	033424	033826	4227	4628	5029	5430	5830	6230	6629	7028	400
9	7426	7825	8223	8620	9017	9414	9811	040207	040602	040998	397
110	041393	041787	042182	042576	042969	043362	043755	044148	044540	044932	393
1	5323	5714	6105	6495	6885	7275	7664	8053	8442	8830	390
2	9218	9606	9993	050380	050766	051153	051538	051924	052309	052694	386
3	053078	053463	053846	4230	4613	4996	5378	5760	6142	6524	383
4	6905	7286	7666	8046	8426	8805	9185	9563	9942	060320	379
5	060698	061075	061452	061829	062206	062582	062958	063333	063709	4083	376
6	4458	4832	5206	5580	5953	6326	6699	7071	7443	7815	373
7	8186	8557	8928	9298	9668	070038	070407	070776	071145	071514	370
8	071882	072250	072617	072985	073352	3718	4085	4451	4816	5182	366
9	5547	5912	6276	6640	7004	7368	7731	8094	8457	8819	363
120	079181	079543	079904	080266	080626	080987	081347	081707	082067	082426	360
1	082785	083144	083503	3861	4219	4576	4934	5291	5647	6004	357
2	6360	6716	7071	7426	7781	8136	8490	8845	9198	9552	355
3	9905	090258	090611	090963	091315	091667	092018	092370	092721	093071	352
4	093422	3772	4122	4471	4820	5169	5518	5866	6215	6562	349
5	6910	7257	7604	7951	8298	8644	8990	9335	9681	100026	346
6	100371	100715	101059	101403	101747	102091	102434	102777	103119	3462	343
7	3804	4146	4487	4828	5169	5510	5851	6191	6531	6871	341
8	7210	7549	7888	8227	8565	8903	9241	9579	9916	110253	338
9	110590	110926	111263	111599	111934	112270	112605	112940	113275	3609	335
130	113943	114277	114611	114944	115278	115611	115943	116276	116608	116940	333
1	7271	7603	7934	8265	8595	8926	9256	9586	9915	120245	330
2	120574	120903	121231	121560	121888	122216	122544	122871	123198	3525	328
3	3852	4178	4504	4830	5156	5481	5806	6131	6456	6781	325
4	7105	7429	7753	8076	8399	8722	9045	9368	9690	130012	323
5	130334	130655	130977	131298	131619	131939	132260	132580	132900	3219	321
6	3539	3858	4177	4496	4814	5133	5451	5769	6086	6403	318
7	6721	7037	7354	7671	7987	8303	8618	8934	9249	9564	316
8	9879	140194	140508	140822	141136	141450	141763	142076	142389	142702	314
9	143015	3327	3639	3951	4263	4574	4885	5196	5507	5818	311
140	146128	146438	146748	147058	147367	147676	147985	148294	148603	148911	309
1	9219	9527	9835	150142	150449	150756	151063	151370	151676	151982	307
2	152288	152594	152900	3205	3510	3815	4120	4424	4728	5032	305
3	5336	5640	5943	6246	6549	6852	7154	7457	7759	8061	303
4	8362	8664	8965	9266	9567	9868	160168	160469	160769	161068	301
5	161368	161667	161967	162266	162564	162863	3161	3460	3758	4055	299
6	4353	4650	4947	5244	5541	5838	6134	6430	6726	7022	297
7	7317	7613	7908	8203	8497	8792	9086	9380	9674	9968	295
8	170262	170555	170848	171141	171434	171726	172019	172311	172603	172895	293
9	3186	3478	3769	4060	4351	4641	4932	5222	5512	5802	291
150	176091	176381	176670	176959	177248	177536	177825	178113	178401	178689	289
1	8977	9264	9552	9839	180126	180413	180699	180986	181272	181558	287
2	181844	182129	182415	182700	2985	3270	3555	3839	4123	4407	285
3	4691	4975	5259	5542	5825	6108	6391	6674	6956	7239	283
4	7521	7803	8084	8366	8647	8928	9209	9490	9771	190051	281
5	190332	190612	190892	191171	191451	191730	192010	192289	192567	2846	279
6	3125	3403	3681	3959	4237	4514	4792	5069	5346	5623	278
7	5900	6176	6453	6729	7005	7281	7556	7832	8107	8382	276
8	8657	8932	9206	9481	9755	200029	200303	200577	200850	201124	274
9	201397	201670	201943	202216	202488	2761	3033	3305	3577	3848	272
No.	0	1	2	3	4	5	6	7	8	9	Diff.

No.	0	1	2	3	4	5	6	7	8	9	Diff.
160	204120	204391	204663	204931	205204	205475	205746	206016	206286	206556	271
1	6826	7096	7365	7634	7904	8173	8441	8710	8979	9247	269
2	9515	9783	210051	210319	210586	210853	211121	211388	211654	211921	267
3	212188	212454	2720	2986	3252	3518	3783	4049	4314	4579	266
4	4844	5109	5373	5638	5902	6166	6430	6694	6957	7221	264
5	7484	7747	8010	8273	8536	8798	9060	9323	9585	9846	262
6	220108	220370	220631	220892	221153	221414	221675	221936	222196	222456	261
7	2716	2976	3236	3496	3755	4015	4274	4533	4792	5051	259
8	5309	5568	5826	6084	6342	6600	6858	7115	7372	7630	258
9	7887	8144	8400	8657	8913	9170	9426	9682	9938	230193	256
170	230449	230704	230960	231215	231470	231724	231979	232234	232488	232742	255
1	2996	3250	3504	3757	4011	4264	4517	4770	5023	5276	253
2	5528	5781	6033	6285	6537	6789	7041	7292	7544	7795	252
3	8046	8297	8548	8799	9049	9299	9550	9800	240050	240300	250
4	240549	240799	241048	241297	241546	241795	242044	242293	2541	2790	249
5	3038	3286	3534	3782	4030	4277	4525	4772	5019	5266	248
6	5513	5759	6006	6252	6499	6745	6991	7237	7482	7728	246
7	7973	8219	8464	8709	8954	9198	9443	9687	9932	250176	245
8	250420	250664	250908	251151	251395	251638	251881	252125	252368	2610	243
9	2853	3096	3338	3580	3822	4064	4306	4548	4790	5031	242
180	255273	255514	255755	255996	256237	256477	256718	256958	257198	257439	241
1	7679	7918	8158	8398	8637	8877	9116	9355	9594	9833	239
2	260071	260310	260548	260787	261025	261263	261501	261739	261976	262214	238
3	2451	2688	2925	3162	3399	3636	3873	4109	4346	4582	237
4	4818	5054	5290	5525	5761	5996	6232	6467	6702	6937	235
5	7172	7406	7641	7875	8110	8344	8578	8812	9046	9279	234
6	9513	9746	9980	270213	270446	270679	270912	271144	271377	271609	233
7	271842	272074	272306	2538	2770	3001	3233	3464	3696	3927	232
8	4158	4389	4620	4850	5081	5311	5542	5772	6002	6232	230
9	6462	6692	6921	7151	7380	7609	7838	8067	8296	8525	229
190	278754	278982	279211	279439	279667	279895	280123	280351	280578	280806	228
1	281033	281261	281488	281715	281942	282169	2396	2622	2849	3075	227
2	3301	3527	3753	3979	4205	4431	4656	4882	5107	5332	226
3	5557	5782	6007	6232	6456	6681	6905	7130	7354	7578	225
4	7802	8026	8249	8473	8696	8920	9143	9366	9589	9812	223
5	290035	290257	290480	290702	290925	291147	291369	291591	291813	292034	222
6	2256	2478	2699	2920	3141	3363	3584	3804	4025	4246	221
7	4466	4687	4907	5127	5347	5567	5787	6007	6226	6446	220
8	6665	6884	7104	7323	7542	7761	7979	8198	8416	8635	219
9	8853	9071	9289	9507	9725	9943	300161	300378	300595	300813	218
200	301030	301247	301464	301681	301898	302114	302331	302547	302764	302980	217
1	3196	3412	3628	3844	4059	4275	4491	4706	4921	5136	216
2	5351	5566	5781	5996	6211	6425	6639	6854	7068	7282	215
3	7496	7710	7924	8137	8351	8564	8778	8991	9204	9417	213
4	9630	9843	310056	310268	310481	310693	310906	311118	311330	311542	212
5	311754	311966	2177	2389	2600	2812	3023	3234	3445	3656	211
6	3867	4078	4289	4499	4710	4920	5130	5340	5551	5760	210
7	5970	6180	6390	6599	6809	7018	7227	7436	7646	7854	209
8	8063	8272	8481	8689	8898	9106	9314	9522	9730	9938	208
9	320146	320354	320562	320769	320977	321184	321391	321598	321805	322012	207
210	322219	322426	322633	322839	323046	323252	323458	323665	323871	324077	206
1	4282	4488	4694	4899	5105	5310	5516	5721	5926	6131	205
2	6336	6541	6745	6950	7155	7359	7563	7767	7972	8176	204
3	8380	8583	8787	8991	9194	9398	9601	9805	330008	330211	203
4	330414	330617	330819	331022	331225	331427	331630	331832	2034	2236	202
5	2438	2640	2842	3044	3246	3447	3649	3850	4051	4253	202
6	4454	4655	4856	5057	5257	5458	5658	5859	6059	6260	201
7	6460	6660	6860	7060	7260	7459	7659	7858	8058	8257	200
8	8456	8656	8855	9054	9253	9451	9650	9849	340047	340246	199
9	340444	340642	340841	341039	341237	341435	341632	341830	2028	2225	198
No.	0	1	2	3	4	5	6	7	8	9	Diff.

No.	0	1	2	3	4	5	6	7	8	9	Diff.
220	342423	342620	342817	343014	343212	343409	343606	343802	343999	344196	197
1	4392	4589	4785	4981	5178	5374	5570	5766	5962	6157	196
2	6353	6549	6744	6939	7135	7330	7525	7720	7915	8110	195
3	8305	8500	8694	8889	9083	9278	9472	9666	9860	350054	194
4	350248	350442	350636	350829	351023	351216	351410	351603	351796	1989	193
5	2183	2375	2568	2761	2954	3147	3339	3532	3724	3916	193
6	4108	4301	4493	4685	4876	5068	5260	5452	5643	5834	192
7	6026	6217	6408	6599	6790	6981	7172	7363	7554	7744	191
8	7935	8125	8316	8506	8696	8886	9076	9266	9456	9646	190
9	9835	360025	360215	360404	360593	360783	360972	361161	361350	361539	189
230	361728	361917	362105	362294	362482	362671	362859	363048	363236	363424	188
1	3612	3800	3988	4176	4363	4551	4739	4926	5113	5301	188
2	5488	5675	5862	6049	6236	6423	6610	6796	6983	7169	187
3	7356	7542	7729	7915	8101	8287	8473	8659	8845	9030	186
4	9216	9401	9587	9772	9958	370143	370328	370513	370698	370883	185
5	371068	371253	371437	371622	371806	1991	2175	2360	2544	2728	184
6	2912	3096	3280	3464	3647	3831	4015	4198	4382	4565	184
7	4748	4932	5115	5298	5481	5664	5846	6029	6212	6394	183
8	6577	6759	6942	7124	7306	7488	7670	7852	8034	8216	182
9	8398	8580	8761	8943	9124	9306	9487	9668	9849	380030	181
240	380211	380392	380573	380754	380934	381115	381296	381476	381656	381837	181
1	2017	2197	2377	2557	2737	2917	3097	3277	3456	3636	180
2	3815	3995	4174	4353	4533	4712	4891	5070	5249	5428	179
3	5606	5785	5964	6142	6321	6499	6677	6856	7034	7212	178
4	7390	7568	7746	7923	8101	8279	8456	8634	8811	8989	178
5	9166	9343	9520	9698	9875	390051	390228	390405	390582	390759	177
6	390935	391112	391288	391464	391641	1817	1993	2169	2345	2521	176
7	2697	2873	3048	3224	3400	3575	3751	3926	4101	4277	176
8	4452	4627	4802	4977	5152	5326	5501	5676	5850	6025	175
9	6199	6374	6548	6722	6896	7071	7245	7419	7592	7766	174
250	397940	398114	398287	398461	398634	398808	398981	399154	399328	399501	173
1	9674	9847	400020	400192	400365	400538	400711	400883	401056	401228	173
2	401401	401573	1745	1917	2089	2261	2433	2605	2777	2949	172
3	3121	3292	3464	3635	3807	3978	4149	4320	4492	4663	171
4	4834	5005	5176	5346	5517	5688	5858	6029	6199	6370	171
5	6540	6710	6881	7051	7221	7391	7561	7731	7901	8070	170
6	8240	8410	8579	8749	8918	9087	9257	9426	9595	9764	169
7	9933	410102	410271	410440	410609	410777	410946	411114	411283	411451	169
8	411620	1788	1956	2124	2293	2461	2629	2796	2964	3132	168
9	3300	3467	3635	3803	3970	4137	4305	4472	4639	4806	167
260	414973	415140	415307	415474	415641	415808	415974	416141	416308	416474	167
1	6641	6807	6973	7139	7306	7472	7638	7804	7970	8135	166
2	8301	8467	8633	8798	8964	9129	9295	9460	9625	9791	165
3	9956	420121	420286	420451	420616	420781	420945	421110	421275	421439	165
4	421604	1768	1933	2097	2261	2426	2590	2754	2918	3082	164
5	3246	3410	3574	3737	3901	4065	4228	4392	4555	4718	164
6	4882	5045	5208	5371	5534	5697	5860	6023	6186	6349	163
7	6511	6674	6836	6999	7161	7324	7486	7648	7811	7973	162
8	8135	8297	8459	8621	8783	8944	9106	9268	9429	9591	162
9	9752	9914	430075	430236	430398	430559	430720	430881	431042	431203	161
270	431364	431525	431685	431846	432007	432167	432328	432488	432649	432809	161
1	2969	3130	3290	3450	3610	3770	3930	4090	4249	4409	160
2	4569	4729	4888	5048	5207	5367	5526	5685	5844	6004	159
3	6163	6322	6481	6640	6799	6957	7116	7275	7433	7592	159
4	7751	7909	8067	8226	8384	8542	8701	8859	9017	9175	158
5	9333	9491	9648	9806	9964	440122	440279	440437	440594	440752	158
6	440909	441066	441224	441381	441538	1695	1852	2009	2166	2323	157
7	2480	2637	2793	2950	3106	3263	3419	3576	3732	3889	157
8	4045	4201	4357	4513	4669	4825	4981	5137	5293	5449	156
9	5604	5760	5915	6071	6226	6382	6537	6692	6848	7003	155
No.	0	1	2	3	4	5	6	7	8	9	Diff.

No	0	1	2	3	4	5	6	7	8	9	Diff.
280	447158	447313	447468	447623	447778	447933	448088	448242	448397	448552	155
1	8706	8861	9015	9170	9324	9478	9633	9787	9941	450095	154
2	450249	450403	450557	450711	450865	451018	451172	451326	451479	1633	154
3	1786	1940	2093	2247	2400	2553	2706	2859	3012	3165	153
4	3318	3471	3624	3777	3930	4082	4235	4387	4540	4692	153
5	4845	4997	5150	5302	5454	5606	5758	5910	6062	6214	152
6	6366	6518	6670	6821	6973	7125	7276	7428	7579	7731	152
7	7882	8033	8184	8336	8487	8638	8789	8940	9091	9242	151
8	9392	9543	9694	9845	9995	460146	460296	460447	460597	460748	151
9	460898	461048	461198	461348	461499	1649	1799	1948	2098	2248	150
290	462398	462548	462697	462847	462997	463146	463296	463445	463594	463744	150
1	3893	4042	4191	4340	4490	4639	4788	4936	5085	5234	149
2	5383	5532	5680	5829	5977	6126	6274	6423	6571	6719	149
3	6868	7016	7164	7312	7460	7608	7756	7904	8052	8200	148
4	8347	8495	8643	8790	8938	9085	9233	9380	9527	9675	148
5	9822	9969	470116	470263	470410	470557	470704	470851	470998	471145	147
6	471292	471438	1585	1732	1878	2025	2171	2318	2464	2610	146
7	2756	2903	3049	3195	3341	3487	3633	3779	3925	4071	146
8	4216	4362	4508	4653	4799	4944	5090	5235	5381	5526	146
9	5671	5816	5962	6107	6252	6397	6542	6687	6832	6976	145
300	477121	477266	477411	477555	477700	477844	477989	478133	478278	478422	145
1	8566	8711	8855	8999	9143	9287	9431	9575	9719	9863	144
2	480007	480151	480294	480438	480582	480725	480869	481012	481156	481299	144
3	1443	1586	1729	1872	2016	2159	2302	2445	2588	2731	143
4	2874	3016	3159	3302	3445	3587	3730	3872	4015	4157	143
5	4300	4442	4585	4727	4869	5011	5153	5295	5437	5579	142
6	5721	5863	6005	6147	6289	6430	6572	6714	6855	6997	142
7	7138	7280	7421	7563	7704	7845	7986	8127	8269	8410	141
8	8551	8692	8833	8974	9114	9255	9396	9537	9677	9818	141
9	9958	490099	490239	490380	490520	490661	490801	490941	491081	491222	140
310	491362	491502	491642	491782	491922	492062	492201	492341	492481	492621	140
1	2760	2900	3040	3179	3319	3458	3597	3737	3876	4015	139
2	4155	4294	4433	4572	4711	4850	4989	5128	5267	5406	139
3	5544	5683	5822	5960	6099	6238	6376	6515	6653	6791	139
4	6930	7068	7206	7344	7483	7621	7759	7897	8035	8173	138
5	8311	8448	8586	8724	8862	8999	9137	9275	9412	9550	138
6	9687	9824	9962	500099	500236	500374	500511	500648	500785	500922	137
7	501059	501196	501383	1470	1607	1744	1880	2017	2154	2291	137
8	2427	2564	2700	2837	2973	3109	3246	3382	3518	3655	136
9	3791	3927	4063	4199	4335	4471	4607	4743	4878	5014	136
320	505150	505286	505421	505557	505693	505828	505964	506099	506234	506370	136
1	6505	6640	6776	6911	7046	7181	7316	7451	7586	7721	135
2	7856	7991	8126	8260	8395	8530	8664	8799	8934	9068	135
3	9203	9337	9471	9606	9740	9874	510009	510143	510277	510411	134
4	510545	510679	510813	510947	511081	511215	1349	1482	1616	1750	134
5	1883	2017	2151	2284	2418	2551	2684	2818	2951	3084	133
6	3218	3351	3484	3617	3750	3883	4016	4149	4282	4415	133
7	4548	4681	4813	4946	5079	5211	5344	5476	5609	5741	133
8	5874	6006	6139	6271	6403	6535	6668	6800	6932	7064	132
9	7196	7328	7460	7592	7724	7855	7987	8119	8251	8382	132
330	518514	518646	518777	518909	519040	519171	519303	519434	519566	519697	131
1	9828	9959	520090	520221	520353	520484	520615	520745	520876	521007	131
2	521138	521269	1400	1530	1661	1792	1922	2053	2183	2314	131
3	2444	2575	2705	2835	2966	3096	3226	3356	3486	3616	130
4	3746	3876	4006	4136	4266	4396	4526	4656	4785	4915	130
5	5045	5174	5304	5434	5563	5693	5822	5951	6081	6210	129
6	6339	6469	6598	6727	6856	6985	7114	7243	7372	7501	129
7	7630	7759	7888	8016	8145	8274	8402	8531	8660	8788	129
8	8917	9045	9174	9302	9430	9559	9687	9815	9943	530072	128
9	530200	530328	530456	530584	530712	530840	530968	531096	531223	1351	128
No.	0	1	2	3	4	5	6	7	8	9	Diff.

No.	0	1	2	3	4	5	6	7	8	9	Diff.
340	531479	531607	531734	531862	531990	532117	532245	532372	532500	532627	128
1	2754	2882	3009	3136	3264	3391	3518	3645	3772	3899	127
2	4026	4153	4280	4407	4534	4661	4787	4914	5041	5167	127
3	5294	5421	5547	5674	5800	5927	6053	6180	6306	6432	126
4	6558	6685	6811	6937	7063	7189	7315	7441	7567	7693	126
5	7819	7945	8071	8197	8322	8448	8574	8699	8825	8951	126
6	9076	9202	9327	9452	9578	9703	9829	9954	540079	540204	125
7	540329	540455	540580	540705	540830	540955	541080	541205	1330	1454	125
8	1579	1704	1829	1953	2078	2203	2327	2452	2576	2701	125
9	2825	2950	3074	3199	3323	3447	3571	3696	3820	3944	124
350	544068	544192	544316	544440	544564	544688	544812	544936	545060	545183	124
1	5307	5431	5555	5678	5802	5925	6049	6172	6296	6419	124
2	6543	6666	6789	6913	7036	7159	7282	7405	7529	7652	123
3	7775	7898	8021	8144	8267	8389	8512	8635	8758	8881	123
4	9003	9126	9249	9371	9494	9616	9739	9861	9984	550106	123
5	550228	550351	550473	550595	550717	550840	550962	551084	551206	1328	122
6	1450	1572	1694	1816	1938	2060	2181	2303	2425	2547	122
7	2668	2790	2911	3033	3155	3276	3398	3519	3640	3762	121
8	3883	4004	4126	4247	4368	4489	4610	4731	4852	4973	121
9	5094	5215	5336	5457	5578	5699	5820	5940	6061	6182	121
360	556303	556423	556544	556664	556785	556905	557026	557146	557267	557387	120
1	7507	7627	7748	7868	7988	8108	8228	8349	8469	8589	120
2	8709	8829	8948	9068	9188	9308	9428	9548	9667	9787	120
3	9907	560026	560146	560265	560385	560504	560624	560743	560863	560982	119
4	561101	1221	1340	1459	1578	1698	1817	1936	2055	2174	119
5	2293	2412	2531	2650	2769	2887	3006	3125	3244	3362	119
6	3481	3600	3718	3837	3955	4074	4192	4311	4429	4548	119
7	4666	4784	4903	5021	5139	5257	5376	5494	5612	5730	118
8	5848	5966	6084	6202	6320	6437	6555	6673	6791	6909	118
9	7026	7144	7262	7379	7497	7614	7732	7849	7967	8084	118
370	568202	568319	568436	568554	568671	568788	568905	569023	569140	569257	117
1	9374	9491	9608	9725	9842	9959	570076	570193	570309	570426	117
2	570543	570660	570776	570893	571010	571126	1243	1359	1476	1592	117
3	1709	1825	1942	2058	2174	2291	2407	2523	2639	2755	116
4	2872	2988	3104	3220	3336	3452	3568	3684	3800	3915	116
5	4031	4147	4263	4379	4494	4610	4726	4841	4957	5072	116
6	5188	5303	5419	5534	5650	5765	5880	5996	6111	6226	115
7	6341	6457	6572	6687	6802	6917	7032	7147	7262	7377	115
8	7492	7607	7722	7836	7951	8066	8181	8295	8410	8525	115
9	8639	8754	8868	8983	9097	9212	9326	9441	9555	9669	114
380	579784	579898	580012	580126	580241	580355	580469	580583	580697	580811	114
1	580925	581039	1153	1267	1381	1495	1608	1722	1836	1950	114
2	2063	2177	2291	2404	2518	2631	2745	2858	2972	3085	114
3	3199	3312	3426	3539	3652	3765	3879	3992	4105	4218	113
4	4331	4444	4557	4670	4783	4896	5009	5122	5235	5348	113
5	5461	5574	5686	5799	5912	6024	6137	6250	6362	6475	113
6	6587	6700	6812	6925	7037	7149	7262	7374	7486	7599	112
7	7711	7823	7935	8047	8160	8272	8384	8496	8608	8720	112
8	8832	8944	9056	9167	9279	9391	9503	9615	9726	9838	112
9	9950	590061	590173	590284	590396	590507	590619	590730	590842	590953	112
390	591065	591176	591287	591399	591510	591621	591732	591843	591955	592066	111
1	2177	2288	2399	2510	2621	2732	2843	2954	3064	3175	111
2	3286	3397	3508	3618	3729	3840	3950	4061	4171	4282	111
3	4393	4503	4614	4724	4834	4945	5055	5165	5276	5386	110
4	5496	5606	5717	5827	5937	6047	6157	6267	6377	6487	110
5	6597	6707	6817	6927	7037	7146	7256	7366	7476	7586	110
6	7695	7805	7914	8024	8134	8243	8353	8462	8572	8681	110
7	8791	8900	9009	9119	9228	9337	9446	9556	9665	9774	109
8	9883	9992	600101	600210	600319	600428	600537	600646	600755	600864	109
9	600973	601082	1191	1299	1408	1517	1625	1734	1843	1951	109
No.	0	1	2	3	4	5	6	7	8	9	Diff.

No.	0	1	2	3	4	5	6	7	8	9	Diff.
400	602060	602169	602277	602386	602494	602603	602711	602819	602928	603036	108
1	3144	3253	3361	3469	3577	3686	3794	3902	4010	4118	108
2	4226	4334	4442	4550	4658	4766	4874	4982	5089	5197	108
3	5305	5413	5521	5628	5736	5844	5951	6059	6166	6274	108
4	6381	6489	6596	6704	6811	6919	7026	7133	7241	7348	107
5	7455	7562	7669	7777	7884	7991	8098	8205	8312	8419	107
6	8526	8633	8740	8847	8954	9061	9167	9274	9381	9488	107
7	9594	9701	9808	9914	610021	610128	610234	610341	610447	610554	107
8	610660	610767	610873	610979	1086	1192	1298	1405	1511	1617	106
9	1723	1829	1936	2042	2148	2254	2360	2466	2572	2678	106
410	612784	612890	612996	613102	613207	613313	613419	613525	613630	613736	106
1	3842	3947	4053	4159	4264	4370	4475	4581	4686	4792	106
2	4897	5003	5108	5213	5319	5424	5529	5634	5740	5845	105
3	5950	6055	6160	6265	6370	6476	6581	6686	6790	6895	105
4	7000	7105	7210	7315	7420	7525	7629	7734	7839	7943	105
5	8048	8153	8257	8362	8466	8571	8676	8780	8884	8989	105
6	9093	9198	9302	9406	9511	9615	9719	9824	9928	620032	104
7	620136	620240	620344	620448	620552	620656	620760	620864	620968	1072	104
8	1176	1280	1384	1488	1592	1695	1799	1903	2007	2110	104
9	2214	2318	2421	2525	2628	2732	2835	2939	3042	3146	104
420	623249	623353	623456	623559	623663	623766	623869	623973	624076	624179	103
1	4282	4385	4488	4591	4695	4798	4901	5004	5107	5210	103
2	5312	5415	5518	5621	5724	5827	5929	6032	6135	6238	103
3	6340	6443	6546	6648	6751	6853	6956	7058	7161	7263	103
4	7366	7468	7571	7673	7775	7878	7980	8082	8185	8287	102
5	8389	8491	8593	8695	8797	8900	9002	9104	9206	9308	102
6	9410	9512	9613	9715	9817	9919	630021	630123	630224	630326	102
7	630428	630530	630631	630733	630835	630936	1038	1139	1241	1342	102
8	1444	1545	1647	1748	1849	1951	2052	2153	2255	2356	101
9	2457	2559	2660	2761	2862	2963	3064	3165	3266	3367	101
430	633468	633569	633670	633771	633872	633973	634074	634175	634276	634376	101
1	4477	4578	4679	4779	4880	4981	5081	5182	5283	5383	101
2	5484	5584	5685	5785	5886	5986	6087	6187	6287	6388	100
3	6488	6588	6688	6789	6889	6989	7089	7189	7290	7390	100
4	7490	7590	7690	7790	7890	7990	8090	8190	8290	8389	100
5	8489	8589	8689	8789	8888	8988	9088	9188	9287	9387	100
6	9486	9586	9686	9785	9885	9984	640084	640183	640283	640382	99
7	640481	640581	640680	640779	640879	640978	1077	1177	1276	1375	99
8	1474	1573	1672	1771	1871	1970	2069	2168	2267	2366	99
9	2465	2563	2662	2761	2860	2959	3058	3156	3255	3354	99
440	643453	643551	643650	643749	643847	643946	644044	644143	644242	644340	98
1	4439	4537	4636	4734	4832	4931	5029	5127	5226	5324	98
2	5422	5521	5619	5717	5815	5913	6011	6110	6208	6306	98
3	6404	6502	6600	6698	6796	6894	6992	7089	7187	7285	98
4	7383	7481	7579	7676	7774	7872	7969	8067	8165	8262	98
5	8360	8458	8555	8653	8750	8848	8945	9043	9140	9237	97
6	9335	9432	9530	9627	9724	9821	9919	650016	650113	650210	97
7	650308	650405	650502	650599	650696	650793	650890	0987	1084	1181	97
8	1278	1375	1472	1569	1666	1762	1859	1956	2053	2150	97
9	2246	2343	2440	2536	2633	2730	2826	2923	3019	3116	97
450	653213	653309	653405	653502	653598	653695	653791	653888	653984	654080	96
1	4177	4273	4369	4465	4562	4658	4754	4850	4946	5042	96
2	5138	5235	5331	5427	5523	5619	5715	5810	5906	6002	96
3	6098	6194	6290	6386	6482	6577	6673	6769	6864	6960	96
4	7056	7152	7247	7343	7438	7534	7629	7725	7820	7916	96
5	8011	8107	8202	8298	8393	8488	8584	8679	8774	8870	95
6	8965	9060	9155	9250	9346	9441	9536	9631	9726	9821	95
7	9916	660011	660106	660201	660296	660391	660486	660581	660676	660771	95
8	660865	0960	1055	1150	1245	1339	1434	1529	1623	1718	95
9	1813	1907	2002	2096	2191	2286	2380	2475	2569	2663	95
No.	0	1	2	3	4	5	6	7	8	9	Diff.

No.	0	1	2	3	4	5	6	7	8	9	Diff.
460	662758	662852	662947	663041	663135	663230	663324	663418	663512	663607	94
1	3701	3795	3889	3983	4078	4172	4266	4360	4454	4548	94
2	4642	4736	4830	4924	5018	5112	5206	5299	5393	5487	94
3	5581	5675	5769	5862	5956	6050	6143	6237	6331	6424	94
4	6518	6612	6705	6799	6892	6986	7079	7173	7266	7360	94
5	7453	7546	7640	7733	7826	7920	8013	8106	8199	8293	93
6	8386	8479	8572	8665	8759	8852	8945	9038	9131	9224	93
7	9317	9410	9503	9596	9689	9782	9875	9967	670060	670153	93
8	670246	670339	670431	670524	670617	670710	670802	670895	0988	1080	93
9	1173	1265	1358	1451	1543	1636	1728	1821	1913	2005	93
470	672098	672190	672283	672375	672467	672560	672652	672744	672836	672929	92
1	3021	3113	3205	3297	3390	3482	3574	3666	3758	3850	92
2	3942	4034	4126	4218	4310	4402	4494	4586	4677	4769	92
3	4861	4953	5045	5137	5228	5320	5412	5503	5595	5687	92
4	5778	5870	5962	6053	6145	6236	6328	6419	6511	6602	92
5	6694	6785	6876	6968	7059	7151	7242	7333	7424	7516	91
6	7607	7698	7789	7881	7972	8063	8154	8245	8336	8427	91
7	8518	8609	8700	8791	8882	8973	9064	9155	9246	9337	91
8	9428	9519	9610	9700	9791	9882	9973	680063	680154	680245	91
9	680336	680426	680517	680607	680698	680789	680879	0970	1060	1151	91
480	681241	681332	681422	681513	681603	681693	681784	681874	681964	682055	90
1	2145	2235	2326	2416	2506	2596	2686	2777	2867	2957	90
2	3047	3137	3227	3317	3407	3497	3587	3677	3767	3857	90
3	3947	4037	4127	4217	4307	4396	4486	4576	4666	4756	90
4	4845	4935	5025	5114	5204	5294	5383	5473	5563	5652	90
5	5742	5831	5921	6010	6100	6189	6279	6368	6458	6547	89
6	6636	6726	6815	6904	6994	7083	7172	7261	7351	7440	89
7	7529	7618	7707	7796	7886	7975	8064	8153	8242	8331	89
8	8420	8509	8598	8687	8776	8865	8953	9042	9131	9220	89
9	9309	9398	9486	9575	9664	9753	9841	9930	690019	690107	89
490	690196	690285	690373	690462	690550	690639	690728	690816	690905	690993	89
1	1081	1170	1258	1347	1435	1524	1612	1700	1789	1877	88
2	1965	2053	2142	2230	2318	2406	2494	2583	2671	2759	88
3	2847	2935	3023	3111	3199	3287	3375	3463	3551	3639	88
4	3727	3815	3903	3991	4078	4166	4254	4342	4430	4517	88
5	4605	4693	4781	4868	4956	5044	5131	5219	5307	5394	88
6	5482	5569	5657	5744	5832	5919	6007	6094	6182	6269	87
7	6356	6444	6531	6618	6706	6793	6880	6968	7055	7142	87
8	7229	7317	7404	7491	7578	7665	7752	7839	7926	8014	87
9	8101	8188	8275	8362	8449	8535	8622	8709	8796	8883	87
500	698970	699057	699144	699231	699317	699404	699491	699578	699664	699751	87
1	9838	9924	700011	700098	700184	700271	700358	700444	700531	700617	87
2	700704	700790	0877	0963	1050	1136	1222	1309	1395	1482	86
3	1568	1654	1741	1827	1913	1999	2086	2172	2258	2344	86
4	2431	2517	2603	2689	2775	2861	2947	3033	3119	3205	86
5	3291	3377	3463	3549	3635	3721	3807	3893	3979	4065	86
6	4151	4236	4322	4408	4494	4579	4665	4751	4837	4922	86
7	5008	5094	5179	5265	5350	5436	5522	5607	5693	5778	86
8	5864	5949	6035	6120	6206	6291	6376	6462	6547	6632	85
9	6718	6803	6888	6974	7059	7144	7229	7315	7400	7485	85
510	707570	707655	707740	707826	707911	707996	708081	708166	708251	708336	85
1	8421	8506	8591	8676	8761	8846	8931	9015	9100	9185	85
2	9270	9355	9440	9524	9609	9694	9779	9863	9948	710033	85
3	710117	710202	710287	710371	710456	710540	710625	710710	710794	0879	85
4	0963	1048	1132	1217	1301	1385	1470	1554	1639	1723	84
5	1807	1892	1976	2060	2144	2229	2313	2397	2481	2566	84
6	2650	2734	2818	2902	2986	3070	3154	3238	3323	3407	84
7	3491	3575	3659	3742	3826	3910	3994	4078	4162	4246	84
8	4330	4414	4497	4581	4665	4749	4833	4916	5000	5084	84
9	5167	5251	5335	5418	5502	5586	5669	5753	5836	5920	84
No	0	1	2	3	4	5	6	7	8	9	Diff.

No.	0	1	2	3	4	5	6	7	8	9	Diff.
520	716003	716087	716170	716254	716337	716421	716504	716588	716671	716754	83
1	6838	6921	7004	7088	7171	7254	7338	7421	7504	7587	83
2	7671	7754	7837	7920	8003	8086	8169	8253	8336	8419	83
3	8502	8586	8668	8751	8834	8917	9000	9083	9165	9248	83
4	9331	9414	9497	9580	9663	9745	9828	9911	9994	720077	83
5	720159	720242	720325	720407	720490	720573	720655	720738	720821	0903	83
6	0986	1068	1151	1233	1316	1398	1481	1563	1646	1728	82
7	1811	1893	1975	2058	2140	2222	2305	2387	2469	2552	82
8	2634	2716	2798	2881	2963	3045	3127	3209	3291	3374	82
9	3456	3538	3620	3702	3784	3866	3948	4030	4112	4194	82
530	724276	724358	724440	724522	724604	724685	724767	724849	724931	725013	82
1	5095	5176	5258	5340	5422	5503	5585	5667	5748	5830	82
2	5912	5993	6075	6156	6238	6320	6401	6483	6564	6646	82
3	6727	6809	6890	6972	7053	7134	7216	7297	7379	7460	81
4	7541	7623	7704	7785	7866	7948	8029	8110	8191	8273	81
5	8354	8435	8516	8597	8678	8759	8841	8922	9003	9084	81
6	9165	9246	9327	9408	9489	9570	9651	9732	9813	9893	81
7	9974	730055	730136	730217	730298	730378	730459	730540	730621	730702	81
8	730782	0863	0944	1024	1105	1186	1266	1347	1428	1508	81
9	1589	1669	1750	1830	1911	1991	2072	2152	2233	2313	81
540	732394	732474	732555	732635	732715	732796	732876	732956	733037	733117	80
1	3197	3278	3358	3138	3518	3598	3679	3759	3839	3919	80
2	3999	4079	4160	4240	4320	4400	4480	4560	4640	4720	80
3	4800	4880	4900	5040	5120	5200	5279	5359	5439	5519	80
4	5599	5679	5759	5838	5918	5998	6078	6157	6237	6317	80
5	6397	6476	6556	6635	6715	6795	6874	6954	7034	7113	80
6	7193	7272	7352	7431	7511	7590	7670	7749	7829	7908	79
7	7987	8067	8146	8225	8305	8384	8463	8543	8622	8701	79
8	8781	8860	8939	9018	9097	9177	9256	9335	9414	9493	79
9	9572	9651	9731	9810	9889	9968	740047	740126	740205	740284	79
550	740363	740442	710521	740600	740678	740757	740836	740915	740994	741073	79
1	1152	1230	1309	1388	1467	1546	1624	1703	1782	1860	79
2	1939	2018	2096	2175	2254	2332	2411	2489	2568	2647	79
3	2725	2804	2882	2961	3039	3118	3196	3275	3353	3431	78
4	3510	3588	3667	3745	3823	3902	3980	4058	4136	4215	78
5	4293	4371	4449	4528	4606	4684	4762	4840	4919	4997	78
6	5075	5153	5231	5309	5387	5465	5543	5621	5699	5777	78
7	5855	5933	6011	6089	6167	6245	6323	6401	6479	6556	78
8	6634	6712	6790	6863	6945	7023	7101	7179	7256	7334	78
9	7412	7489	7567	7645	7722	7800	7878	7955	8033	8110	78
560	748188	748266	748343	748421	748498	748576	748653	748731	748808	748885	77
1	8963	9040	9118	9195	9272	9350	9427	9504	9582	9659	77
2	9736	9814	9891	9968	750045	750123	750200	750277	750354	750431	77
3	750508	750586	750663	750740	0817	0894	0971	1048	1125	1202	77
4	1279	1356	1433	1510	1587	1664	1741	1818	1895	1972	77
5	2048	2125	2202	2279	2356	2433	2509	2586	2663	2740	77
6	2816	2893	2970	3047	3123	3200	3277	3353	3430	3506	77
7	3583	3660	3736	3813	3889	3966	4042	4119	4195	4272	77
8	4348	4425	4501	4578	4654	4730	4807	4883	4960	5036	76
9	5112	5189	5265	5341	5417	5494	5570	5646	5722	5799	76
570	755875	755951	756027	756103	756180	756256	756332	756408	756484	756560	76
1	6636	6712	6788	6864	6940	7016	7092	7168	7244	7320	76
2	7396	7472	7548	7624	7700	7775	7851	7927	8003	8079	76
3	8155	8230	8306	8382	8458	8533	8609	8685	8761	8836	76
4	8912	8988	9063	9139	9214	9290	9366	9441	9517	9592	76
5	9668	9743	9819	9894	9970	760045	760121	760196	760272	760347	75
6	760422	760498	760573	760649	760724	0799	0875	0950	1025	1101	75
7	1176	1251	1326	1402	1477	1552	1627	1702	1778	1853	75
8	1928	2003	2078	2153	2228	2303	2378	2453	2529	2604	75
9	2679	2754	2329	2904	2978	3053	3128	3203	3278	3353	75
No.	0	1	2	3	4	5	6	7	8	9	Diff.

No.	0	1	2	3	4	5	6	7	8	9	Diff.
580	763428	763503	763578	763653	763727	763802	763877	763952	764027	764101	75
1	4176	4251	4326	4400	4475	4550	4624	4699	4774	4848	75
2	4923	4998	5072	5147	5221	5296	5370	5445	5520	5594	75
3	5669	5743	5818	5892	5966	6041	6115	6190	6264	6338	74
4	6413	6487	6562	6636	6710	6785	6859	6933	7007	7082	74
5	7156	7230	7304	7379	7453	7527	7601	7675	7749	7823	74
6	7898	7972	8046	8120	8194	8268	8342	8416	8490	8564	74
7	8638	8712	8786	8860	8934	9008	9082	9156	9230	9303	74
8	9377	9451	9525	9599	9673	9746	9820	9894	9968	770042	74
9	770115	770189	770263	770336	770410	770484	770557	770631	770705	0778	74
590	770852	770926	770999	771073	771146	771220	771293	771367	771440	771514	74
1	1587	1661	1734	1808	1881	1955	2028	2102	2175	2248	73
2	2322	2395	2468	2542	2615	2688	2762	2835	2908	2981	73
3	3055	3128	3201	3274	3348	3421	3494	3567	3640	3713	73
4	3786	3860	3933	4006	4079	4152	4225	4298	4371	4444	73
5	4517	4590	4663	4736	4809	4882	4955	5028	5100	5173	73
6	5246	5319	5392	5465	5538	5610	5683	5756	5829	5902	73
7	5974	6047	6120	6193	6265	6338	6411	6483	6556	6629	73
8	6701	6774	6846	6919	6992	7064	7137	7209	7282	7354	73
9	7427	7499	7572	7644	7717	7789	7862	7934	8006	8079	72
600	778151	778224	778296	778368	778441	778513	778585	778658	778730	778802	72
1	8874	8947	9019	9091	9163	9236	9308	9380	9452	9524	72
2	9596	9669	9741	9813	9885	9957	780029	780101	780173	780245	72
3	780317	780389	780461	780533	780605	780677	0749	0821	0893	0965	72
4	1037	1109	1181	1253	1324	1396	1468	1540	1612	1684	72
5	1755	1827	1899	1971	2042	2114	2186	2258	2329	2401	72
6	2473	2544	2616	2688	2759	2831	2902	2974	3046	3117	72
7	3189	3260	3332	3403	3475	3546	3618	3689	3761	3832	71
8	3904	3975	4046	4118	4189	4261	4332	4403	4475	4546	71
9	4617	4689	4760	4831	4902	4974	5045	5116	5187	5259	71
610	785330	785401	785472	785543	785615	785686	785757	785828	785899	785970	71
1	6041	6112	6183	6254	6325	6396	6467	6538	6609	6680	71
2	6751	6822	6893	6964	7035	7106	7177	7248	7319	7390	71
3	7460	7531	7602	7673	7744	7815	7885	7956	8027	8098	71
4	8168	8239	8310	8381	8451	8522	8593	8663	8734	8804	71
5	8875	8946	9016	9087	9157	9228	9299	9369	9440	9510	71
6	9581	9651	9722	9792	9863	9933	790004	790074	790144	790215	70
7	790285	790356	790426	790496	790567	790637	0707	0778	0848	0918	70
8	0988	1059	1129	1199	1269	1340	1410	1480	1550	1620	70
9	1691	1761	1831	1901	1971	2041	2111	2181	2252	2322	70
620	792392	792462	792532	792602	792672	792742	792812	792882	792952	793022	70
1	3092	3162	3231	3301	3371	3441	3511	3581	3651	3721	70
2	3790	3860	3930	4000	4070	4139	4209	4279	4349	4418	70
3	4488	4558	4627	4697	4767	4836	4906	4976	5045	5115	70
4	5185	5254	5324	5393	5463	5532	5602	5672	5741	5811	70
5	5880	5949	6019	6088	6158	6227	6297	6366	6436	6505	69
6	6574	6644	6713	6782	6852	6921	6990	7060	7129	7198	69
7	7268	7337	7406	7475	7545	7614	7683	7752	7821	7890	69
8	7960	8029	8098	8167	8236	8305	8374	8443	8513	8582	69
9	8651	8720	8789	8858	8927	8996	9065	9134	9203	9272	69
630	799341	799409	799478	799547	799616	799685	799754	799823	799892	799961	69
1	800029	800098	800167	800236	800305	800373	800442	800511	800580	800648	69
2	0717	0786	0854	0923	0992	1061	1129	1198	1266	1335	69
3	1404	1472	1541	1609	1678	1747	1815	1884	1952	2021	69
4	2089	2158	2226	2295	2363	2432	2500	2568	2637	2705	68
5	2774	2842	2910	2979	3047	3116	3184	3252	3321	3389	68
6	3457	3525	3594	3662	3730	3798	3867	3935	4003	4071	68
7	4139	4208	4276	4344	4412	4480	4548	4616	4685	4753	68
8	4821	4889	4957	5025	5093	5161	5229	5297	5365	5433	68
9	5501	5569	5637	5705	5773	5841	5908	5976	6044	6112	68
No.	0	1	2	3	4	5	6	7	8	9	Diff.

No.	0	1	2	3	4	5	6	7	8	9	Diff.
640	806180	806248	806316	806384	806451	806519	806587	806655	806723	806790	68
1	6858	6926	6994	7061	7129	7197	7264	7332	7400	7467	68
2	7535	7603	7670	7738	7806	7873	7941	8008	8076	8143	68
3	8211	8279	8346	8414	8481	8549	8616	8684	8751	8818	67
4	8886	8953	9021	9088	9156	9223	9290	9358	9425	9492	67
5	9560	9627	9694	9762	9829	9896	9964	810031	810098	810165	67
6	810233	810300	810367	810434	810501	810569	810636	0703	0770	0837	67
7	0904	0971	1039	1106	1173	1240	1307	1374	1441	1508	67
8	1575	1642	1709	1776	1843	1910	1977	2044	2111	2178	67
9	2245	2312	2379	2445	2512	2579	2646	2713	2780	2847	67
650	812913	812980	813047	813114	813181	813247	813314	813381	813448	813514	67
1	3581	3648	3714	3781	3848	3914	3981	4048	4114	4181	67
2	4248	4314	4381	4447	4514	4581	4647	4714	4780	4847	67
3	4913	4980	5046	5113	5179	5246	5312	5378	5445	5511	66
4	5578	5644	5711	5777	5843	5910	5976	6042	6109	6175	66
5	6241	6308	6374	6440	6506	6573	6639	6705	6771	6838	66
6	6904	6970	7036	7102	7169	7235	7301	7367	7433	7499	66
7	7565	7631	7698	7764	7830	7896	7962	8028	8094	8160	66
8	8226	8292	8358	8424	8490	8556	8622	8688	8754	8820	66
9	8885	8951	9017	9083	9149	9215	9281	9346	9412	9478	66
660	819544	819610	819676	819741	819807	819873	819939	820004	820070	820136	66
1	820201	820267	820333	820399	820464	820530	820595	0661	0727	0792	66
2	0858	0924	0989	1055	1120	1186	1251	1317	1382	1448	66
3	1514	1579	1645	1710	1775	1841	1906	1972	2037	2103	65
4	2168	2233	2299	2364	2430	2495	2560	2626	2691	2756	65
5	2822	2887	2952	3018	3083	3148	3213	3279	3344	3409	65
6	3474	3539	3605	3670	3735	3800	3865	3930	3996	4061	65
7	4126	4191	4256	4321	4386	4451	4516	4581	4646	4711	65
8	4776	4841	4906	4971	5036	5101	5166	5231	5296	5361	65
9	5426	5491	5556	5621	5686	5751	5815	5880	5945	6010	65
670	826075	826140	826204	826269	826334	826399	826464	826528	826593	826658	65
1	6723	6787	6852	6917	6981	7046	7111	7175	7240	7305	65
2	7369	7434	7499	7563	7628	7692	7757	7821	7886	7951	65
3	8015	8080	8144	8209	8273	8338	8402	8467	8531	8595	64
4	8660	8724	8789	8853	8918	8982	9046	9111	9175	9239	64
5	9304	9368	9132	9497	9561	9625	9690	9754	9818	9882	64
6	9947	830011	830075	830139	830204	830268	830332	830396	830460	830525	64
7	830589	0653	0717	0781	0845	0909	0973	1037	1102	1166	64
8	1230	1294	1358	1422	1486	1550	1614	1678	1742	1806	64
9	1870	1934	1998	2062	2126	2189	2253	2317	2381	2445	64
680	832509	832573	832637	832700	832764	832828	832892	832956	833020	833083	64
1	3147	3211	3275	3338	3402	3466	3530	3593	3657	3721	64
2	3784	3848	3912	3975	4039	4103	4166	4230	4294	4357	64
3	4421	4484	4548	4611	4675	4739	4802	4866	4929	4993	64
4	5056	5120	5183	5247	5310	5373	5437	5500	5564	5627	63
5	5691	5754	5817	5881	5944	6007	6071	6134	6197	6261	63
6	6324	6387	6451	6514	6577	6641	6704	6767	6830	6894	63
7	6957	7020	7083	7146	7210	7273	7336	7399	7462	7525	63
8	7588	7652	7715	7778	7841	7904	7967	8030	8093	8156	63
9	8219	8232	8345	8408	8471	8534	8597	8660	8723	8786	63
690	838849	838912	838975	839038	839101	839164	839227	839289	839352	839415	63
1	9478	9541	9604	9667	9729	9792	9855	9918	9981	840043	63
2	840106	840169	840232	840294	840357	840420	840482	840545	840608	0671	63
3	0733	0796	0859	0921	0984	1046	1109	1172	1234	1297	63
4	1359	1422	1485	1547	1610	1672	1735	1797	1860	1922	63
5	1985	2047	2110	2172	2235	2297	2360	2422	2184	2547	62
6	2609	2672	2734	2796	2859	2921	2983	3046	3108	3170	62
7	3233	3295	3357	3420	3482	3544	3606	3669	3731	3793	62
8	3855	3918	3980	4042	4104	4166	4229	4291	4353	4415	62
9	4477	4539	4601	4664	4726	4788	4850	4912	4974	5036	62
No.	0	1	2	3	4	5	6	7	8	9	Diff.

No.	0	1	2	3	4	5	6	7	8	9	Diff.
700	845098	845160	845222	845284	845346	845408	845470	845532	845594	845656	62
1	5718	5780	5842	5904	5966	6028	6090	6151	6213	6275	62
2	6337	6399	6461	6523	6585	6646	6708	6770	6832	6894	62
3	6955	7017	7079	7141	7202	7264	7326	7388	7449	7511	62
4	7573	7634	7696	7758	7819	7881	7943	8004	8066	8128	62
5	8189	8251	8312	8374	8435	8497	8559	8620	8682	8743	62
6	8805	8866	8928	8989	9051	9112	9174	9235	9297	9358	61
7	9419	9481	9542	9604	9665	9726	9788	9849	9911	9972	61
8	850033	850095	850156	850217	850279	850340	850401	850462	850524	850585	61
9	0646	0707	0769	0830	0891	0952	1014	1075	1136	1197	61
710	851258	851320	851381	851442	851503	851564	851625	851686	851747	851809	61
1	1870	1931	1992	2053	2114	2175	2236	2297	2358	2419	61
2	2480	2541	2602	2663	2724	2785	2846	2907	2968	3029	61
3	3090	3150	3211	3272	3333	3394	3455	3516	3577	3637	61
4	3698	3759	3820	3881	3941	4002	4063	4124	4185	4245	61
5	4306	4367	4428	4488	4549	4610	4670	4731	4792	4852	61
6	4913	4974	5034	5095	5156	5216	5277	5337	5398	5459	61
7	5519	5580	5640	5701	5761	5822	5882	5943	6003	6064	61
8	6124	6185	6245	6306	6366	6427	6487	6548	6608	6668	60
9	6729	6789	6850	6910	6970	7031	7091	7152	7212	7272	60
720	857332	857393	857453	857513	857574	857634	857694	857755	857815	857875	60
1	7935	7995	8056	8116	8176	8236	8297	8357	8417	8477	60
2	8537	8597	8657	8718	8778	8838	8898	8958	9018	9078	60
3	9138	9198	9258	9318	9379	9439	9499	9559	9619	9679	60
4	9739	9799	9859	9918	9978	860038	860098	860158	860218	860278	60
5	860338	860398	860458	860518	860578	0637	0697	0757	0817	0877	60
6	0937	0996	1056	1116	1176	1236	1295	1355	1415	1475	60
7	1534	1594	1654	1714	1773	1833	1893	1952	2012	2072	60
8	2131	2191	2251	2310	2370	2430	2489	2549	2608	2668	60
9	2728	2787	2847	2906	2966	3025	3085	3144	3204	3263	60
730	863323	863382	863442	863501	863561	863620	863680	863739	863799	863858	59
1	3917	3977	4036	4096	4155	4214	4274	4333	4392	4452	59
2	4511	4570	4630	4689	4748	4808	4867	4926	4985	5045	59
3	5104	5163	5222	5282	5341	5400	5459	5519	5578	5637	59
4	5696	5755	5814	5874	5933	5992	6051	6110	6169	6228	59
5	6287	6316	6405	6465	6524	6583	6642	6701	6760	6819	59
6	6878	6937	6996	7055	7114	7173	7232	7291	7350	7409	59
7	7467	7526	7585	7644	7703	7762	7821	7880	7939	7998	59
8	8056	8115	8174	8233	8292	8350	8409	8468	8527	8586	59
9	8644	8703	8762	8821	8879	8938	8997	9056	9114	9173	59
740	869232	869290	869349	869408	869466	869525	869584	869642	869701	869760	59
1	9818	9877	9935	9994	870053	870111	870170	870228	870287	870345	59
2	870404	870462	870521	870579	0638	0696	0755	0813	0872	0930	58
3	0989	1047	1106	1164	1223	1281	1339	1398	1456	1515	58
4	1573	1631	1690	1748	1806	1865	1923	1981	2040	2098	58
5	2156	2215	2273	2331	2389	2448	2506	2564	2622	2681	58
6	2739	2797	2855	2913	2972	3030	3088	3146	3204	3262	58
7	3321	3379	3437	3495	3553	3611	3669	3727	3785	3844	58
8	3902	3960	4018	4076	4134	4192	4250	4308	4366	4424	58
9	4482	4540	4598	4656	4714	4772	4830	4888	4945	5003	58
750	875061	875119	875177	875235	875293	875351	875409	875466	875524	875582	58
1	5640	5698	5756	5813	5871	5929	5987	6045	6102	6160	58
2	6218	6276	6333	6391	6449	6507	6564	6622	6680	6737	58
3	6795	6853	6910	6968	7026	7083	7141	7199	7256	7314	58
4	7371	7429	7487	7544	7602	7659	7717	7774	7832	7889	58
5	7947	8004	8062	8119	8177	8234	8292	8349	8407	8464	57
6	8522	8579	8637	8694	8752	8809	8866	8924	8981	9039	57
7	9096	9153	9211	9268	9325	9383	9440	9497	9555	9612	57
8	9669	9726	9784	9841	9898	9956	880013	880070	880127	880185	57
9	880242	880299	880356	880413	880471	880528	0585	0642	0699	0756	57
No.	0	1	2	3	4	5	6	7	8	9	Diff.

No.	0	1	2	3	4	5	6	7	8	9	Diff.
760	880814	880871	880928	880985	881042	881099	881156	881213	881271	881328	57
1	1385	1442	1499	1556	1613	1670	1727	1784	1841	1898	57
2	1955	2012	2069	2126	2183	2240	2297	2354	2411	2468	57
3	2525	2581	2638	2695	2752	2809	2866	2923	2980	3037	57
4	3093	3150	3207	3264	3321	3377	3434	3491	3548	3605	57
5	3661	3718	3775	3832	3888	3945	4002	4059	4115	4172	57
6	4229	4285	4342	4399	4455	4512	4569	4625	4682	4739	57
7	4795	4852	4909	4965	5022	5078	5135	5192	5248	5305	57
8	5361	5418	5474	5531	5587	5644	5700	5757	5813	5870	57
9	5926	5983	6039	6096	6152	6209	6265	6321	6378	6434	56
770	886491	886547	886604	886660	886716	886773	886829	886885	886942	886998	56
1	7054	7111	7167	7223	7280	7336	7392	7449	7505	7561	56
2	7617	7674	7730	7786	7842	7898	7955	8011	8067	8123	56
3	8179	8236	8292	8348	8404	8460	8516	8573	8629	8685	56
4	8741	8797	8853	8909	8965	9021	9077	9134	9190	9246	56
5	9302	9358	9414	9470	9526	9582	9638	9694	9750	9806	56
6	9862	9918	9974	890030	890086	890141	890197	890253	890309	890365	56
7	890421	890477	890533	0589	0645	0700	0756	0812	0868	0924	56
8	0980	1035	1091	1147	1203	1259	1314	1370	1426	1482	56
9	1537	1593	1649	1705	1760	1816	1872	1928	1983	2039	56
780	892095	892150	892206	892262	892317	892373	892429	892484	892540	892595	56
1	2651	2707	2762	2818	2873	2929	2985	3040	3096	3151	56
2	3207	3262	3318	3373	3429	3484	3540	3595	3651	3706	56
3	3762	3817	3873	3928	3984	4039	4094	4150	4205	4261	55
4	4316	4371	4427	4482	4538	4593	4648	4704	4759	4814	55
5	4870	4925	4980	5036	5091	5146	5201	5257	5312	5367	55
6	5423	5478	5533	5588	5644	5699	5754	5809	5864	5920	55
7	5975	6030	6085	6140	6195	6251	6306	6361	6416	6471	55
8	6526	6581	6636	6692	6747	6802	6857	6912	6967	7022	55
9	7077	7132	7187	7242	7297	7352	7407	7462	7517	7572	55
790	897627	897682	897737	897792	897847	897902	897957	898012	898067	898122	55
1	8176	8231	8286	8341	8396	8451	8506	8561	8615	8670	55
2	8725	8780	8835	8890	8944	8999	9054	9109	9164	9218	55
3	9273	9328	9383	9437	9492	9547	9602	9656	9711	9766	55
4	9821	9875	9930	9985	900039	900094	900149	900203	900258	900312	55
5	900367	900422	900476	900531	0586	0640	0695	0749	0804	0859	55
6	0913	0968	1022	1077	1131	1186	1240	1295	1349	1404	55
7	1458	1513	1567	1622	1676	1731	1785	1840	1894	1948	54
8	2003	2057	2112	2166	2221	2275	2329	2384	2438	2492	54
9	2547	2601	2655	2710	2764	2818	2873	2927	2981	3036	54
800	903090	903144	903199	903253	903307	903361	903416	903470	903524	903578	54
1	3633	3687	3741	3795	3849	3904	3958	4012	4066	4120	54
2	4174	4229	4283	4337	4391	4445	4499	4553	4607	4661	54
3	4716	4770	4824	4878	4932	4986	5040	5094	5148	5202	54
4	5256	5310	5364	5418	5472	5526	5580	5634	5688	5742	54
5	5796	5850	5904	5958	6012	6066	6119	6173	6227	6281	54
6	6335	6389	6443	6497	6551	6604	6658	6712	6766	6820	54
7	6874	6927	6981	7035	7089	7143	7196	7250	7304	7358	54
8	7411	7465	7519	7573	7626	7680	7734	7787	7841	7895	54
9	7949	8002	8056	8110	8163	8217	8270	8324	8378	8431	54
810	908485	908539	908592	908646	908699	908753	908807	908860	908914	908967	54
1	9021	9074	9128	9181	9235	9289	9342	9396	9449	9503	54
2	9556	9610	9663	9716	9770	9823	9877	9930	9984	910037	53
3	910091	910144	910197	910251	910304	910358	910411	910464	910518	0571	53
4	0624	0678	0731	0784	0838	0891	0944	0998	1051	1104	53
5	1158	1211	1264	1317	1371	1424	1477	1530	1584	1637	53
6	1690	1743	1797	1850	1903	1956	2009	2063	2116	2169	53
7	2222	2275	2328	2381	2435	2488	2541	2594	2647	2700	53
8	2753	2806	2859	2913	2966	3019	3072	3125	3178	3231	53
9	3284	3337	3390	3443	3496	3549	3602	3655	3708	3761	53
No.	0	1	2	3	4	5	6	7	8	9	Diff.

No.	0	1	2	3	4	5	6	7	8	9	Diff.
820	913814	913867	913920	913973	914026	914079	914132	914184	914237	914290	53
1	4343	4396	4449	4502	4555	4608	4660	4713	4766	4819	53
2	4872	4925	4977	5030	5083	5136	5189	5241	5294	5347	53
3	5400	5453	5505	5558	5611	5664	5716	5769	5822	5875	53
4	5927	5980	6033	6085	6138	6191	6243	6296	6349	6401	53
5	6454	6507	6559	6612	6664	6717	6770	6822	6875	6927	53
6	6980	7033	7085	7138	7190	7243	7295	7348	7400	7453	53
7	7506	7558	7611	7663	7716	7768	7820	7873	7925	7978	52
8	8030	8083	8135	8188	8240	8293	8345	8397	8450	8502	52
9	8555	8607	8659	8712	8764	8816	8869	8921	8973	9026	52
830	919078	919130	919183	919235	919287	919340	919392	919444	919496	919549	52
1	9601	9653	9706	9758	9810	9862	9914	9967	920019	920071	52
2	920123	920176	920228	920280	920332	920384	920436	920489	0541	0593	52
3	0645	0697	0749	0801	0853	0906	0958	1010	1062	1114	52
4	1166	1218	1270	1322	1374	1426	1478	1530	1582	1634	52
5	1686	1738	1790	1842	1894	1946	1998	2050	2102	2154	52
6	2206	2258	2310	2362	2414	2466	2518	2570	2622	2674	52
7	2725	2777	2829	2881	2933	2985	3037	3089	3140	3192	52
8	3244	3296	3348	3399	3451	3503	3555	3607	3658	3710	52
9	3762	3814	3865	3917	3969	4021	4072	4124	4176	4228	52
840	924279	924331	924383	924434	924486	924538	924589	924641	924693	924744	52
1	4796	4848	4899	4951	5003	5054	5106	5157	5209	5261	52
2	5312	5364	5415	5467	5518	5570	5621	5673	5725	5776	52
3	5828	5879	5931	5982	6034	6085	6137	6188	6240	6291	51
4	6342	6394	6445	6497	6548	6600	6651	6702	6754	6805	51
5	6857	6908	6959	7011	7062	7114	7165	7216	7268	7319	51
6	7370	7422	7473	7524	7576	7627	7678	7730	7781	7832	51
7	7883	7935	7986	8037	8088	8140	8191	8242	8293	8345	51
8	8396	8447	8498	8549	8601	8652	8703	8754	8805	8857	51
9	8908	8959	9010	9061	9112	9163	9215	9266	9317	9368	51
850	929419	929470	929521	929572	929623	929674	929725	929776	929827	929879	51
1	9930	9981	930032	930083	930134	930185	930236	930287	930338	930389	51
2	930440	930491	0542	0592	0643	0694	0745	0796	0847	0898	51
3	0949	1000	1051	1102	1153	1204	1254	1305	1356	1407	51
4	1458	1509	1560	1610	1661	1712	1763	1814	1865	1915	51
5	1966	2017	2068	2118	2169	2220	2271	2322	2372	2423	51
6	2474	2524	2575	2626	2677	2727	2778	2829	2879	2930	51
7	2981	3031	3082	3133	3183	3234	3285	3335	3386	3437	51
8	3487	3538	3589	3639	3690	3740	3791	3841	3892	3943	51
9	3993	4044	4094	4145	4195	4246	4296	4347	4397	4448	51
860	934498	934549	934599	934650	934700	934751	934801	934852	934902	934953	50
1	5003	5054	5104	5154	5205	5255	5306	5356	5406	5457	50
2	5507	5558	5608	5658	5709	5759	5809	5860	5910	5960	50
3	6011	6061	6111	6162	6212	6262	6313	6363	6413	6463	50
4	6514	6564	6614	6665	6715	6765	6815	6865	6916	6966	50
5	7016	7066	7117	7167	7217	7267	7317	7367	7418	7468	50
6	7518	7568	7618	7668	7718	7769	7819	7869	7919	7969	50
7	8019	8069	8119	8169	8219	8269	8320	8370	8420	8470	50
8	8520	8570	8620	8670	8720	8770	8820	8870	8920	8970	50
9	9020	9070	9120	9170	9220	9270	9320	9369	9419	9469	50
870	939519	939569	939619	939669	939719	939769	939819	939869	939918	939968	50
1	940018	940068	940118	940168	940218	940267	940317	940367	940417	940467	50
2	0516	0566	0616	0666	0716	0765	0815	0865	0915	0964	50
3	1014	1064	1114	1163	1213	1263	1313	1362	1412	1462	50
4	1511	1561	1611	1660	1710	1760	1809	1859	1909	1958	50
5	2008	2058	2107	2157	2207	2256	2306	2355	2405	2455	50
6	2504	2554	2603	2653	2702	2752	2801	2851	2901	2950	50
7	3000	3049	3099	3148	3198	3247	3297	3346	3396	3445	49
8	3495	3544	3593	3643	3692	3742	3791	3841	3890	3939	49
9	3989	4038	4088	4137	4186	4236	4285	4335	4384	4433	49
No.	0	1	2	3	4	5	6	7	8	9	Diff.

No.	0	1	2	3	4	5	6	7	8	9	Diff.
880	944483	944532	944581	944631	944680	944729	944779	944828	944877	944927	49
1	4976	5025	5074	5124	5173	5222	5272	5321	5370	5419	49
2	5469	5518	5567	5616	5665	5715	5764	5813	5862	5912	49
3	5961	6010	6059	6108	6157	6207	6256	6305	6354	6403	49
4	6452	6501	6551	6600	6649	6698	6747	6796	6845	6894	49
5	6943	6992	7041	7090	7140	7189	7238	7287	7336	7385	49
6	7434	7483	7532	7581	7630	7679	7728	7777	7826	7875	49
7	7924	7973	8022	8070	8119	8168	8217	8266	8315	8364	49
8	8413	8462	8511	8560	8609	8657	8706	8755	8804	8853	49
9	8902	8951	8999	9048	9097	9146	9195	9244	9292	9341	49
890	949390	949439	949488	949536	949585	949634	949683	949731	949780	949829	49
1	9878	9926	9975	950024	950073	950121	950170	950219	950267	950316	49
2	950365	950414	950462	0511	0560	0608	0657	0706	0754	0803	49
3	0851	0900	0949	0997	1046	1095	114	1192	1240	1289	49
4	1338	1386	1435	1483	1532	1580	1629	1677	1726	1775	49
5	1823	1872	1920	1969	2017	2066	2114	2163	2211	2260	48
6	2308	2356	2405	2453	2502	2550	2599	2647	2696	2744	48
7	2792	2841	2889	2938	2986	3034	3083	3131	3180	3228	48
8	3276	3325	3373	3421	3470	3518	3566	3615	3663	3711	48
9	3760	3808	3856	3905	3953	4001	4049	4098	4146	4194	48
900	954243	954291	954339	954387	954435	954484	954532	954580	954628	954677	48
1	4725	4773	4821	4869	4918	4966	5014	5062	5110	5158	48
2	5207	5255	5303	5351	5399	5447	5495	5543	5592	5640	48
3	5688	5736	5784	5832	5880	5928	5976	6024	6072	6120	48
4	6168	6216	6265	6313	6361	6409	6457	6505	6553	6601	48
5	6649	6697	6745	6793	6840	6888	6936	6984	7032	7080	48
6	7128	7176	7224	7272	7320	7368	7416	7464	7512	7559	48
7	7607	7655	7703	7751	7799	7847	7894	7942	7990	8038	48
8	8086	8134	8181	8229	8277	8325	8373	8421	8468	8516	48
9	8564	8612	8659	8707	8755	8803	8850	8898	8946	8994	48
910	959041	959089	959137	959185	959232	959280	959328	959375	959423	959471	48
1	9518	9566	9614	9661	9709	9757	9804	9852	9900	9947	48
2	9995	960042	960090	960138	960185	960233	960280	960328	960376	960423	48
3	960471	0518	0566	0613	0661	0709	0756	0804	0851	0899	48
4	0946	0994	1041	1089	1136	1184	1231	1279	1326	1374	48
5	1421	1469	1516	1563	1611	1658	1706	1753	1801	1848	47
6	1895	1943	1990	2038	2085	2132	2180	2227	2275	2322	47
7	2369	2417	2464	2511	2559	2606	2653	2701	2748	2795	47
8	2843	2890	2937	2985	3032	3079	3126	3174	3221	3268	47
9	3316	3363	3410	3457	3504	3552	3599	3646	3693	3741	47
920	963788	963835	963882	963929	963977	964024	964071	964118	964165	964212	47
1	4260	4307	4354	4401	4448	4495	4542	4590	4637	4684	47
2	4731	4778	4825	4872	4919	4966	5013	5061	5108	5155	47
3	5202	5249	5296	5343	5390	5437	5484	5531	5578	5625	47
4	5672	5719	5766	5813	5860	5907	5954	6001	6048	6095	47
5	6142	6189	6236	6283	6329	6376	6423	6470	6517	6564	47
6	6611	6658	6705	6752	6799	6845	6892	6939	6986	7033	47
7	7080	7127	7173	7220	7267	7314	7361	7408	7454	7501	47
8	7548	7595	7642	7688	7735	7782	7829	7875	7922	7969	47
9	8016	8062	8109	8156	8203	8249	8296	8343	8390	8436	47
930	968483	968530	968576	968623	968670	968716	968763	968810	968856	968903	47
1	8950	8996	9043	9090	9136	9183	9229	9276	9323	9369	47
2	9416	9463	9509	9556	9602	9649	9695	9742	9789	9835	47
3	9882	9928	9975	970021	970068	970114	970161	970207	970254	970300	47
4	970347	970393	970440	0486	0533	0579	0626	0672	0719	0765	46
5	0812	0858	0904	0951	0997	1044	1090	1137	1183	1229	46
6	1276	1322	1369	1415	1461	1508	1554	1601	1647	1693	46
7	1740	1786	1832	1879	1925	1971	2018	2064	2110	2157	46
8	2203	2249	2295	2342	2388	2434	2481	2527	2573	2619	46
9	2666	2712	2758	2804	2851	2897	2943	2989	3035	3082	46
No.	0	1	2	3	4	5	6	7	8	9	Diff.

No.	0	1	2	3	4	5	6	7	8	9	Diff.
940	973128	973174	973220	973266	973313	973359	973405	973451	973497	973543	46
1	3590	3636	3682	3728	3774	3820	3866	3913	3959	4005	46
2	4051	4097	4143	4189	4235	4281	4327	4374	4420	4466	46
3	4512	4558	4604	4650	4696	4742	4788	4834	4880	4926	46
4	4972	5018	5064	5110	5156	5202	5248	5294	5340	5386	46
5	5432	5478	5524	5570	5616	5662	5707	5753	5799	5845	46
6	5891	5937	5983	6029	6075	6121	6167	6212	6258	6304	46
7	6350	6396	6442	6488	6533	6579	6625	6671	6717	6763	46
8	6808	6854	6900	6946	6992	7037	7083	7129	7175	7220	46
9	7266	7312	7358	7403	7449	7495	7541	7586	7632	7678	46
950	977724	977769	977815	977861	977906	977952	977998	978043	978089	[illegible]	46
1	8181	8226	8272	8317	8363	8409	8454	8500	8546	8591	46
2	8637	8683	8728	8774	8819	8865	8911	8956	9002	9047	46
3	9093	9138	9184	9230	9275	9321	9366	9412	9457	[illegible]	46
4	9548	9594	9639	9685	9730	9776	9821	9867	9912	9958	46
5	980003	980049	980094	980140	980185	980231	980276	980322	980367	980412	45
6	0458	0503	0549	0594	0640	0685	0730	0776	0821	[illegible]	45
7	0912	0957	1003	1048	1093	1139	1184	1229	1275	1320	45
8	1366	1411	1456	1501	1547	1592	1637	1683	1728	1773	45
9	1819	1864	1909	1954	2000	2045	2090	2135	2181	[illegible]	45
960	982271	982316	982362	982407	982452	982497	982543	982588	982633	982678	45
1	2723	2769	2814	2859	2904	2949	2994	3040	3085	[illegible]	45
2	3175	3220	3265	3310	3356	3401	3446	3491	3536	3581	45
3	3626	3671	3716	3762	3807	3852	3897	3942	3987	4032	45
4	4077	4122	4167	4212	4257	4302	4347	4392	4437	[illegible]	45
5	4527	4572	4617	4662	4707	4752	4797	4842	4887	4932	45
6	4977	5022	5067	5112	5157	5202	5247	5292	5337	5382	45
7	5426	5471	5516	5561	5606	5651	5696	5741	5786	[illegible]	45
8	5875	5920	5965	6010	6055	6100	6144	6189	6234	6279	45
9	6324	6369	6413	6458	6503	6548	6593	6637	6682	6727	45
970	986772	986817	986861	986906	986951	986996	987040	987085	987130	987175	45
1	7219	7264	7309	7353	7398	7443	7488	7532	7577	7622	45
2	7666	7711	7756	7800	7845	7890	7934	7979	8024	[illegible]	45
3	8113	8157	8202	8247	8291	8336	8381	8425	8470	8514	45
4	8559	8604	8648	8693	8737	8782	8826	8871	8916	8960	45
5	9005	9049	9094	9138	9183	9227	9272	9316	9361	[illegible]	45
6	9450	9494	9539	9583	9628	9672	9717	9761	9806	[illegible]	44
7	9895	9939	9983	990028	990072	990117	990161	990206	990250	990294	44
8	990339	990383	990428	0472	0516	0561	0605	0650	0694	[illegible]	44
9	0783	0827	0871	0916	0960	1004	1049	1093	1137	[illegible]	44
980	991226	991270	991315	991359	991403	991448	991492	991536	991580	991625	44
1	1669	1713	1758	1802	1846	1890	1935	1979	2023	2067	44
2	2111	2156	2200	2244	2288	2333	2377	2421	2465	2509	44
3	2554	2598	2642	2686	2730	2774	2819	2863	2907	[illegible]	44
4	2995	3039	3083	3127	3172	3216	3260	3304	3348	[illegible]	44
5	3436	3480	3524	3568	3613	3657	3701	3745	3789	3833	44
6	3877	3921	3965	4009	4053	4097	4141	4185	4229	[illegible]	44
7	4317	4361	4405	4449	4493	4537	4581	4625	4669	[illegible]	44
8	4757	4801	4845	4889	4933	4977	5021	5065	5108	5152	44
9	5196	5240	5284	5328	5372	5416	5460	5504	5547	5591	44
990	995635	995679	995723	995767	995811	995854	995898	995942	995986	996030	44
1	6074	6117	6161	6205	6249	6293	6337	6380	6424	6468	44
2	6512	6555	6599	6643	6687	6731	6774	6818	6862	[illegible]	44
3	6949	6993	7037	7080	7124	7168	7212	7255	7299	7343	44
4	7386	7430	7474	7517	7561	7605	7648	7692	7736	7779	44
5	7823	7867	7910	7954	7998	8041	8085	8129	8172	[illegible]	44
6	8259	8303	8347	8390	8434	8477	8521	8564	8608	[illegible]	44
7	8695	8739	8782	8826	8869	8913	8956	9000	9043	9087	44
8	9131	9174	9218	9261	9305	9348	9392	9435	9479	[illegible]	44
9	9565	9609	9652	9696	9739	9783	9826	9870	9913	[illegible]	43
No.	0	1	2	3	4	5	6	7	8	9	Diff.

TABLE XIII.

LOGARITHMIC SINES, COSINES, TANGENTS,

AND

COTANGENTS.

NOTE.

The table here given extends to minutes only. The usual method of extending such a table to seconds, by proportional parts of the difference between two consecutive logarithms, is accurate enough for most purposes, especially if the angle is not very small. When the angle is very small, and great accuracy is required, the following method may be used for sines, tangents, and cotangents.

I. Suppose it were required to find the logarithmic sine of 5′ 24″. By the ordinary method we should have

$$\begin{array}{lr} \text{log. sin. } 5' & = 7.162696 \\ \text{diff. for } 24'' & = \quad 31673 \\ \hline \text{log. sin. } 5'\ 24'' & = 7.194369 \end{array}$$

The more accurate method is founded on the proposition in Trigonometry, that the sines or tangents of very small angles are proportional to the angles themselves. In the present case, therefore, we have sin. 5′ : sin. 5′ 24′ $= 5' : 5'\ 24'' = 300'' : 324''$. Hence sin. 5′ 24′ $= \frac{324 \text{ sin. } 5'}{300}$, or log. sin. 5′ 24″ = log. sin. 5′ + log. 324 — log. 300. The difference for 24″ will therefore, be the difference between the logarithm of 324 and the logarithm of 300. The operation will stand thus: —

$$\begin{array}{lr} \text{log. } 324 & = 2.510545 \\ \text{log. } 300 & = 2.477121 \\ \hline \text{diff. for } 24 & = \quad 33424 \\ \text{log. sin. } 5' & = 7.162696 \\ \hline \text{log. sin. } 5'\ 24'' & = 7.196120 \end{array}$$

Comparing this value with that given in tables that extend to seconds, we find it exact even to the last figure

II. Given log. sin. A = 7.004438 to find A. The sine next less than this in the table is sin. 3′ = 6.940847. Now we have sin. 3′ : sin. A = 3 : A. Therefore, $A = \frac{3 \text{ sin. } A}{\text{sin. } 3'}$, or log. A = log. 3 + log. sin. A - log. sin. 3′. Hence it appears, that, to find the logarithm of A in minutes, we must add to the logarithm of 3 the difference between log. sin. A and log. sin. $3'$.

$$\begin{array}{lr} \text{log. sin. } A = & 7.004438 \\ \text{log. sin. } 3' = & 6.940847 \\ \hline & 63591 \\ \text{log. } 3 \quad = & 0.477121 \\ \hline A = 3.473 & 0.540712 \end{array}$$

or $A = 3'\ 28.38''$. By the common method we should have found $A = 3'\ 30.54''$.

The same method applies to tangents and cotangents, except that in the case of cotangents the differences are to be subtracted.

*** The radius of this table is unity, and the characteristics **9, 8, 7, and 6 stand respectively for —1, —2, —3, and —4.**

0° 179°

M.	Sine.	D. 1″.	Cosine.	D. 1″.	Tang.	D. 1″.	Cotang.	M.
0	Inf. neg.		0.000000	.00	Inf. neg.		Infinite.	60
1	6.463726	5017.17	.000000	.00	6.463726	5017.17	3.536274	59
2	.764756	2934.85	.000000	.00	.764756	2934.85	.235244	58
3	.940847	2082.31	.000000	.00	.940847	2082.31	.059153	57
4	7.065786	1615.17	.000000	.00	7.065786	1615.17	2.934214	56
5	.162696	1319.69	.000000	.00	.162696	1319.69	.837304	55
6	.241877	1115.78	9.999999	.00	.241878	1115.78	.758122	54
7	.308824	966.53	.999999	.00	.308825	966.54	.691175	53
8	.366816	852.54	.999999	.01	.366817	852.55	.633183	52
9	.417968	762.62	.999999	.01	.417970	762.63	.582030	51
10	7.463726	689.88	9.999998	.01	7.463727	689.88	2.536273	50
11	.505118	629.81	.999998	.01	.505120	629.81	.494880	49
12	.542906	579.37	.999997	.01	.542909	579.37	.457091	48
13	.577668	536.41	.999997	.01	.577672	536.42	.422328	47
14	.609853	499.38	.999996	.01	.609857	499.39	.390143	46
15	.639816	467.14	.999996	.01	.639820	467.15	.360180	45
16	.667845	438.81	.999995	.01	.667849	438.82	.332151	44
17	.694173	413.72	.999995	.01	.694179	413.73	.305821	43
18	.718997	391.35	.999994	.01	.719003	391.36	.280997	42
19	.742478	371.27	.999993	.01	.742484	371.28	.257516	41
20	7.764754	353.15	9.999993	.01	7.764761	353.16	2.235239	40
21	.785943	336.72	.999992	.01	.785951	336.73	.214049	39
22	.806146	321.75	.999991	.01	.806155	321.76	.193845	38
23	.825451	308.05	.999990	.01	.825460	308.07	.174540	37
24	.843934	295.47	.999989	.02	.843944	295.49	.156056	36
25	.861662	283.88	.999989	.02	.861674	283.90	.138326	35
26	.878695	273.17	.999988	.02	.878708	273.18	.121292	34
27	.895085	263.23	.999987	.02	.895099	263.25	.104901	33
28	.910879	253.99	.999986	.02	.910894	254.01	.089106	32
29	.926119	245.38	.999985	.02	.926134	245.40	.073866	31
30	7.940842	237.33	9.999983	.02	7.940858	237.35	2.059142	30
31	.955082	229.80	.999982	.02	.955100	229.82	.044900	29
32	.968870	222.73	.999981	.02	.968889	222.75	.031111	28
33	.982233	216.08	.999980	.02	.982253	216.10	.017747	27
34	.995198	209.81	.999979	.02	.995219	209.83	.004781	26
35	8.007787	203.90	.999977	.02	8.007809	203.92	1.992191	25
36	.020021	198.31	.999976	.02	.020044	198.33	.979956	24
37	.031919	193.02	.999975	.02	.031945	193.05	.968055	23
38	.043501	188.01	.999973	.02	.043527	188.03	.956473	22
39	.054781	183.25	.999972	.02	.054809	183.27	.945191	21
40	8.065776	178.72	9.999971	.02	8.065806	178.75	1.934194	20
41	.076500	174.42	.999969	.03	.076531	174.44	.923469	19
42	.086965	170.31	.999968	.03	.086997	170.34	.913003	18
43	.097183	166.39	.999966	.03	.097217	166.42	.902783	17
44	.107167	162.65	.999964	.03	.107203	162.68	.892797	16
45	.116926	159.08	.999963	.03	.116963	159.11	.883037	15
46	.126471	155.66	.999961	.03	.126510	155.69	.873490	14
47	.135810	152.38	.999959	.03	.135851	152.41	.864149	13
48	.144953	149.24	.999958	.03	.144996	149.27	.855004	12
49	.153907	146.22	.999956	.03	.153952	146.25	.846048	11
50	8.162681	143.33	9.999954	.03	8.162727	143.36	1.837273	10
51	.171280	140.54	.999952	.03	.171328	140.57	.828672	9
52	.179713	137.86	.999950	.03	.179763	137.90	.820237	8
53	.187985	135.29	.999948	.03	.188036	135.32	.811964	7
54	.196102	132.80	.999946	.03	.196156	132.84	.803844	6
55	.204070	130.41	.999944	.03	.204126	130.44	.795874	5
56	.211895	128.10	.999942	.03	.211953	128.14	.788047	4
57	.219581	125.87	.999940	.04	.219641	125.91	.780359	3
58	.227134	123.72	.999938	.04	.227195	123.76	.772805	2
59	.234557	121.64	.999936	.04	.234621	121.68	.765379	1
60	.241855		.999934		.241921		.758079	0
M.	Cosine.	D. 1″.	Sine.	D. 1″.	Cotang.	D. 1″.	Tang.	M.

90° 89°

M.	Sine.	D. 1″.	Cosine.	D. 1″.	Tang.	D. 1″.	Cotang.	M.
0	8.241855	119.63	9.999934	.04	8.241921	119.67	1.758079	60
1	.249033	117.69	.999932	.04	.249102	117.72	.750898	59
2	.256094	115.80	.999929	.04	.256165	115.84	.743835	58
3	.263042	113.98	.999927	.04	.263115	114.02	.736885	57
4	.269881	112.21	.999925	.04	.269956	112.25	.730044	56
5	.276614	110.50	.999922	.04	.276691	110.54	.723309	55
6	.283243	108.83	.999920	.04	.283323	108.87	.716677	54
7	.289773	107.22	.999918	.04	.289856	107.26	.710144	53
8	.296207	105.66	.999915	.04	.296292	105.70	.703708	52
9	.302546	104.13	.999913	.04	.302634	104.18	.697366	51
10	8.308794	102.66	9.999910	.04	8.308884	102.70	1.691116	50
11	.314954	101.22	.999907	.04	.315046	101.26	.684954	49
12	.321027	99.82	.999905	.04	.321122	99.87	.678878	48
13	.327016	98.47	.999902	.05	.327114	98.51	.672886	47
14	.332924	97.14	.999899	.05	.333025	97.19	.666975	46
15	.338753	95.86	.999897	.05	.338856	95.90	.661144	45
16	.344504	94.60	.999894	.05	.344610	94.65	.655390	44
17	.350181	93.38	.999891	.05	.350289	93.43	.649711	43
18	.355783	92.19	.999888	.05	.355895	92.24	.644105	42
19	.361315	91.03	.999885	.05	.361430	91.08	.638570	41
20	8.366777	89.90	9.999882	.05	8.366895	89.95	1.633105	40
21	.372171	88.80	.999879	.05	.372292	88.85	.627708	39
22	.377499	87.72	.999876	.05	.377622	87.77	.622378	38
23	.382762	86.67	.999873	.05	.382889	86.72	.617111	37
24	.387962	85.64	.999870	.05	.388092	85.70	.611908	36
25	.393101	84.64	.999867	.05	.393234	84.69	.606766	35
26	.398179	83.66	.999864	.05	.398315	83.71	.601685	34
27	.403199	82.71	.999861	.05	.403338	82.76	.596662	33
28	.408161	81.77	.999858	.05	.408304	81.82	.591696	32
29	.413068	80.86	.999854	.05	.413213	80.91	.586787	31
30	8.417919	79.96	9.999851	.06	8.418068	80.02	1.581932	30
31	.422717	79.09	.999848	.06	.422869	79.14	.577131	29
32	.427462	78.23	.999844	.06	.427618	78.29	.572382	28
33	.432156	77.40	.999841	.06	.432315	77.45	.567685	27
34	.436800	76.58	.999838	.06	.436962	76.63	.563038	26
35	.441394	75.77	.999834	.06	.441560	75.83	.558440	25
36	.445941	74.99	.999831	.06	.446110	75.05	.553890	24
37	.450440	74.22	.999827	.06	.450613	74.28	.549387	23
38	.454893	73.47	.999824	.06	.455070	73.53	.544930	22
39	.459301	72.73	.999820	.06	.459481	72.79	.540519	21
40	8.463665	72.00	9.999816	.06	8.463849	72.06	1.536151	20
41	.467935	71.29	.999813	.06	.468172	71.35	.531828	19
42	.472263	70.60	.999809	.06	.472454	70.66	.527546	18
43	.476498	69.91	.999805	.06	.476693	69.98	.523307	17
44	.480693	69.24	.999801	.06	.480892	69.31	.519108	16
45	.484848	68.59	.999797	.06	.485050	68.65	.514950	15
46	.488963	67.94	.999794	.07	.489170	68.01	.510830	14
47	.493040	67.31	.999790	.07	.493250	67.38	.506750	13
48	.497078	66.69	.999786	.07	.497293	66.76	.502707	12
49	.501080	66.08	.999782	.07	.501298	66.15	.498702	11
50	8.505045	65.48	9.999778	.07	8.505267	65.55	1.494733	10
51	.508974	64.89	.999774	.07	.509200	64.96	.490800	9
52	.512867	64.32	.999769	.07	.513098	64.39	.486902	8
53	.516726	63.75	.999765	.07	.516961	63.82	.483039	7
54	.520551	63.19	.999761	.07	.520790	63.26	.479210	6
55	.524343	62.65	.999757	.07	.524586	62.72	.475414	5
56	.528102	62.11	.999753	.07	.528349	62.18	.471651	4
57	.531828	61.58	.999748	.07	.532080	61.65	.467920	3
58	.535523	61.06	.999744	.07	.535779	61.13	.464221	2
59	.539186	60.55	.999740	.07	.539447	60.62	.460553	1
60	.542819		.999735		.543084		.456916	0
M.	Cosine.	D. 1″.	Sine.	D. 1″.	Cotang.	D. 1″.	Tang.	M.

M.	Sine.	D. 1″.	Cosine.	D. 1″.	Tang.	D. 1″.	Cotang.	M.
0	8.542819	60.04	9.999735	.07	8.543084	60.12	1.456916	60
1	.546422	59.55	.999731	.07	.546691	59.62	.453309	59
2	.549995	59.06	.999726	.08	.550268	59.14	.449732	58
3	.553539	58.58	.999722	.08	.553817	58.66	.446183	57
4	.557054	58.11	.999717	.08	.557336	58.19	.442664	56
5	.560540	57.65	.999713	.08	.560828	57.73	.439172	55
6	.563999	57.19	.999708	.08	.564291	57.27	.435709	54
7	.567431	56.74	.999704	.08	.567727	56.82	.432273	53
8	.570836	56.30	.999699	.08	.571137	56.38	.428863	52
9	.574214	55.87	.999694	.08	.574520	55.95	.425480	51
10	8.577566	55.44	9.999689	.08	8.577877	55.52	1.422123	50
11	.580892	55.02	.999685	.08	.581208	55.10	.418792	49
12	.584193	54.60	.999680	.08	.584514	54.68	.415486	48
13	.587469	54.19	.999675	.08	.587795	54.27	.412205	47
14	.590721	53.79	.999670	.08	.591051	53.87	.408949	46
15	.593948	53.39	.999665	.08	.594283	53.47	.405717	45
16	.597152	53.00	.999660	.08	.597492	53.08	.402508	44
17	.600332	52.61	.999655	.08	.600677	52.70	.399323	43
18	.603489	52.23	.999650	.08	.603839	52.32	.396161	42
19	.606623	51.86	.999645	.09	.606978	51.94	.393022	41
20	8.609734	51.49	9.999640	.09	8.610094	51.58	1.389906	40
21	.612823	51.12	.999635	.09	.613189	51.21	.386811	39
22	.615891	50.77	.999629	.09	.616262	50.85	.383738	38
23	.618937	50.41	.999624	.09	.619313	50.50	380687	37
24	.621962	50.06	.999619	.09	.622343	50.15	377657	36
25	.624965	49.72	.999614	.09	.625352	49.81	374648	35
26	.627948	49.38	.999608	.09	.628340	49.47	.371660	34
27	.630911	49.04	.999603	.09	.631308	49.13	.368692	33
28	.633854	48.71	.999597	.09	.634256	48.80	.365744	32
29	.636776	48.39	.999592	.09	.637184	48.48	.362816	31
30	8.639680	48.06	9.999586	.09	8.640093	48.16	1.359907	30
31	.642563	47.75	.999581	.09	.642982	47.84	.357018	29
32	.645428	47.43	.999575	.09	.645853	47.53	.354147	28
33	.648274	47.12	.999570	.09	.648704	47.22	.351296	27
34	.651102	46.82	.999564	.09	.651537	46.91	.348463	26
35	.653911	46.52	.999558	.10	.654352	46.61	.345648	25
36	.656702	46.22	.999553	.10	.657149	46.31	.342851	24
37	.659475	45.93	.999547	.10	.659928	46.02	.340072	23
38	.662230	45.63	.999541	.10	.662689	45.73	.337311	22
39	.664968	45.35	.999535	.10	.665433	45.45	.334567	21
40	8.667689	45.07	9.999529	.10	8.668160	45.16	1.331840	20
41	.670393	44.79	.999524	.10	.670870	44.88	.329130	19
42	.673080	44.51	.999518	.10	.673563	44.61	.326437	18
43	.675751	44.24	.999512	.10	.676239	44.34	.323761	17
44	.678405	43.97	.999506	.10	.678900	44.07	.321100	16
45	.681043	43.70	.999500	.10	.681544	43.80	.318456	15
46	.683665	43.44	.999493	.10	.684172	43.54	.315828	14
47	.686272	43.18	.999487	.10	.686784	43.28	.313216	13
48	.688863	42.92	.999481	.10	.689381	43.03	.310619	12
49	.691438	42.67	.999475	.10	.691963	42.77	.308037	11
50	8.693998	42.42	9.999469	.10	8.694529	42.52	1.305471	10
51	.696543	42.17	.999463	.11	.697081	42.28	.302919	9
52	.699073	41.93	.999456	.11	.699617	42.03	.300383	8
53	.701589	41.68	.999450	.11	.702139	41.79	.297861	7
54	.704090	41.44	.999443	.11	.704646	41.55	.295354	6
55	.706577	41.21	.999437	.11	.707140	41.32	.292860	5
56	.709049	40.97	.999431	.11	.709618	41.08	.290382	4
57	.711507	40.74	.999424	.11	.712083	40.85	.287917	3
58	.713952	40.51	.999418	.11	.714534	40.62	.285466	2
59	.716383	40.29	.999411	.11	.716972	40.40	.283028	1
60	.718800		.999404		.719396		.280604	0
M.	Cosine.	D. 1″.	Sine.	D. 1″.	Cotang.	D. 1″.	Tang.	M.

3° | 176°

M.	Sine.	D. 1".	Cosine.	D. 1".	Tang.	D. 1".	Cotang.	M.
0	8.718800	40.06	9.999404	.11	8.719396	40.17	1.280604	60
1	.721204	39.84	.999398	.11	.721806	39.95	.278194	59
2	.723595	39.62	.999391	.11	.724204	39.74	.275796	58
3	.725972	39.41	.999384	.11	.726588	39.52	.273412	57
4	.728337	39.19	.999378	.11	.728959	39.31	.271041	56
5	.730688	38.98	.999371	.11	.731317	39.10	.268683	55
6	.733027	38.77	.999364	.11	.733663	38.89	.266337	54
7	.735354	38.57	.999357	.11	.735996	38.68	.264004	53
8	.737667	38.36	.999350	.12	.738317	38.48	.261683	52
9	.739969	38.16	.999343	.12	.740626	38.27	.259374	51
10	8.742259	37.96	9.999336	.12	8.742922	38.07	1.257078	50
11	.744536	37.76	.999329	.12	.745207	37.88	.254793	49
12	.746802	37.56	.999322	.12	.747479	37.68	.252521	48
13	.749055	37.37	.999315	.12	.749740	37.49	.250260	47
14	.751297	37.17	.999308	.12	.751989	37.29	.248011	46
15	.753528	36.98	.999301	.12	.754227	37.10	.245773	45
16	.755747	36.80	.999294	.12	.756453	36.92	.243547	44
17	.757955	36.61	.999287	.12	.758668	36.73	.241332	43
18	.760151	36.42	.999279	.12	.760872	36.55	.239128	42
19	.762337	36.24	.999272	.12	.763065	36.36	.236935	41
20	8.764511	36.06	9.999265	.12	8.765246	36.18	1.234754	40
21	.766675	35.88	.999257	.12	.767417	36.00	.232583	39
22	.768828	35.70	.999250	.12	.769578	35.83	.230422	38
23	.770970	35.53	.999242	.12	.771727	35.65	.228273	37
24	.773101	35.35	.999235	.13	.773866	35.48	.226134	36
25	.775223	35.18	.999227	.13	.775995	35.31	.224005	35
26	.777333	35.01	.999220	.13	.778114	35.14	.221886	34
27	.779434	34.81	.999212	.13	.780222	34.97	.219778	33
28	.781524	34.67	.999205	.13	.782320	34.80	.2176 0	32
29	.783605	34.51	.999197	.13	.784408	34.64	.215592	31
30	8.785675	34.34	9.999189	.13	8.786486	34.47	1.213514	30
31	.787736	34.18	.999181	.13	.788554	34.31	.211446	29
32	.789787	34.02	.999174	.13	.790613	34.15	.209387	28
33	.791828	33.86	.999166	.13	.792662	33.99	.207338	27
34	.793859	33.70	.999158	.13	.794701	33.83	.205299	26
35	.795881	33.54	.999150	.13	.796731	33.68	.203269	25
36	.797894	33.39	.999142	.13	.798752	33.52	.201248	24
37	.799897	33.23	.999134	.13	.800763	33.37	.199237	23
38	.801892	33.08	.999126	.13	.802765	33.22	.197235	22
39	.803876	32.93	.999118	.13	.804758	33.07	.195242	21
40	8.805852	32.78	9.999110	.14	8.806742	32.92	1.193258	20
41	.807819	32.63	.999102	.14	.808717	32.77	.191283	19
42	.809777	32.49	.999094	.14	.810683	32.62	.189317	18
43	.811726	32.34	.999086	.14	.812641	32.48	.187359	17
44	.813667	32.20	.999077	.14	.814589	32.33	.185411	16
45	.815599	32.05	.999069	.14	.816529	32.19	.183471	15
46	.817522	31.91	.999061	.14	.818461	32.05	.181539	14
47	.819436	31.77	.999053	.14	.820384	31.91	.179616	13
48	.821313	31.63	.999044	.14	.822298	31.77	.177702	12
49	.823240	31.49	.999036	.14	.824205	31.63	.175795	11
50	8.825130	31.36	9.999027	.14	8.826103	31.50	1.173897	10
51	.827011	31.22	.999019	.14	.827992	31.36	.172008	9
52	.828884	31.08	.999010	.14	.829874	31.23	.170126	8
53	.830749	30.95	.999002	.14	.831748	31.09	.168252	7
54	.832607	30.82	.998993	.14	.833613	30.96	.166387	6
55	.834456	30.69	.998984	.14	.835471	30.83	.164529	5
56	.836297	30.56	.998976	.15	.837321	30.70	.162679	4
57	.838130	30.43	.998967	.15	.839163	30.57	.160837	3
58	.839956	30.30	.998958	.15	.840998	30.45	.159002	2
59	.841774	30.17	.998950	.15	.842825	30.32	.157175	1
60	.843585		.998941		.844644		.155356	0
M.	Cosine.	D. 1".	Sine.	D. 1".	Cotang.	D. 1".	Tang.	M.

4° | 175°

M.	Sine.	D. 1″.	Cosine.	D. 1″.	Tang.	D. 1″.	Cotang.	M.
0	8.843585	30.05	9.998941	.15	8.844644	30.20	1.155356	60
1	.845387	29.92	.998932	.15	.846455	30.07	.153545	59
2	.847183	29.80	.998923	.15	.848260	29.95	.151740	58
3	.848971	29.68	.998914	.15	.850057	29.83	.149943	57
4	.850751	29.55	.998905	.15	.851846	29.70	.148154	56
5	.852525	29.43	.998896	.15	.853628	29.58	.146372	55
6	.854291	29.31	.998887	.15	.855403	29.46	.144597	54
7	.856049	29.19	.998878	.15	.857171	29.35	.142829	53
8	.857801	29.08	.998869	.15	.858932	29.23	.141068	52
9	.859546	28.96	.998860	.15	.860686	29.11	.139314	51
10	8.861283	28.84	9.998851	.15	8.862433	29.00	1.137567	50
11	.863014	28.73	.998841	.15	.864173	28.88	.135827	49
12	.864738	28.61	.998832	.15	.865906	28.77	.134094	48
13	.866455	28.50	.998823	.16	.867632	28.66	.132368	47
14	.868165	28.39	.998813	.16	.869351	28.55	.130649	46
15	.869868	28.28	.998804	.16	.871064	28.43	.128936	45
16	.871565	28.17	.998795	.16	.872770	28.32	.127230	44
17	.873255	28.06	.998785	.16	.874469	28.22	.125531	43
18	.874938	27.95	.998776	.16	.876162	28.11	.123838	42
19	.876615	27.84	.998766	.16	.877849	28.00	.122151	41
20	8.878285	27.73	9.998757	.16	8.879529	27.89	1.120471	40
21	.879949	27.63	.998747	.16	.881202	27.79	.118798	39
22	.881607	27.52	.998738	.16	.882869	27.68	.117131	38
23	.883258	27.42	.998728	.16	.884530	27.58	.115470	37
24	.884903	27.31	.998718	.16	.886185	27.47	.113815	36
25	.886542	27.21	.998708	.16	.887833	27.37	.112167	35
26	.888174	27.11	.998699	.16	.889476	27.27	.110524	34
27	.889801	27.00	.998689	.16	.891112	27.17	.108888	33
28	.891421	26.90	.998679	.16	.892742	27.07	.107258	32
29	.893035	26.80	.998669	.17	.894366	26.97	.105634	31
30	8.894643	26.70	9.998659	.17	8.895984	26.87	1.104016	30
31	.896246	26.60	.998649	.17	.897596	26.77	.102404	29
32	.897842	26.51	.998639	.17	.899203	26.67	.100797	28
33	.899432	26.41	.998629	.17	.900803	26.58	.099197	27
34	.901017	26.31	.998619	.17	.902398	26.48	.097602	26
35	.902596	26.22	.998609	.17	.903987	26.39	.096013	25
36	.904169	26.12	.998599	.17	.905570	26.29	.094430	24
37	.905736	26.03	.998589	.17	.907147	26.20	.092853	23
38	.907297	25.93	.998578	.17	.908719	26.10	.091281	22
39	.908853	25.84	.998568	.17	.910285	26.01	.089715	21
40	8.910404	25.75	9.998558	.17	8.911846	25.92	1.088154	20
41	.911949	25.66	.998548	.17	.913401	25.83	.086599	19
42	.913488	25.56	.998537	.17	.914951	25.74	.085049	18
43	.915022	25.47	.998527	.17	.916495	25.65	.083505	17
44	916550	25.38	.998516	.17	.918034	25.56	.081966	16
45	918073	25.29	.998506	.18	.919568	25.47	.080432	15
46	.919591	25.21	.998495	.18	.921096	25.38	.078904	14
47	.921103	25.12	.998485	.18	.922619	25.29	.077381	13
48	.922610	25.03	.998474	.18	.924136	25.21	.075864	12
49	.924112	24.94	.998464	.18	.925649	25.12	.074351	11
50	8.925609	24.86	9.998453	.18	8.927156	25.04	1.072844	10
51	.927100	24.77	.998442	18	.928658	24.95	.071342	9
52	.928587	24.69	.998431	.18	.930155	24.87	.069845	8
53	.930068	24.60	.998421	18	.931647	24.78	.068353	7
54	.931544	24.52	.998410	.18	.933134	24.70	.066866	6
55	.933015	24.43	.998399	.18	.934616	24.62	.065384	5
56	.934481	24.35	.998388	.18	.936093	24.53	.063907	4
57	.935942	24.27	.998377	.18	.937565	24.45	.062435	3
58	.937398	24.19	.998366	.18	.939032	24.37	.060968	2
59	.938850	24.11	.998355	.18	.940494	24.29	.059506	1
60	.940296		.998344		.941952		.058048	0
M.	Cosine.	D. 1″.	Sine.	D. 1″.	Cotang.	D. 1″.	Tang.	M.

94° | 85°

M.	Sine.	D. 1″.	Cosine.	D. 1″.	Tang.	D. 1″.	Cotang.	M.
0	8.940296	24.03	9.998344	.18	8.941952	24.21	1.058048	60
1	.941738	23.95	.998333	.19	.943404	24.13	.056596	59
2	.943174	23.87	.998322	.19	.944852	24.05	.055148	58
3	.944606	23.79	.998311	.19	.946295	23.97	.053705	57
4	.946034	23.71	.998300	.19	.947734	23.90	.052266	56
5	.947456	23.63	.998289	.19	.949168	23.82	.050832	55
6	.948874	23.55	.998277	.19	.950597	23.74	.049403	54
7	.950287	23.48	.998266	.19	.952021	23.67	.047979	53
8	.951696	23.40	.998255	.19	.953441	23.59	.046559	52
9	.953100	23.32	.998243	.19	.954856	23.51	.045144	51
10	8.954499	23.25	9.998232	.19	8.956267	23.44	1.043733	50
11	.955894	23.17	.998220	.19	.957674	23.36	.042326	49
12	.957284	23.10	.998209	.19	.959075	23.29	.040925	48
13	.958670	23.02	.998197	.19	.960473	23.22	.039527	47
14	.960052	22.95	.998186	.19	.961866	23.14	.038134	46
15	.961429	22.88	.998174	.19	.963255	23.07	.036745	45
16	.962801	22.81	.998163	.19	.964639	23.00	.035361	44
17	.964170	22.73	.998151	.20	.966019	22.93	.033981	43
18	.965534	22.66	.998139	.20	.967394	22.86	.032606	42
19	.966893	22.59	.998128	.20	.968766	22.79	.031234	41
20	8.968249	22.52	9.998116	.20	8.970133	22.72	1.029867	40
21	.969600	22.45	.998104	.20	.971496	22.65	.028504	39
22	.970947	22.38	.998092	.20	.972855	22.58	.027145	38
23	.972289	22.31	.998080	.20	.974209	22.51	.025791	37
24	.973628	22.24	.998068	.20	.975560	22.44	.024440	36
25	.974962	22.17	.998056	.20	.976906	22.37	.023094	35
26	.976293	22.10	.998044	.20	.978248	22.30	.021752	34
27	.977619	22.03	.998032	.20	.979586	22.24	.020414	33
28	.978941	21.97	.998020	.20	.980921	22.17	.019079	32
29	.980259	21.90	.998008	.20	.982251	22.10	.017749	31
30	8.981573	21.83	9.997996	.20	8.983577	22.04	1.016423	30
31	.982883	21.77	.997984	.20	.984899	21.97	.015101	29
32	.984189	21.70	.997972	.20	.986217	21.91	.013783	28
33	.985491	21.64	.997959	.20	.987532	21.84	.012468	27
34	.986789	21.57	.997947	.21	.988842	21.78	.011158	26
35	.988083	21.51	.997935	.21	.990149	21.71	.009851	25
36	.989374	21.44	.997922	.21	.991451	21.65	.008549	24
37	.990660	21.38	.997910	.21	.992750	21.59	.007250	23
38	.991943	21.31	.997897	.21	.994045	21.52	.005955	22
39	.993222	21.25	.997885	.21	.995337	21.46	.004663	21
40	8.994497	21.19	9.997872	21	8.996624	21.40	1.003376	20
41	.995768	21.12	.997860	.21	.997908	21.34	.002092	19
42	.997036	21.06	.997847	.21	.999188	21.27	.000812	18
43	.998299	21.00	.997835	.21	9.000465	21.21	0.999535	17
44	.999560	20.94	.997822	.21	.001738	21.15	.998262	16
45	9.000816	20.88	.997809	.21	.003007	21.09	.996993	15
46	.002069	20.82	.997797	.21	.004272	21.03	.995728	14
47	.003318	20.76	.997784	.21	.005534	20.97	.994466	13
48	.004563	20.70	.997771	.21	.006792	20.91	.993208	12
49	.005805	20.64	.997758	.21	.008047	20.85	.991953	11
50	9.007044	20.58	9.997745	.22	9.009298	20.80	0.990702	10
51	.008278	20.52	.997732	.22	.010546	20.74	.989454	9
52	.009510	20.46	.997719	.22	.011790	20.68	.988210	8
53	.010737	20.40	.997706	.22	.013031	20.62	.986969	7
54	.011962	20.35	.997693	.22	.014268	20.56	.985732	6
55	.013182	20.29	.997680	.22	.015502	20.51	.984498	5
56	.014400	20.23	.997667	.22	.016732	20.45	.983268	4
57	.015613	20.17	.997654	.22	.017959	20.39	.982041	3
58	.016824	20.12	.997641	.22	.019183	20.34	.980817	2
59	.018031	20.06	.997628	.22	.020403	20.28	.979597	1
60	.019235		.997614		.021620		.978380	0
M.	Cosine.	D. 1″.	Sine.	D. 1″.	Cotang.	D. 1″.	Tang.	M.

6° 173°

M.	Sine.	D. 1″.	Cosine.	D. 1″.	Tang.	D. 1″.	Cotang.	M.
0	9.019235	20.00	9.997614	.22	9.021620	20.23	0.978380	60
1	.020435	19.95	.997601	.22	.022834	20.17	.977166	59
2	.021632	19.89	.997588	.22	.024044	20.12	.975956	58
3	.022825	19.84	.997574	.22	.025251	20.06	.974749	57
4	.024016	19.78	.997561	.22	.026455	20.01	.973545	56
5	.025203	19.73	.997547	.22	.027655	19.95	.972345	55
6	.026386	19.67	.997534	.23	.028852	19.90	.971148	54
7	.027567	19.62	.997520	.23	.030046	19.85	.969954	53
8	.028744	19.57	.997507	.23	.031237	19.79	.968763	52
9	.029918	19.51	.997493	.23	.032425	19.74	.967575	51
10	9.031089	19.46	9.997480	.23	9.033609	19.69	0.966391	50
11	.032257	19.41	.997466	.23	.034791	19.64	.965209	49
12	.033421	19.36	.997452	.23	.035969	19.58	.964031	48
13	.034582	19.30	.997439	.23	.037144	19.53	.962856	47
14	.035741	19.25	.997425	.23	.038316	19.48	.961684	46
15	.036896	19.20	.997411	.23	.039485	19.43	.960515	45
16	.038048	19.15	.997397	.23	.040651	19.38	.959349	44
17	.039197	19.10	.997383	.23	.041813	19.33	.958187	43
18	.040342	19.05	.997369	.23	.042973	19.28	.957027	42
19	.041485	19.00	.997355	.23	.044130	19.23	.955870	41
20	9.042625	18.95	9.997341	.23	9.045284	19.18	0.954716	40
21	.043762	18.90	.997327	.23	.046434	19.13	.953566	39
22	.044895	18.85	.997313	.24	.047582	19.08	.952418	38
23	.046026	18.80	.997299	.24	.048727	19.03	.951273	37
24	.047154	18.75	.997285	.24	.049869	18.98	.950131	36
25	.048279	18.70	.997271	.24	.051008	18.93	.948992	35
26	.049400	18.65	.997257	.24	.052144	18.89	.947856	34
27	.050519	18.60	.997242	.24	.053277	18.84	.946723	33
28	.051635	18.55	.997228	.24	.054407	18.79	.945593	32
29	.052749	18.50	.997214	.24	.055535	18.74	.944465	31
30	9.053859	18.46	9.997199	.24	9.056659	18.70	0.943341	30
31	.054966	18.41	.997185	.24	.057781	18.65	.942219	29
32	.056071	18.36	.997170	.24	.058900	18.60	.941100	28
33	.057172	18.31	.997156	.24	.060016	18.56	.939984	27
34	.058271	18.27	.997141	.24	.061130	18.51	.938870	26
35	.059367	18.22	.997127	.24	.062240	18.46	.937760	25
36	.060460	18.17	.997112	.24	.063348	18.42	.936652	24
37	.061551	18.13	.997098	.24	.064453	18.37	.935547	23
38	.062639	18.08	.997083	.25	.065556	18.33	.934444	22
39	.063724	18.04	.997068	.25	.066655	18.28	.933345	21
40	9.064806	17.99	9.997053	.25	9.067752	18.24	0.932248	20
41	.065885	17.95	.997039	.25	.068846	18.19	.931154	19
42	.066962	17.90	.997024	.25	.069938	18.15	.930062	18
43	.068036	17.86	.997009	.25	.071027	18.10	.928973	17
44	.069107	17.81	.996994	.25	.072113	18.06	.927887	16
45	.070176	17.77	.996979	.25	.073197	18.02	.926803	15
46	.071242	17.72	.996964	.25	.074278	17.97	.925722	14
47	.072306	17.68	.996949	.25	.075356	17.93	.924644	13
48	.073366	17.64	.996934	.25	.076432	17.89	.923568	12
49	.074424	17.59	.996919	.25	.077505	17.84	.922495	11
50	9.075480	17.55	9.996904	.25	9.078576	17.80	0.921424	10
51	.076533	17.51	.996889	.25	.079644	17.76	.920356	9
52	.077583	17.46	.996874	.25	.080710	17.72	.919290	8
53	.078631	17.42	.996858	.25	.081773	17.67	.918227	7
54	.079676	17.38	.996843	.26	.082833	17.63	.917167	6
55	.080719	17.34	.996828	.26	.083891	17.59	.916109	5
56	.081759	17.29	.996812	.26	.084947	17.55	.915053	4
57	.082797	17.25	.996797	.26	.086000	17.51	.914000	3
58	.083832	17.21	.996782	.26	.087050	17.47	.912950	2
59	.084864	17.17	.996766	.26	.088098	17.43	.911902	1
60	.085894		.996751		.089144		.910856	0
M.	Cosine.	D. 1″.	Sine.	D. 1″.	Cotang.	D. 1″.	Tang.	M.

M.	Sine.	D. 1″.	Cosine.	D. 1″.	Tang.	D. 1″.	Cotang.	M.
0	9.085894	17.13	9.996751	.26	9.089144	17.39	0.910856	60
1	.086922	17.09	.996735	.26	.090187	17.35	.909813	59
2	.087947	17.05	.996720	.26	.091228	17.31	.908772	58
3	.088970	17.00	.996704	.26	.092266	17.27	.907734	57
4	.089990	16.96	.996688	.26	.093302	17.23	.906698	56
5	.091008	16.92	.996673	.26	.094336	17.19	.905664	55
6	.092024	16.88	.996657	.26	.095367	17.15	.904633	54
7	.093037	16.84	.996641	.26	.096395	17.11	.903605	53
8	.094047	16.80	.996625	.26	.097422	17.07	.902578	52
9	.095056	16.76	.996610	.26	.098446	17.03	.901554	51
10	9.096062	16.73	9.996594	.27	9.099468	16.99	0.900532	50
11	.097065	16.69	.996578	.27	.100487	16.95	.899513	49
12	.098066	16.65	.996562	.27	.101504	16.91	.898496	48
13	.099065	16.61	.996546	.27	.102519	16.88	.897481	47
14	.100062	16.57	.996530	.27	.103532	16.84	.896468	46
15	.101056	16.53	.996514	.27	.104542	16.80	.895458	45
16	.102048	16.49	.996498	.27	.105550	16.76	.894450	44
17	.103037	16.46	.996482	.27	.106556	16.72	.893444	43
18	.104025	16.42	.996465	.27	.107559	16.69	.892441	42
19	.105010	16.38	.996449	.27	.108560	16.65	.891440	41
20	9.105992	16.34	9.996433	.27	9.109559	16.61	0.890441	40
21	.106973	16.30	.996417	.27	.110556	16.58	.889444	39
22	.107951	16.27	.996400	.27	.111551	16.54	.888449	38
23	.108927	16.23	.996384	.27	.112543	16.50	.887457	37
24	.109901	16.19	.996368	.27	.113533	16.47	.886467	36
25	.110873	16.16	.996351	.27	.114521	16.43	.885479	35
26	.111842	16.12	996335	.28	.115507	16.39	.884493	34
27	.112809	16.08	.996318	.28	.116491	16.36	.883509	33
28	.113774	16.05	.996302	.28	.117472	16.32	.882528	32
29	.114737	16.01	.996285	.28	.118452	16.29	.881548	31
30	9.115698	15.98	9.996269	.28	9.119429	16.25	0.880571	30
31	.116656	15.94	.996252	.28	.120404	16.22	.879596	29
32	.117613	15.90	.996235	.28	.121377	16.18	.878623	28
33	.118567	15.87	.996219	.28	.122348	16.15	.877652	27
34	.119519	15.83	.996202	.28	.123317	16.11	.876683	26
35	.120469	15.80	.996185	.28	.124284	16.08	.875716	25
36	.121417	15.76	.996168	.28	.125249	16.04	.874751	24
37	.122362	15.73	.996151	.28	.126211	16.01	.873789	23
38	.123306	15.69	.996134	.28	.127172	15.98	.872828	22
39	.124248	15.66	.996117	.28	.128130	15.94	.871870	21
40	9.125187	15.62	9.996100	.28	9.129087	15.91	0.870913	20
41	.126125	15.59	.996083	.28	.130041	15.87	.869959	19
42	.127060	15.56	.996066	.28	.130994	15.84	.869006	18
43	.127993	15.52	.996049	.29	.131944	15.81	.868056	17
44	.128925	15.49	.996032	.29	.132893	15.77	.867107	16
45	.129854	15.45	.996015	.29	.133839	15.74	.866161	15
46	.130781	15.42	.995998	.29	.134784	15.71	.865216	14
47	.131706	15.39	.995980	.29	.135726	15.68	.864274	13
48	.132630	15.35	.995963	.29	.136667	15.64	.863333	12
49	.133551	15.32	.995946	.29	.137605	15.61	.862395	11
50	9.134470	15.29	9.995928	.29	9.138542	15.58	0.861458	10
51	.135387	15.26	.995911	.29	.139476	15.55	.860524	9
52	.136303	15.22	.995894	.29	.140409	15.51	.859591	8
53	.137216	15.19	.995876	.29	.141340	15.48	.858660	7
54	.138128	15.16	.995859	.29	.142269	15.45	.857731	6
55	.139037	15.13	.995841	.29	.143196	15.42	.856804	5
56	.139944	15.09	.995823	.29	.144121	15.39	.855879	4
57	.140850	15.06	.995806	.29	.145044	15.36	.854956	3
58	.141754	15.03	.995788	.29	.145966	15.32	.854034	2
59	142655	15.00	.995771	.30	.146885	15.29	.853115	1
60	.143555		.995753		.147803		.852197	0
M.	Cosine.	D. 1″.	Sine.	D. 1″.	Cotang.	D. 1″.	Tang.	M.

8° 171°

M.	Sine.	D. 1″.	Cosine.	D. 1′	Tang.	D. 1″.	Cotang.	M.
0	9.143555	14.97	9.995753	.30	9.147803	15.26	0.852197	60
1	.144453	14.93	.995735	.30	.148718	15.23	.851282	59
2	.145349	14.90	.995717	.30	.149632	15.20	.850368	58
3	.146243	14.87	.995699	.30	.150544	15.17	.849456	57
4	.147136	14.84	.995681	.30	.151454	15.14	.848546	56
5	.148026	14.81	.995664	.30	.152363	15.11	.847637	55
6	.148915	14.78	.995646	.30	.153269	15.08	.846731	54
7	.149802	14.75	.995628	.30	.154174	15.05	.845826	53
8	.150686	14.72	.995610	.30	.155077	15.02	.844923	52
9	.151569	14.69	.995591	.30	.155978	14.99	.844022	51
10	9.152451	14.66	9.995573	.30	9.156877	14.96	0.843123	50
11	.153330	14.63	.995555	.30	.157775	14.93	.842225	49
12	.154208	14.60	.995537	.30	.158671	14.90	.841329	48
13	.155083	14.57	.995519	.30	.159565	14.87	.840435	47
14	.155957	14.54	.995501	.30	.160457	14.84	.839543	46
15	.156830	14.51	.995482	.31	.161347	14.81	.838653	45
16	.157700	14.48	.995464	.31	.162236	14.78	.837764	44
17	.158569	14.45	.995446	.31	.163123	14.75	.836877	43
18	.159435	14.42	.995427	.31	.164008	14.73	.835992	42
19	.160301	14.39	.995409	.31	.164892	14.70	.835108	41
20	9.161164	14.36	9.995390	.31	9.165774	14.67	0.834226	40
21	.162025	14.33	.995372	.31	.166654	14.64	.833346	39
22	.162885	14.30	.995353	.31	.167532	14.61	.832468	38
23	.163743	14.27	.995334	.31	.168409	14.58	.831591	37
24	.164600	14.24	.995316	.31	.169284	14.56	.830716	36
25	.165454	14.22	.995297	.31	.170157	14.53	.829843	35
26	.166307	14.19	.995278	.31	.171029	14.50	.828971	34
27	.167159	14.16	.995260	.31	.171899	14.47	.828101	33
28	.168008	14.13	.995241	.31	.172767	14.44	.827233	32
29	.168856	14.10	.995222	.31	.173634	14.42	.826366	31
30	9.169702	14.07	9.995203	.31	9.174499	14.39	0.825501	30
31	.170547	14.05	.995184	.32	.175362	14.36	.824638	29
32	.171389	14.02	.995165	.32	.176224	14.33	.823776	28
33	.172230	13.99	.995146	.32	.177084	14.31	.822916	27
34	.173070	13.96	.995127	.32	.177942	14.28	.822058	26
35	.173908	13.94	.995108	.32	.178799	14.25	.821201	25
36	.174744	13.91	.995089	.32	.179655	14.23	.820345	24
37	.175578	13.88	.995070	.32	.180508	14.20	.819492	23
38	.176411	13.85	.995051	.32	.181360	14.17	.818640	22
39	.177242	13.83	.995032	.32	.182211	14.15	.817789	21
40	9.178072	13.80	9.995013	.32	9.183059	14.12	0.816941	20
41	.178900	13.77	.994993	.32	.183907	14.09	.816093	19
42	.179726	13.75	.994974	.32	.184752	14.07	.815248	18
43	.180551	13.72	.994955	.32	.185597	14.04	.814403	17
44	.181374	13.69	.994935	.32	.186439	14.02	.813561	16
45	.182196	13.67	.994916	.32	.187280	13.99	.812720	15
46	.183016	13.64	.994896	.33	.188120	13.97	.811880	14
47	.183834	13.61	.994877	.33	.188958	13.94	.811042	13
48	.184651	13.59	.994857	.33	.189794	13.91	.810206	12
49	.185466	13.56	.994838	.33	.190629	13.89	.809371	11
50	9.186280	13.54	9.994818	.33	9.191462	13.86	0.808538	10
51	.187092	13.51	.994798	.33	.192294	13.84	.807706	9
52	.187903	13.48	.994779	.33	.193124	13.81	.806876	8
53	.188712	13.46	.994759	.33	.193953	13.79	.806047	7
54	.189519	13.43	.994739	.33	.194780	13.76	.805220	6
55	.190325	13.41	.994720	.33	.195606	13.74	.804394	5
56	.191130	13.38	994700	.33	.196430	13.71	.803570	4
57	.191933	13.36	.994680	.33	.197253	13.69	.802747	3
58	.192734	13.33	.994660	.33	.198074	13.66	.801926	2
59	.193534	13.31	.994640	.33	.198894	13.64	.801106	1
60	.194332		.994620		199713		.800287	0
M.	Cosine.	D. 1″.	Sine.	D. 1″.	Cotang.	D. 1″.	Tang.	M.

98° 81°

9° | 170°

M	Sine.	D. 1″.	Cosine.	D. 1″.	Tang.	D. 1″.	Cotang.	M.
0	9.194332	13.28	9.994620	.33	9.199713	13.62	0.800287	60
1	.195129	13.26	.994600	.33	.200529	13.59	.799471	59
2	.195925	13.23	.994580	.34	.201345	13.57	.798655	58
3	.196719	13.21	.994560	.34	.202159	13.54	.797841	57
4	.197511	13.18	.994540	.34	.202971	13.52	.797029	56
5	.198302	13.16	.994519	.34	.203782	13.49	.796218	55
6	.199091	13.13	.994499	.34	.204592	13.47	.795408	54
7	.199879	13.11	.994479	.34	.205400	13.45	.794600	53
8	.200666	13.08	.994459	.34	.206207	13.42	.793793	52
9	.201451	13.06	.994438	.34	.207013	13.40	.792987	51
10	9.202234	13.04	9.994418	.34	9.207817	13.38	0.792183	50
11	.203017	13.01	.994398	.34	.208619	13.35	.791381	49
12	.203797	12.99	.994377	.34	.209420	13.33	.790580	48
13	.204577	12.96	.994357	.34	.210220	13.31	.789780	47
14	.205354	12.94	.994336	.34	.211018	13.28	.788982	46
15	.206131	12.92	.994316	.34	.211815	13.26	.788185	45
16	.206906	12.89	.994295	.34	.212611	13.24	.787389	44
17	.207679	12.87	.994274	.34	.213405	13.21	.786595	43
18	.208452	12.85	.994254	.35	.214193	13.19	.785802	42
19	.209222	12.82	.994233	.35	.214989	13.17	.785011	41
20	9.209992	12.80	9.994212	.35	9.215780	13.15	0.784220	40
21	.210760	12.78	.994191	.35	.216568	13.12	.783432	39
22	.211526	12.75	.994171	.35	.217356	13.10	.782644	38
23	.212291	12.73	.994150	.35	.218142	13.08	.781858	37
24	.213055	12.71	.994129	.35	.218926	13.06	.781074	36
25	.213818	12.68	.994108	.35	.219710	13.03	.780290	35
26	.214579	12.66	.994087	.35	.220492	13.01	.779508	34
27	.215338	12.64	.994066	.35	.221272	12.99	.778728	33
28	.216097	12.62	.994045	.35	.222052	12.97	.777948	32
29	.216854	12.59	.994024	.35	.222830	12.95	.777170	31
30	9.217609	12.57	9.994003	.35	9.223607	12.92	0.776393	30
31	.218363	12.55	.993982	.35	.224382	12.90	.775618	29
32	.219116	12.53	.993960	.35	.225156	12.88	.774844	28
33	.219868	12.50	.993939	.35	.225929	12.86	.774071	27
34	.220618	12.48	.993918	.36	.226700	12.84	.773300	26
35	.221367	12.46	.993897	.36	.227471	12.82	.772529	25
36	.222115	12.44	.993875	.36	.228239	12.79	.771761	24
37	.222861	12.42	.993854	.36	.229007	12.77	.770993	23
38	.223606	12.39	.993832	.36	.229773	12.75	.770227	22
39	.224349	12.37	.993811	.36	.230539	12.73	.769461	21
40	9.225092	12.35	9.993789	.36	9.231302	12.71	0.768698	20
41	.225833	12.33	.993768	.36	.232065	12.69	.767935	19
42	.226573	12.31	.993746	.36	.232826	12.67	.767174	18
43	.227311	12.29	.993725	.36	.233586	12.65	.766414	17
44	.228048	12.26	.993703	.36	.234345	12.63	.765655	16
45	.228784	12.24	.993681	.36	.235103	12.60	.764897	15
46	.229518	12.22	.993660	.36	.235859	12.58	.764141	14
47	.230252	12.20	.993638	.36	.236614	12.56	.763386	13
48	.230984	12.18	.993616	.36	.237368	12.54	.762632	12
49	.231715	12.16	.993594	.36	.238120	12.52	.761880	11
50	9.232444	12.14	9.993572	.37	9.238872	12.50	0.761128	10
51	.233172	12.12	.993550	.37	.239622	12.48	.760378	9
52	.233899	12.10	.993528	.37	.240371	12.46	.759629	8
53	.234625	12.07	.993506	.37	.241118	12.44	.758882	7
54	.235349	12.05	.993484	.37	.241865	12.42	.758135	6
55	.236073	12.03	.993462	.37	.242610	12.40	.757390	5
56	.236795	12.01	.993440	.37	.243354	12.38	.756646	4
57	.237515	11.99	.993418	.37	.244097	12.36	.755903	3
58	.238235	11.97	.993396	.37	.244839	12.34	.755161	2
59	.238953	11.95	.993374	.37	.245579	12.32	.754421	1
60	.239670		.993351		.216319		.753681	0
M.	Cosine.	D. 1″.	Sine.	D. 1″.	Cotang.	D. 1″.	Tang.	M.

10° 169°

M.	Sine.	D. 1″.	Cosine.	D. 1″.	Tang.	D. 1″.	Cotang.	M
0	9.239670	11.93	9.993351	.37	9.246319	12.30	0.753681	60
1	.240386	11.91	.993329	.37	.247057	12.28	.752943	59
2	.241101	11.89	.993307	.37	.247794	12.26	.752206	58
3	.241814	11.87	.993284	.37	.248530	12.24	.751470	57
4	.242526	11.85	.993262	.37	.249264	12.22	.750736	56
5	.243237	11.83	.993240	.37	.249998	12.20	.750002	55
6	.243947	11.81	.993217	.38	.250730	12.18	.749270	54
7	.244656	11.79	.993195	.38	.251461	12.17	.748539	53
8	.245363	11.77	.993172	.38	.252191	12.15	.747809	52
9	.246069	11.75	.993149	.38	.252920	12.13	.747080	51
10	9.246775	11.73	9.993127	.38	9.253648	12.11	0.746352	50
11	.247478	11.71	.993104	.38	.254374	12.09	.745626	49
12	.248181	11.69	.993081	.38	.255100	12.07	.744900	48
13	.248883	11.67	.993059	.38	.255824	12.05	.744176	47
14	.249583	11.65	.993036	.38	.256547	12.03	.743453	46
15	.250282	11.63	.993013	.38	.257269	12.01	.742731	45
16	.250980	11.61	.992990	.38	.257990	12.00	.742010	44
17	.251677	11.59	.992967	.38	.258710	11.98	.741290	43
18	.252373	11.58	.992944	.38	.259429	11.96	.740571	42
19	.253067	11.56	.992921	.38	.260146	11.94	.739854	41
20	9.253761	11.54	9.992898	.38	9.260863	11.92	0.739137	40
21	.254453	11.52	.992875	.38	.261578	11.90	.738422	39
22	.255144	11.50	.992852	.39	.262292	11.89	.737708	38
23	.255834	11.48	.992829	.39	.263005	11.87	.736995	37
24	.256523	11.46	.992806	.39	.263717	11.85	.736283	36
25	.257211	11.44	.992783	.39	.264428	11.83	.735572	35
26	.257898	11.42	.992759	.39	.265138	11.81	.734862	34
27	.258583	11.41	.992736	.39	.265847	11.79	.734153	33
28	.259268	11.39	.992713	.39	.266555	11.78	.733445	32
29	.259951	11.37	.992690	.39	.267261	11.76	.732739	31
30	9.260633	11.35	9.992666	.39	9.267967	11.74	0.732033	30
31	.261314	11.33	.992643	.39	.268671	11.72	.731329	29
32	.261994	11.31	.992619	.39	.269375	11.70	.730625	28
33	.262673	11.30	.992596	.39	.270077	11.69	.729923	27
34	.263351	11.28	.992572	.39	.270779	11.67	.729221	26
35	.264027	11.26	.992549	.39	.271479	11.65	.728521	25
36	.264703	11.24	.992525	.39	.272178	11.64	.727822	24
37	.265377	11.22	.992501	.39	.272876	11.62	.727124	23
38	.266051	11.20	.992478	.40	.273573	11.60	.726427	22
39	.266723	11.19	.992454	.40	.274269	11.58	.725731	21
40	9.267395	11.17	9.992430	.40	9.274964	11.57	0.725036	20
41	.268065	11.15	.992406	.40	.275658	11.55	.724342	19
42	.268734	11.13	.992382	.40	.276351	11.53	.723649	18
43	.269402	11.12	.992359	.40	.277043	11.51	.722957	17
44	.270069	11.10	.992335	.40	.277734	11.50	.722266	16
45	.270735	11.08	.992311	.40	.278424	11.48	.721576	15
46	.271400	11.06	.992287	.40	.279113	11.46	.720887	14
47	.272064	11.05	.992263	.40	.279801	11.45	.720199	13
48	.272726	11.03	.992239	.40	.280488	11.43	.719512	12
49	.273388	11.01	.992214	.40	.281174	11.41	.718826	11
50	9.274049	10.99	9.992190	.40	9.281858	11.40	0.718142	10
51	.274708	10.98	.992166	.40	.282542	11.38	.717458	9
52	.275367	10.96	.992142	.40	.283225	11.36	.716775	8
53	.276025	10.94	.992118	.41	.283907	11.35	716093	7
54	.276681	10.92	.992093	.41	.284588	11.33	.715412	6
55	.277337	10.91	.992069	.41	.285268	11.31	.714732	5
56	.277991	10.89	.992044	.41	.285947	11.30	.714053	4
57	.278645	10.87	.992020	.41	.286624	11.28	.713376	3
58	.279297	10.86	.991996	.41	.287301	11.26	.712699	2
59	.279948	10.84	.991971	.41	.287977	11.25	.712023	1
60	.280599		.991947		.288652		.711348	0
M.	Cosine.	D. 1″.	Sine.	D. 1″.	Cotang.	D. 1″.	Tang.	M.

100° 79°

M.	Sine.	D. 1".	Cosine.	D. 1".	Tang.	D. 1".	Cotang.	M.
0	9.280599	10.82	9.991947	.41	9.288652	11.23	0.711348	60
1	.281248	10.81	.991922	.41	.289326	11.22	.710674	59
2	.281897	10.79	.991897	.41	.289999	11.20	.710001	58
3	.282544	10.77	.991873	.41	.290671	11.18	.709329	57
4	.283190	10.76	.991848	.41	.291342	11.17	.708658	56
5	.283836	10.74	.991823	.41	.292013	11.15	.707987	55
6	.284480	10.72	.991799	.41	.292682	11.14	.707318	54
7	.285124	10.71	.991774	.41	.293350	11.12	.706650	53
8	.285766	10.69	.991749	.41	.294017	11.11	.705983	52
9	.286408	10.67	.991724	42	.294684	11.09	.705316	51
10	9.287048	10.66	9.991699	.42	9.295349	11.07	0.704651	50
11	.287688	10.64	.991674	.42	.296013	11.06	.703987	49
12	.288326	10.63	.991649	.42	.296677	11.04	.703323	48
13	.288964	10.61	.991624	42	.297339	11.03	.702661	47
14	.289600	10.59	.991599	.42	.298001	11.01	.701999	46
15	.290236	10.58	.991574	.42	.298662	11.00	.701338	45
16	.290870	10.56	.991549	.42	.299322	10.98	.700678	44
17	.291504	10.55	.991524	.42	.299980	10.97	.700020	43
18	.292137	10.53	.991498	.42	.300638	10.95	.699362	42
19	.292768	10.51	.991473	.42	.301295	10.93	.698705	41
20	9.293399	10.50	9.991448	.42	9.301951	10.92	0.698049	40
21	.294029	10.48	.991422	.42	.302607	10.90	.697393	39
22	.294658	10.47	.991397	.42	.303261	10.89	.696739	38
23	.295286	10.45	.991372	.42	.303914	10.87	.696086	37
24	.295913	10.43	.991346	.42	.304567	10.86	.695433	36
25	.296539	10.42	.991321	.43	.305218	10.84	.694782	35
26	.297164	10.40	.991295	.43	.305869	10.83	.694131	34
27	.297788	10.39	.991270	.43	.306519	10.81	.693481	33
28	.298412	10.37	.991244	.43	.307168	10.80	.692832	32
29	.299034	10.36	.991218	.43	.307816	10.78	.692184	31
30	9.299655	10.34	9.991193	.43	9.308463	10.77	0.691537	30
31	.300276	10.33	.991167	.43	.309109	10.76	.690891	29
32	.300895	10.31	.991141	.43	.309754	10.74	.690246	28
33	.301514	10.30	.991115	.43	.310399	10.73	.689601	27
34	.302132	10.28	.991090	.43	.311042	10.71	.688958	26
35	.302748	10.26	.991064	.43	.311685	10.70	.688315	25
36	.303364	10.25	.991038	.43	.312327	10.68	.687673	24
37	.303979	10.23	.991012	.43	.312968	10.67	.687032	23
38	.304593	10.22	.990986	.43	.313608	10.65	.686392	22
39	.305207	10.20	.990960	.43	.314247	10.64	.685753	21
40	9.305819	10.19	9.990934	.44	9.314885	10.62	0.685115	20
41	.306430	10.17	.990908	.44	.315523	10.61	.684477	19
42	.307041	10.16	.990882	.44	.316159	10.60	.683841	18
43	.307650	10.14	.990855	.44	.316795	10.58	.683205	17
44	.308259	10.13	.990829	.44	.317430	10.57	.682570	16
45	.308867	10.12	.990803	.44	.318064	10.55	.681936	15
46	.309474	10.10	.990777	.44	.318697	10.54	.681303	14
47	.310080	10.09	.990750	.44	.319330	10.53	.680670	13
48	.310685	10.07	.990724	.44	.319961	10.51	.680039	12
49	.311289	10.06	.990697	.44	.320592	10.50	.679408	11
50	9.311893	10.04	9.990671	.44	9.321222	10.48	0.678778	10
51	.312495	10.03	.990645	.44	.321851	10.47	.678149	9
52	.313097	10.01	.990618	.44	.322479	10.46	.677521	8
53	.313698	10.00	.990591	.44	.323106	10.44	.676894	7
54	.314297	9.98	.990565	.44	.323733	10.43	.676267	6
55	.314897	9.97	.990538	.44	.324358	10.41	.675642	5
56	.315495	9.96	.990511	.45	.324983	10.40	.675017	4
57	.316092	9.94	.990485	.45	.325607	10.39	.674393	3
58	.316689	9.93	.990458	.45	.326231	10.37	.673769	2
59	.317284	9.91	.990431	.45	.326853	10.36	.673147	1
60	.317879		.990404		.327475		.672525	0
M.	Cosine.	D. 1".	Sine.	D. 1".	Cotang.	D. 1".	Tang.	M.

12° **167°**

M.	Sine.	D. 1″.	Cosine.	D. 1″.	Tang.	D. 1″.	Cotang.	M.
0	9.317879	9.90	9.990404	.45	9.327475	10.35	0.672525	60
1	.318473	9.88	.990378	.45	.328095	10.33	.671905	59
2	.319066	9.87	.990351	.45	.328715	10.32	.671285	58
3	.319658	9.86	.990324	.45	.329334	10.31	.670666	57
4	.320249	9.84	.990297	.45	.329953	10.29	.670047	56
5	.320840	9.83	.990270	.45	.330570	10.28	.669430	55
6	.321430	9.81	.990243	.45	.331187	10.27	.668813	54
7	.322019	9.80	.990215	.45	.331803	10.25	.668197	53
8	.322607	9.79	.990188	.45	.332418	10.24	.667582	52
9	.323194	9.77	.990161	.45	.333033	10.23	.666967	51
10	9.323780	9.76	9.990134	.45	9.333646	10.21	0.666354	50
11	.324366	9.75	.990107	.45	.334259	10.20	.665741	49
12	.324950	9.73	.990079	.46	.334871	10.19	.665129	48
13	.325534	9.72	.990052	.46	.335482	10.17	.664518	47
14	.326117	9.70	.990025	.46	.336093	10.16	.663907	46
15	.326700	9.69	.989997	.46	.336702	10.15	.663298	45
16	.327281	9.68	.989970	.46	.337311	10.14	.662689	44
17	.327862	9.66	.989942	.46	.337919	10.12	.662081	43
18	.328442	9.65	.989915	.46	.338527	10.11	.661473	42
19	.329021	9.64	.989887	.46	.339133	10.10	.660867	41
20	9.329599	9.62	9.989860	.46	9.339739	10.08	0.660261	40
21	.330176	9.61	.989832	.46	.340344	10.07	.659656	39
22	.330753	9.60	.989804	.46	.340948	10.06	.659052	38
23	.331329	9.58	.989777	.46	.341552	10.05	.658448	37
24	.331903	9.57	.989749	.46	.342155	10.03	.657845	36
25	.332478	9.56	.989721	.46	.342757	10.02	.657243	35
26	.333051	9.54	.989693	.46	.343358	10.01	.656642	34
27	.333624	9.53	.989665	.47	.343958	10.00	.656042	33
28	.334195	9.52	.989637	.47	.344558	9.98	.655442	32
29	.334767	9.50	.989610	.47	.345157	9.97	.654843	31
30	9.355337	9.49	9.989582	.47	9.345755	9.96	0.654245	30
31	.335906	9.48	.989553	.47	.346353	9.95	.653647	29
32	.336475	9.46	.989525	.47	.346949	9.93	.653051	28
33	.337043	9.45	.989497	.47	.347545	9.92	.652455	27
34	.337610	9.44	.989469	.47	.348141	9.91	.651859	26
35	.338176	9.43	.989441	.47	.348735	9.90	.651265	25
36	.338742	9.41	.989413	.47	.349329	9.88	.650671	24
37	.339307	9.40	.989385	.47	.349922	9.87	.650078	23
38	.339871	9.39	.989356	.47	.350514	9.86	.649486	22
39	.340434	9.37	.989328	.47	.351106	9.85	.648894	21
40	9.340996	9.36	9.989300	.47	9.351697	9.84	0.648303	20
41	.341558	9.35	.989271	.47	.352287	9.82	.647713	19
42	.342119	9.34	.989243	.47	.352876	9.81	.647124	18
43	.342679	9.32	.989214	.48	.353465	9.80	.646535	17
44	.343239	9.31	.989186	.48	.354053	9.79	.645947	16
45	.343797	9.30	.989157	.48	.354640	9.78	.645360	15
46	.344355	9.29	.989128	.48	.355227	9.76	.644773	14
47	.344912	9.27	.989100	.48	.355813	9.75	.644187	13
48	.345469	9.26	.989071	.48	.356398	9.74	.643602	12
49	.346024	9.25	.989042	.48	.356982	9.73	.643018	11
50	9.346579	9.24	9.989014	.48	9.357566	9.72	0.642434	10
51	.347134	9.22	.988985	.48	.358149	9.70	.641851	9
52	.347687	9.21	.988956	.48	.358731	9.69	.641269	8
53	.348240	9.20	.988927	.48	.359313	9.68	.640687	7
54	.348792	9.19	.988898	.48	.359893	9.67	.640107	6
55	.349343	9.17	.988869	.48	.360474	9.66	.639526	5
56	.349893	9.16	.988840	.48	.361053	9.65	.638947	4
57	.350443	9.15	.988811	.48	.361632	9.63	.638368	3
58	.350992	9.14	.988782	.49	.362210	9.62	.637790	2
59	.351540	9.13	.988753	.49	.362787	9.61	.637213	1
60	.352088		.988724		.363364		.636636	0
M.	**Cosine.**	**D. 1″.**	**Sine.**	**D. 1″.**	**Cotang.**	**D. 1″.**	**Tang.**	**M.**

102° **77°**

13° 166°

M.	Sine.	D. 1″.	Cosine.	D. 1″	Tang.	D. 1″.	Cotang.	M.
0	9.352088	9.11	9.988724	.49	9.363364	9.60	0.636636	60
1	.352635	9.10	.988695	.49	.363940	9.59	.636060	59
2	.353181	9.09	.988666	.49	.364515	9.58	.635485	58
3	.353726	9.08	.988636	.49	.365090	9.57	.634910	57
4	.354271	9.07	.988607	.49	.365664	9.55	.634336	56
5	.354815	9.05	.988578	.49	.366237	9.54	.633763	55
6	.355358	9.04	.988548	.49	.366810	9.53	.633190	54
7	.355901	9.03	.988519	.49	.367382	9.52	.632618	53
8	.356443	9.02	.988489	.49	.367953	9.51	.632047	52
9	.356984	9.01	.988460	.49	.368524	9.50	.631476	51
10	9.357524	8.99	9.988430	.49	9.369094	9.49	0.630906	50
11	.358064	8.98	.988401	.49	.369663	9.48	.630337	49
12	.358603	8.97	.988371	.49	.370232	9.47	.629768	48
13	.359141	8.96	.988342	.50	.370799	9.45	.629201	47
14	.359678	8.95	.988312	.50	.371367	9.44	.628633	46
15	.360215	8.94	.988282	.50	.371933	9.43	.628067	45
16	.360752	8.92	.988252	.50	.372499	9.42	.627501	44
17	.361287	8.91	.988223	.50	.373064	9.41	.626936	43
18	.361822	8.90	.988193	.50	.373629	9.40	.626371	42
19	.362356	8.89	.988163	.50	.374193	9.39	.625807	41
20	9.362889	8.88	9.988133	.50	9.374756	9.38	0.625244	40
21	.363422	8.87	.988103	.50	.375319	9.37	.624681	39
22	.363954	8.86	.988073	.50	.375881	9.36	.624119	38
23	.364485	8.84	.988043	.50	.376442	9.35	.623558	37
24	.365016	8.83	.988013	.50	.377003	9.33	.622997	36
25	.365546	8.82	.987983	.50	.377563	9.32	.622437	35
26	.366075	8.81	.987953	.50	.378122	9.31	.621878	34
27	.366604	8.80	.987922	.50	.378681	9.30	.621319	33
28	.367131	8.79	.987892	.50	.379239	9.29	.620761	32
29	.367659	8.78	.987862	.51	.379797	9.28	.620203	31
30	9.368185	8.76	9.987832	.51	9.380354	9.27	0.619646	30
31	.368711	8.75	.987801	.51	.380910	9.26	.619090	29
32	.369236	8.74	.987771	.51	.381466	9.25	.618534	28
33	369761	8.73	.987740	.51	.382020	9.24	.617980	27
34	.370285	8.72	.987710	.51	.382575	9.23	.617425	26
35	.370808	8.71	.987679	.51	.383129	9.22	.616871	25
36	.371330	8.70	.987649	.51	.383682	9.21	.616318	24
37	.371852	8.69	.987618	.51	.384234	9.20	.615766	23
38	.372373	8.68	.987588	.51	.384786	9.19	.615214	22
39	.372894	8.66	.987557	.51	.385337	9.18	.614663	21
40	9.373414	8.65	9.987526	.51	9.385888	9.17	0.614112	20
41	.373933	8.64	.987496	.51	.386438	9.16	.613562	19
42	.374452	8.63	.987465	.51	.386987	9.15	.613013	18
43	.374970	8.62	.987434	.51	.387536	9.14	.612464	17
44	.375487	8.61	.987403	.51	.388084	9.12	.611916	16
45	.376003	8.60	.987372	.52	.388631	9.11	.611369	15
46	.376519	8.59	.987341	.52	.389178	9.10	.610822	14
47	.377035	8.58	.987310	.52	.389724	9.09	.610276	13
48	.377549	8.57	.987279	.52	.390270	9.08	.609730	12
49	.378063	8.56	.987248	.52	.390815	9.07	.609185	11
50	9.378577	8.55	9.987217	.52	9.391360	9.06	0.608640	10
51	.379089	8.53	.987186	.52	.391903	9.05	.608097	9
52	.379601	8.52	.987155	.52	.392447	9.04	.607553	8
53	.380113	8.51	.987124	.52	.392989	9.03	.607011	7
54	.380624	8.50	.987092	.52	.393531	9.02	.606469	6
55	.381134	8.49	.987061	.52	.394073	9.01	.605927	5
56	.381643	8.48	.987030	.52	.394614	9.00	.605386	4
57	.382152	8.47	.986998	52	.395154	8.99	.604846	3
58	.382661	8.46	.986967	.52	.395694	8.98	.604306	2
59	.383168	8.45	.986936	.52	.396233	8.97	.603767	1
60	.383675		.986904		.396771		.603229	0
M.	Cosine.	D. 1″.	Sine.	D. 1″.	Cotang.	D. 1″.	Tang.	M.

103° 76°

14° 165°

M.	Sine.	D. 1''.	Cosine.	D. 1''.	Tang.	D. 1''.	Cotang	M.
0	9.383675	8.44	9.986904	.53	9.396771	8.96	0.603229	60
1	.384182	8.43	.986873	.53	.397309	8.96	.602691	59
2	.384687	8.42	.986841	.53	.397846	8.95	.602154	58
3	.385192	8.41	.986809	.53	.398383	8.94	.601617	57
4	.385697	8.40	.986778	.53	.398919	8.93	.601081	56
5	.386201	8.39	.986746	.53	.399455	8.92	.600545	55
6	.386704	8.38	.986714	.53	.399990	8.91	.600010	54
7	.387207	8.37	.986683	.53	.400524	8.90	.599476	53
8	.387709	8.36	.986651	.53	.401058	8.89	.598942	52
9	.388210	8.35	.986619	.53	.401591	8.88	.598409	51
10	9.388711	8.34	9.986587	.53	9.402124	8.87	0.597876	50
11	.389211	8.33	.986555	.53	.402656	8.86	.597344	49
12	.389711	8.32	.986523	.53	.403187	8.85	.596813	48
13	.390210	8.31	.986491	.53	.403718	8.84	.596282	47
14	.390708	8.30	.986459	.53	.404249	8.83	.595751	46
15	.391206	8.29	.986427	.54	.404778	8.82	.595222	45
16	.391703	8.28	.986395	.54	.405308	8.81	.594692	44
17	.392199	8.27	.986363	.54	.405836	8.80	.594164	43
18	.392695	8.26	.986331	.54	.406364	8.79	.593636	42
19	.393191	8.25	.986299	.54	.406892	8.78	.593108	41
20	9.393685	8.24	9.986266	.54	9.407419	8.77	0.592581	40
21	.394179	8.23	.986234	.54	.407945	8.76	.592055	39
22	.394673	8.22	.986202	.54	.408471	8.75	.591529	38
23	.395166	8.21	.986169	.54	.408996	8.75	.591004	37
24	.395658	8.20	.986137	.54	.409521	8.74	.590479	36
25	.396150	8.19	.986104	.54	.410045	8.73	.589955	35
26	.396641	8.18	.986072	.54	.410569	8.72	.589431	34
27	.397132	8.17	.986039	.54	.411092	8.71	.588908	33
28	.397621	8.16	.986007	.54	.411615	8.70	.588385	32
29	.398111	8.15	.985974	.54	.412137	8.69	.587863	31
30	9.398600	8.14	9.985942	.54	9.412658	8.68	0.587342	30
31	.399088	8.13	.985909	.55	.413179	8.67	.586821	29
32	.399575	8.12	.985876	.55	.413699	8.66	.586301	28
33	.400062	8.11	.985843	.55	.414219	8.65	.585781	27
34	.400549	8.10	.985811	.55	.414738	8.65	.585262	26
35	.401035	8.09	.985778	.55	.415257	8.64	.584743	25
36	.401520	8.08	.985745	.55	.415775	8.63	.584225	24
37	.402005	8.07	.985712	.55	.416293	8.62	.583707	23
38	.402489	8.06	.985679	.55	.416810	8.61	.583190	22
39	.402972	8.05	.985646	.55	.417326	8.60	.582674	21
40	9.403455	8.04	9.985613	.55	9.417842	8.59	0.582158	20
41	.403938	8.03	.985580	.55	.418358	8.58	.581642	19
42	.404420	8.02	.985547	.55	.418873	8.57	.581127	18
43	.404901	8.01	.985514	.55	.419387	8.56	.580613	17
44	.405382	8.00	.985480	.55	.419901	8.56	.580099	16
45	.405862	7.99	.985447	.55	.420415	8.55	.579585	15
46	.406341	7.98	.985414	.56	.420927	8.54	.579073	14
47	.406820	7.97	.985381	.56	.421440	8.53	.578560	13
48	.407299	7.96	.985347	.56	421952	8.52	.578048	12
49	.407777	7.96	.985314	.56	.422463	8.51	.577537	11
50	9.408254	7.95	9.985280	.56	9.422974	8.50	0.577026	10
51	.408731	7.94	.985247	.56	.423484	8.49	.576516	9
52	.409207	7.93	.985213	.56	.423993	8.49	.576007	8
53	.409682	7.92	.985180	.56	.424503	8.48	.575497	7
54	.410157	7.91	.985146	.56	.425011	8.47	.574989	6
55	.410632	7.90	.985113	.56	.425519	8.46	.574481	5
56	.411106	7.89	.985079	.56	.426027	8.45	.573973	4
57	.411579	7.88	.985045	.56	.426534	8.44	.573466	3
58	.412052	7.87	.985011	.56	.427041	8.43	.572959	2
59	.412524	7.86	.984978	.56	.427547	8.43	.572453	1
60	.412996		.984944		.428052		.571948	0
M.	Cosine.	D. 1''.	Sine.	D. 1''.	Cotang.	D. 1''.	Tang.	M.

104° 75°

M.	Sine.	D. 1″.	Cosine.	D. 1″.	Tang.	D. 1″.	Cotang.	M.
0	9.412996	7.85	9.984944	.56	9.428052	8.42	0.571948	60
1	.413467	7.84	.984910	.57	.428558	8.41	.571442	59
2	.413938	7.84	.984876	.57	.429062	8.40	.570938	58
3	.414408	7.83	.984842	.57	.429566	8.39	.570434	57
4	.414878	7.82	.984808	.57	.430070	8.38	.569930	56
5	.415347	7.81	.984774	.57	.430573	8.38	.569427	55
6	.415815	7.80	.984740	.57	.431075	8.37	.568925	54
7	.416283	7.79	.984706	.57	.431577	8.36	.568423	53
8	.416751	7.78	.984672	.57	.432079	8.35	.567921	52
9	.417217	7.77	.984638	.57	.432580	8.34	.567420	51
10	9.417684	7.76	9.984603	.57	9.433080	8.33	0.566920	50
11	.418150	7.75	.984569	.57	.433580	8.33	.566420	49
12	.418615	7.75	.984535	.57	.434080	8.32	.565920	48
13	.419079	7.74	.984500	.57	.434579	8.31	.565421	47
14	.419544	7.73	.984466	.57	.435078	8.30	.564922	46
15	.420007	7.72	.984432	.57	.435576	8.29	.564424	45
16	.420470	7.71	.984397	.58	.436073	8.28	.563927	44
17	.420933	7.70	.984363	.58	.436570	8.28	.563430	43
18	.421395	7.69	.984328	.58	.437067	8.27	.562933	42
19	.421857	7.68	.984294	.58	.437563	8.26	.562437	41
20	9.422318	7.67	9.984259	.58	9.438059	8.25	0.561941	40
21	.422778	7.67	.984224	.58	.438554	8.24	.561446	39
22	.423238	7.66	.984190	.58	.439048	8.24	.560952	38
23	.423697	7.65	.984155	.58	.439543	8.23	.560457	37
24	.424156	7.64	.984120	.58	.440036	8.22	.559964	36
25	.424615	7.63	.984085	.58	.440529	8.21	.559471	35
26	.425073	7.62	.984050	.58	.441022	8.20	.558978	34
27	.425530	7.61	.984015	.58	.441514	8.20	.558486	33
28	.425987	7.61	.983981	.58	.442006	8.19	.557994	32
29	.426443	7.60	.983946	.58	.442497	8.18	.557503	31
30	9.426899	7.59	9.983911	.58	9.442988	8.17	0.557012	30
31	.427354	7.58	.983875	.58	.443479	8.16	.556521	29
32	.427809	7.57	.983840	.59	.443968	8.16	.556032	28
33	.428263	7.56	.983805	.59	.444458	8.15	.555542	27
34	.428717	7.55	.983770	.59	.444947	8.14	.555053	26
35	.429170	7.55	.983735	.59	.445435	8.13	.554565	25
36	.429623	7.53	.983700	.59	.445923	8.13	.554077	24
37	.430075	7.52	.983664	.59	.446411	8.12	.553589	23
38	.430527	7.52	.983629	.59	.446898	8.11	.553102	22
39	.430978	7.51	.983594	.59	.447384	8.10	.552616	21
40	9.431429	7.50	9.983558	59	9.447870	8.09	0.552130	20
41	.431879	7.49	.983523	.59	.448356	8.09	.551644	19
42	.432329	7.49	.983487	.59	.448841	8.08	.551159	18
43	.432778	7.48	.983452	.59	.449326	8.07	.550674	17
44	.433226	7.47	.983416	.59	.449810	8.06	.550190	16
45	.433675	7.46	.983381	.59	.450294	8.06	.549706	15
46	.434122	7.45	.983345	.59	.450777	8.05	.549223	14
47	.434569	7.44	.983309	.60	.451260	8.04	.548740	13
48	.435016	7.44	.983273	.60	.451743	8.03	.548257	12
49	.435462	7.43	.983238	.60	.452225	8.03	.547775	11
50	9.435908	7.42	9.983202	.60	9.452706	8.02	0.547294	10
51	.436353	7.41	.983166	.60	.453187	8.01	.546813	9
52	.436798	7.40	.983130	.60	.453668	8.00	.546332	8
53	.437242	7.40	.983094	.60	.454148	8.00	.545852	7
54	.437686	7.39	.983058	.60	.454628	7.99	.545372	6
55	.438129	7.38	.983022	.60	.455107	7.98	.544893	5
56	.438572	7.37	.982986	.60	.455586	7.97	.544414	4
57	.439014	7.36	.982950	.60	.456064	7.97	.543936	3
58	.439456	7.36	.982914	.60	.456542	7.96	.543458	2
59	.439897	7.35	.982878	.60	.457019	7.95	.542981	1
60	.440338		.982842		.457496		.542504	0
M.	Cosine.	D. 1″.	Sine.	D. 1″.	Cotang.	D. 1″.	Tang.	M.

M.	Sine.	D. 1″.	Cosine.	D. 1″.	Tang.	D. 1″.	Cotang.	M.
0	9.440338	7.34	9.982842	.60	9.457496	7.94	0.542504	60
1	.440778	7.33	.982805	.60	.457973	7.94	.542027	59
2	.441218	7.32	.982769	.61	.458449	7.93	.541551	58
3	.441658	7.31	.982733	.61	.458925	7.92	.541075	57
4	.442096	7.31	.982696	.61	.459400	7.91	.540600	56
5	.442535	7.30	.982660	.61	.459875	7.91	.540125	55
6	.442973	7.29	.982624	.61	.460349	7.90	.539651	54
7	.443410	7.28	.982587	.61	.460823	7.89	.539177	53
8	.443847	7.27	.982551	.61	.461297	7.88	.538703	52
9	.444284	7.27	.982514	.61	.461770	7.88	.538230	51
10	9.444720	7.26	9.982477	.61	9.462242	7.87	0.537758	50
11	.445155	7.25	.982441	.61	.462715	7.86	.537285	49
12	.445590	7.24	.982404	.61	.463186	7.86	.536814	48
13	.446025	7.24	.982367	.61	.463658	7.85	.536342	47
14	.446459	7.23	.982331	.61	.464128	7.84	.535872	46
15	.446893	7.22	.982294	.61	.464599	7.83	.535401	45
16	.447326	7.21	.982257	.61	.465069	7.83	.534931	44
17	.447759	7.20	.982220	.62	.465539	7.82	.534461	43
18	.448191	7.20	.982183	.62	.466008	7.81	.533992	42
19	.448623	7.19	.982146	.62	.466477	7.81	.533523	41
20	9.449054	7.18	9.982109	.62	9.466945	7.80	0.533055	40
21	.449485	7.17	.982072	.62	.467413	7.79	.532587	39
22	.449915	7.17	.982035	.62	.467880	7.78	.532120	38
23	.450345	7.16	.981998	.62	.468347	7.78	.531653	37
24	.450775	7.15	.981961	.62	.468814	7.77	.531186	36
25	.451204	7.14	.981924	.62	.469280	7.76	.530720	35
26	.451632	7.13	.981886	.62	.469746	7.76	.530254	34
27	.452060	7.13	.981849	.62	.470211	7.75	.529789	33
28	.452488	7.12	.981812	62	.470676	7.74	.529324	32
29	.452915	7.11	.981774	62	.471141	7.74	.528859	31
30	9.453342	7.10	9.981737	.62	9.471605	7.73	0.528395	30
31	.453768	7.10	.981700	.62	.472069	7.72	.527931	29
32	.454194	7.09	.981662	.63	.472532	7.71	.527468	28
33	.454619	7.08	.981625	.63	.472995	7.71	.527005	27
34	.455044	7.07	.981587	.63	.473457	7.70	.526543	26
35	.455469	7.07	.981549	.63	.473919	7.69	.526081	25
36	.455893	7.06	.981512	.63	.474381	7.69	.525619	24
37	.456316	7.05	.981474	.63	.474842	7.68	.525158	23
38	.456739	7.04	.981436	.63	.475303	7.67	.524697	22
39	.457162	7.04	.981399	.63	.475763	7.67	.524237	21
40	9.457584	7.03	9.981361	.63	9.476223	7.66	0.523777	20
41	.458006	7.02	.981323	.63	.476683	7.65	.523317	19
42	.458427	7.01	.981285	.63	.477142	7.65	.522858	18
43	.458848	7.01	.981247	.63	.477601	7.64	.522399	17
44	.459268	7.00	.981209	.63	.478059	7.63	.521941	16
45	.459688	6.99	.981171	.63	.478517	7.63	.521483	15
46	.460108	6.98	.981133	.63	.478975	7.62	.521025	14
47	.460527	6.98	.981095	.64	.479432	7.61	.520568	13
48	.460946	6.97	.981057	.64	.479889	7.61	.520111	12
49	.461364	6.96	.981019	.64	.480345	7.60	.519655	11
50	9.461782	6.96	9.980981	.64	9.480801	7.59	0.519199	10
51	.462199	6.95	.980942	.64	.481257	7.59	.518743	9
52	.462616	6.94	.980904	.64	.481712	7.58	.518288	8
53	.463032	6.93	.980866	.64	.482167	7.57	.517833	7
54	.463448	6.93	.980827	.64	.482621	7.57	.517379	6
55	.463864	6.92	.980789	.64	.483075	7.56	.516925	5
56	.464279	6.91	.980750	.64	.483529	7.55	.516471	4
57	.464694	6.90	.980712	.64	.483982	7.55	.516018	3
58	.465108	6.90	.980673	.64	.484435	7.54	.515565	2
59	.465522	6.89	.980635	.64	.484887	7.53	.515113	1
60	.465935		.980596		.485339		.514661	0
M.	Cosine.	D. 1″.	Sine	D. 1″.	Cotang.	D. 1″.	Tang.	M.

17° **162°**

M.	Sine.	D 1″.	Cosine.	D. 1″.	Tang.	D. 1″.	Cotang.	M.
0	9.465935	6.88	9.980596	.64	9.485339	7.53	0.514661	60
1	.466348	6.88	.980558	.64	.485791	7.52	.514209	59
2	.466761	6.87	.980519	.65	.486242	7.51	.513758	58
3	.467173	6.86	.980480	.65	.486693	7.51	.513307	57
4	.467585	6.85	.980442	.65	.487143	7.50	.512857	56
5	.467996	6.85	.980403	.65	.487593	7.50	.512407	55
6	.468407	6.84	.980364	.65	.488043	7.49	.511957	54
7	.468817	6.83	.980325	.65	.488492	7.48	.511508	53
8	.469227	6.83	.980286	.65	.488941	7.48	.511059	52
9	.469637	6.82	.980247	.65	.489390	7.47	.510610	51
10	9.470046	6.81	9.980208	.65	9.489838	7.46	0.510162	50
11	.470455	6.81	.980169	.65	.490286	7.46	.509714	49
12	.470863	6.80	.980130	.65	.490733	7.45	.509267	48
13	.471271	6.79	.980091	.65	.491180	7.44	.508820	47
14	.471679	6.78	.980052	.65	.491627	7.44	.508373	46
15	.472086	6.78	.980012	.65	.492073	7.43	.507927	45
16	.472492	6.77	.979973	.65	.492519	7.43	.507481	44
17	.472898	6.76	.979934	.66	.492965	7.42	.507035	43
18	.473304	6.76	.979895	.66	.493410	7.41	.506590	42
19	.473710	6.75	.979855	.66	.493854	7.41	.506146	41
20	9.474115	6.74	9.979816	.66	9.494299	7.40	0.505701	40
21	.474519	6.74	.979776	.66	.494743	7.39	.505257	39
22	.474923	6.73	.979737	.66	.495186	7.39	.504814	38
23	.475327	6.72	.979697	.66	.495630	7.38	.504370	37
24	.475730	6.72	.979658	.66	.496073	7.38	.503927	36
25	.476133	6.71	.979618	.66	.496515	7.37	.503485	35
26	.476536	6.70	.979579	.66	.496957	7.36	.503043	34
27	.476938	6.69	.979539	.66	497399	7.36	.502601	33
28	.477340	6.69	.979499	.66	.497841	7.35	.502159	32
29	.477741	6.68	.979459	.66	.498282	7.34	.501718	31
30	9.478142	6.67	9.979420	.66	9.498722	7.34	0.501278	30
31	.478542	6.67	.979380	.66	.499163	7.33	.500837	29
32	.478942	6.66	.979340	.67	.499603	7.33	.500397	28
33	.479342	6.65	.979300	.67	.500042	7.32	.499958	27
34	.479741	6.65	.979260	.67	.500481	7.31	.499519	26
35	.480140	6.64	.979220	.67	.500920	7.31	.499080	25
36	.480539	6.63	.979180	.67	.501359	7.30	.498641	24
37	.480937	6.63	.979140	.67	.501797	7.30	.498203	23
38	.481334	6.62	.979100	.67	.502235	7.29	.497765	22
39	.481731	6.61	.979059	.67	502672	7.28	.497328	21
40	9.482128	6.61	9.979019	.67	9.503109	7.28	0.496891	20
41	.482525	6.60	.978979	.67	503546	7.27	.496454	19
42	.482921	6.59	.978939	.67	503982	7.27	.496018	18
43	.483316	6.59	.978898	.67	.504418	7.26	.495582	17
44	.483712	6.58	.978858	.67	.504854	7.25	.495146	16
45	.484107	6.57	.978817	.67	.505289	7.25	.494711	15
46	.484501	6.57	.978777	.67	.505724	7.24	.494276	14
47	.484895	6.56	.978737	.68	.506159	7.24	.493841	13
48	.485289	6.55	.978696	.68	.506593	7.23	.493407	12
49	.485682	6.55	.978655	.68	.507027	7.23	.492973	11
50	9.486075	6.54	9.978615	.68	9.507460	7.22	0.492540	10
51	.486467	6.54	.978574	.68	.507893	7.21	.492107	9
52	.486860	6.53	.978533	.68	.508326	7.21	.491674	8
53	.487251	6.52	.978493	.68	.508759	7.20	.491241	7
54	.487643	6.52	.978452	.68	.509191	7.20	.490809	6
55	.488034	6.51	.978411	.68	.509622	7.19	.490378	5
56	.488424	6.50	.978370	.68	.510054	7.18	.489946	4
57	.488814	6.50	.978329	.68	.510485	7.18	.489515	3
58	.489204	6.49	.978288	.68	.510916	7.17	.489084	2
59	.489593	6.49	.978247	.68	.511346	7.17	.488654	1
60	.489982		.978206		.511776		.488224	0
M.	Cosine.	D. 1″.	Sine	D. 1″.	Cotang.	D. 1″.	Tang.	M.

107° **72°**

M.	Sine.	D. 1″.	Cosine.	D. 1″.	Tang.	D. 1″.	Cotang.	M.
0	9.489982	6.48	9.978206	.68	9.511776	7.16	0.488224	60
1	.490371	6.47	.978165	.69	.512206	7.16	.487794	59
2	.490759	6.46	.978124	.69	.512635	7.15	.487365	58
3	.491147	6.46	.978083	.69	.513064	7.14	.486936	57
4	.491535	6.45	.978042	.69	.513493	7.14	.486507	56
5	.491922	6.45	.978001	.69	.513921	7.13	.486079	55
6	.492308	6.44	.977959	.69	.514349	7.13	.485651	54
7	.492695	6.43	.977918	.69	.514777	7.12	.485223	53
8	.493081	6.43	.977877	.69	.515204	7.12	.484796	52
9	.493466	6.42	.977835	.69	.515631	7.11	.484369	51
10	9.493851	6.41	9.977794	.69	9.516057	7.10	0.483943	50
11	.494236	6.41	.977752	.69	.516484	7.10	.483516	49
12	.494621	6.40	.977711	.69	.516910	7.09	.483090	48
13	.495005	6.39	.977669	.69	.517335	7.09	.482665	47
14	.495388	6.39	.977628	.69	.517761	7.08	.482239	46
15	.495772	6.38	.977586	.69	.518186	7.08	.481814	45
16	.496154	6.38	.977544	.70	.518610	7.07	.481390	44
17	.496537	6.37	.977503	.70	.519034	7.07	.480966	43
18	.496919	6.36	.977461	.70	.519458	7.06	.480542	42
19	.497301	6.36	.977419	.70	.519882	7.05	.480118	41
20	9.497682	6.35	9.977377	.70	9.520305	7.05	0.479695	40
21	.498064	6.34	.977335	.70	.520728	7.04	.479272	39
22	.498444	6.34	.977293	.70	.521151	7.04	.478849	38
23	.498825	6.33	.977251	.70	.521573	7.03	.478427	37
24	.499204	6.33	.977209	.70	.521995	7.03	.478005	36
25	.499584	6.32	.977167	.70	.522417	7.02	.477583	35
26	.499963	6.31	.977125	.70	.522838	7.02	.477162	34
27	.500342	6.31	.977083	.70	.523259	7.01	.476741	33
28	.500721	6.30	.977041	.70	.523680	7.01	.476320	32
29	.501099	6.30	.976999	.70	.524100	7.00	.475900	31
30	9.501476	6.29	9.976957	.70	9.524520	6.99	0.475480	30
31	.501854	6.28	.976914	.71	.524940	6.99	.475060	29
32	.502231	6.28	.976872	.71	.525359	6.98	.474641	28
33	.502607	6.27	.976830	.71	.525778	6.98	.474222	27
34	.502984	6.27	.976787	.71	.526197	6.97	.473803	26
35	.503360	6.26	.976745	.71	.526615	6.97	.473385	25
36	.503735	6.25	.976702	.71	.527033	6.96	.472967	24
37	.504110	6.25	.976660	.71	.527451	6.96	.472549	23
38	.504485	6.24	.976617	.71	.527868	6.95	.472132	22
39	.504860	6.24	.976574	.71	.528285	6.95	.471715	21
40	9.505234	6.23	9.976532	.71	9.528702	6.94	0.471298	20
41	.505608	6.22	.976489	.71	.529119	6.94	.470881	19
42	.505981	6.22	.976446	.71	.529535	6.93	.470465	18
43	.506354	6.21	.976404	.71	.529951	6.93	.470049	17
44	.506727	6.21	.976361	.71	.530366	6.92	.469634	16
45	.507099	6.20	.976318	.72	.530781	6.91	.469219	15
46	.507471	6.19	.976275	.72	.531196	6.91	.468804	14
47	.507843	6.19	.976232	.72	.531611	6.90	.468389	13
48	.508214	6.18	.976189	.72	.532025	6.90	.467975	12
49	.508585	6.18	.976146	.72	.532439	6.89	.467561	11
50	9.508956	6.17	9.976103	.72	9.532853	6.89	0.467147	10
51	.509326	6.16	.976060	.72	.533266	6.88	.466734	9
52	.509696	6.16	.976017	.72	.533679	6.88	.466321	8
53	.510065	6.15	.975974	.72	.534092	6.87	.465908	7
54	.510434	6.15	.975930	.72	.534504	6.87	.465496	6
55	.510803	6.14	.975887	.72	.534916	6.86	.465084	5
56	.511172	6.14	.975844	.72	.535328	6.86	.464672	4
57	.511540	6.13	.975800	.72	.535739	6.85	.464261	3
58	.511907	6.12	.975757	.72	.536150	6.85	.463850	2
59	.512275	6.12	.975714	.72	.536561	6.84	.463439	1
60	.512642		.975670		.536972		.463028	0
M.	Cosine.	D. 1″.	Sine.	D. 1″.	Cotang.	D. 1″.	Tang.	M.

19° **160°**

M.	Sine.	D. 1″.	Cosine.	D. 1″	Tang.	D. 1″.	Cotang.	M.
0	9.512642	6.11	9.975670	.73	9.536972	6.84	0.463028	60
1	.513009	6.11	.975627	.73	.537382	6.83	.462618	59
2	.513375	6.10	.975583	.73	.537792	6.83	.462208	58
3	.513741	6.09	.975539	.73	.538202	6.82	.461798	57
4	.514107	6.09	.975496	.73	.538611	6.82	.461389	56
5	.514472	6.08	.975452	.73	.539020	6.81	.460980	55
6	.514837	6.08	.975408	.73	.539429	6.81	.460571	54
7	.515202	6.07	.975365	.73	.539837	6.80	.460163	53
8	.515566	6.07	.975321	.73	.540245	6.80	.459755	52
9	.515930	6.06	.975277	.73	.540653	6.79	.459347	51
10	9.516294	6.05	9.975233	.73	9.541061	6.79	0.458939	50
11	.516657	6.05	.975189	.73	.541468	6.78	.458532	49
12	.517020	6.04	.975145	.73	.541875	6.78	.458125	48
13	.517382	6.04	.975101	.73	.542281	6.77	.457719	47
14	.517745	6.03	.975057	.73	.542688	6.77	.457312	46
15	.518107	6.03	.975013	.74	.543094	6.76	.456906	45
16	.518468	6.02	.974969	.74	.543499	6.76	.456501	44
17	.518829	6.02	.974925	.74	.543905	6.75	.456095	43
18	.519190	6.01	.974880	.74	.544310	6.75	.455690	42
19	.519551	6.00	.974836	.74	.544715	6.74	.455285	41
20	9.519911	6.00	9.974792	.74	9.545119	6.74	0.454881	40
21	.520271	5.99	.974748	.74	.545524	6.73	.454476	39
22	.520631	5.99	.974703	.74	.545928	6.73	.454072	38
23	.520990	5.98	.974659	.74	.546331	6.72	.453669	37
24	.521349	5.98	.974614	.74	.546735	6.72	.453265	36
25	.521707	5.97	.974570	.74	.547138	6.71	.452862	35
26	.522066	5.97	.974525	.74	.547540	6.71	.452460	34
27	.522424	5.96	.974481	.74	.547943	6.70	.452057	33
28	.522781	5.95	.974436	.74	.548345	6.70	.451655	32
29	.523138	5.95	.974391	.75	.548747	6.69	.451253	31
30	9.523495	5.94	9.974347	.75	9.549149	6.69	0.450851	30
31	.523852	5.94	.974302	.75	.549550	6.68	.450450	29
32	.524208	5.93	.974257	.75	.549951	6.68	.450049	28
33	.524564	5.93	.974212	.75	.550352	6.67	.449648	27
34	.524920	5.92	.974167	.75	.550752	6.67	.449248	26
35	.525275	5.92	.974122	.75	.551153	6.67	.448847	25
36	.525630	5.91	.974077	.75	.551552	6.66	.448448	24
37	.525984	5.90	.974032	.75	.551952	6.66	.448048	23
38	.526339	5.90	.973987	.75	.552351	6.65	.447649	22
39	.526693	5.89	.973942	.75	.552750	6.65	.447250	21
40	9.527046	5.89	9.973897	.75	9.553149	6.64	0.446851	20
41	.527400	5.88	.973852	.75	.553548	6.64	.446452	19
42	.527753	5.88	.973807	.75	.553946	6.63	.446054	18
43	.528105	5.87	.973761	.75	.554344	6.63	.445656	17
44	.528458	5.87	.973716	.76	.554741	6.62	.445259	16
45	.528810	5.86	.973671	.76	.555139	6.62	.444861	15
46	.529161	5.86	.973625	.76	.555536	6.61	.444464	14
47	.529513	5.85	.973580	.76	.555933	6.61	.444067	13
48	.529864	5.85	.973535	.76	.556329	6.60	.443671	12
49	.530215	5.84	.973489	.76	.556725	6.60	.443275	11
50	9.530565	5.83	9.973444	.76	9.557121	6.59	0.442879	10
51	.530915	5.83	.973398	.76	.557517	6.59	.442483	9
52	.531265	5.82	.973352	.76	.557913	6.59	.442087	8
53	.531614	5.82	.973307	.76	.558308	6.58	.441692	7
54	.531963	5.81	.973261	.76	.558703	6.58	.441297	6
55	.532312	5.81	.973215	.76	.559097	6.57	.440903	5
56	.532661	5.80	.973169	.76	.559491	6.57	.440509	4
57	.533009	5.80	.973124	.76	.559885	6.56	.440115	3
58	.533357	5.79	.973078	.77	.560279	6.56	.439721	2
59	.533704	5.79	.973032	.77	.560673	6.55	.439327	1
60	.534052		.972986		.561066		.438934	0
M.	Cosine.	D. 1″.	Sine.	D. 1″.	Cotang.	D. 1″.	Tang.	M.

109° **70°**

20° **159°**

M.	Sine.	D. 1″.	Cosine.	D. 1″.	Tang.	D. 1″.	Cotang.	M.
0	9.534052	5.78	9.972986	.77	9.561066	6.55	0.438934	60
1	.534399	5.78	.972940	.77	.561459	6.54	.438541	59
2	.534745	5.77	.972894	.77	.561851	6.54	.438149	58
3	.535092	5.77	.972848	.77	.562244	6.54	.437756	57
4	.535438	5.76	.972802	.77	.562636	6.53	.437364	56
5	.535783	5.76	.972755	.77	.563028	6.53	.436972	55
6	.536129	5.75	.972709	.77	.563419	6.52	.436581	54
7	.536474	5.75	.972663	.77	.563811	6.52	.436189	53
8	.536818	5.74	.972617	.77	.564202	6.51	.435798	52
9	.537163	5.74	.972570	.77	.564593	6.51	.435407	51
10	9.537507	5.73	9.972524	.77	9.564983	6.50	0.435017	50
11	.537851	5.73	.972478	.77	.565373	6.50	.434627	49
12	.538194	5.72	.972431	.78	.565763	6.50	.434237	48
13	.538538	5.71	.972385	.78	.566153	6.49	.433847	47
14	.538880	5.71	.972338	.78	.566542	6.49	.433458	46
15	.539223	5.70	.972291	.78	.566932	6.48	.433068	45
16	.539565	5.70	.972245	.78	.567320	6.48	.432680	44
17	.539907	5.69	.972198	.78	.567709	6.47	.432291	43
18	.540249	5.69	.972151	.78	.568098	6.47	.431902	42
19	.540590	5.68	.972105	.78	.568486	6.46	.431514	41
20	9.540931	5.68	9.972058	.78	9.568873	6.46	0.431127	40
21	.541272	5.67	.972011	.78	.569261	6.46	.430739	39
22	.541613	5.67	.971964	.78	.569648	6.45	.430352	38
23	.541953	5.66	.971917	.78	.570035	6.45	.429965	37
24	.542293	5.66	.971870	.78	.570422	6.44	.429578	36
25	.542632	5.65	.971823	.78	.570809	6.44	.429191	35
26	.542971	5.65	.971776	.78	.571195	6.43	.428805	34
27	.543310	5.64	.971729	.79	.571581	6.43	.428419	33
28	.543649	5.64	.971682	.79	.571967	6.43	.428033	32
29	.543987	5.63	.971635	.79	.572352	6.42	.427648	31
30	9.544325	5.63	9.971588	.79	9.572738	6.42	0.427262	30
31	.544663	5.62	.971540	.79	.573123	6.41	.426877	29
32	.545000	5.62	.971493	.79	.573507	6.41	.426493	28
33	545338	5.61	.971446	.79	.573892	6.40	.426108	27
34	.545674	5.61	.971398	.79	.574276	6.40	.425724	26
35	.546011	5.60	.971351	.79	.574660	6.40	.425340	25
36	.546347	5.60	.971303	.79	.575044	6.39	.424956	24
37	.546683	5.59	.971256	.79	.575427	6.39	.424573	23
38	.547019	5.59	.971208	.79	.575810	6.38	.424190	22
39	.547354	5.58	.971161	.79	.576193	6.38	.423807	21
40	9.547689	5.58	9.971113	.79	9.576576	6.37	0.423424	20
41	.548024	5.57	.971066	.80	.576959	6.37	.423041	19
42	.548359	5.57	.971018	.80	.577341	6.37	.422659	18
43	.548693	5.56	.970970	.80	.577723	6.36	.422277	17
44	.549027	5.56	.970922	.80	.578104	6.36	.421896	16
45	.549360	5.55	.970874	.80	.578486	6.35	.421514	15
46	.549693	5.55	.970827	.80	.578867	6.35	.421133	14
47	.550026	5.55	.970779	.80	.579248	6.34	.420752	13
48	.550359	5.54	.970731	.80	.579629	6.34	.420371	12
49	.550692	5.54	.970683	.80	.580009	6.34	.419991	11
50	9.551024	5.53	9.970635	.80	9.580389	6.33	0.419611	10
51	.551356	5.53	.970586	.80	.580769	6.33	.419231	9
52	.551687	5.52	.970538	.80	.581149	6.32	.418851	8
53	.552018	5.52	.970490	.80	.581528	6.32	.418472	7
54	.552349	5.51	.970442	.80	.581907	6.32	.418093	6
55	.552680	5.51	.970394	.81	.582286	6.31	.417714	5
56	.553010	5.50	.970345	.81	.582665	6.31	.417335	4
57	.553341	5.50	.970297	.81	.583044	6.30	.416956	3
58	.553670	5.49	.970249	.81	.583422	6.30	.416578	2
59	.554000	5.49	.970200	.81	.583800	6.30	.416200	1
60	.554329		.970152		.584177		.415823	0
M.	Cosine.	D. 1″.	Sine.	D. 1″.	Cotang.	D. 1″.	Tang.	M.

110° **69°**

M.	Sine.	D. 1".	Cosine.	D. 1".	Tang.	D. 1".	Cotang.	M.
0	9.554329	5.48	9.970152	.81	9.584177	6.29	0.415823	60
1	.554658	5.48	.970103	.81	.584555	6.29	.415445	59
2	.554987	5.47	.970055	.81	.584932	6.28	.415068	58
3	.555315	5.47	.970006	.81	.585309	6.28	.414691	57
4	.555643	5.46	.969957	.81	.585686	6.28	.414314	56
5	.555971	5.46	.969909	.81	.586062	6.27	.413938	55
6	.556299	5.45	.969860	.81	.586439	6.27	.413561	54
7	.556626	5.45	.969811	.81	.586815	6.26	.413185	53
8	.556953	5.44	.969762	.81	.587190	6.26	.412810	52
9	.557280	5.44	.969714	.81	.587566	6.26	.412434	51
10	9.557606	5.44	9.969665	.82	9.587941	6.25	0.412059	50
11	.557932	5.43	.969616	.82	.588316	6.25	.411684	49
12	.558258	5.43	.969567	.82	.588691	6.24	.411309	48
13	.558583	5.42	.969518	.82	.589066	6.24	.410934	47
14	.558909	5.42	.969469	.82	.589440	6.24	.410560	46
15	.559234	5.41	.969420	.82	.589814	6.23	.410186	45
16	.559558	5.41	.969370	.82	.590188	6.23	.409812	44
17	.559883	5.40	.969321	.82	.590562	6.22	.409438	43
18	.560207	5.40	.969272	.82	.590935	6.22	.409065	42
19	.560531	5.39	.969223	.82	.591308	6.22	.408692	41
20	9.560855	5.39	9.969173	.82	9.591681	6.21	0.408319	40
21	.561178	5.38	.969124	.82	.592054	6.21	.407946	39
22	.561501	5.38	.969075	.82	.592426	6.20	.407574	38
23	.561824	5.37	.969025	.82	.592799	6.20	.407201	37
24	.562146	5.37	.968976	.83	.593171	6.20	.406829	36
25	.562468	5.37	.968926	.83	.593542	6.19	.406458	35
26	.562790	5.36	.968877	.83	.593914	6.19	.406086	34
27	.563112	5.36	.968827	.83	.594285	6.18	.405715	33
28	.563433	5.35	.968777	.83	.594656	6.18	.405344	32
29	.563755	5.35	.968728	.83	.595027	6.18	.404973	31
30	9.564075	5.34	9.968678	.83	9.595398	6.17	0.404602	30
31	.564396	5.34	.968628	.83	.595768	6.17	.404232	29
32	.564716	5.33	.968578	.83	.596138	6.16	.403862	28
33	.565036	5.33	.968528	.83	.596508	6.16	.403492	27
34	.565356	5.32	.968479	.83	.596878	6.16	.403122	26
35	.565676	5.32	.968429	.83	.597247	6.15	.402753	25
36	.565995	5.32	.968379	.83	.597616	6.15	.402384	24
37	.566314	5.31	.968329	.83	.597985	6.15	.402015	23
38	.566632	5.31	.968278	.84	.598354	6.14	.401646	22
39	.566951	5.30	.968228	.84	.598722	6.14	.401278	21
40	9.567269	5.30	9.968178	84	9.599091	6.13	0.400909	20
41	.567587	5.29	.968128	.84	.599459	6.13	.400541	19
42	.567904	5.29	.968078	.84	.599827	6.13	.400173	18
43	.568222	5.28	.968027	.84	.600194	6.12	.399806	17
44	.568539	5.28	.967977	.84	.600562	6.12	.399438	16
45	.568856	5.28	.967927	.84	.600929	6.12	.399071	15
46	.569172	5.27	.967876	.84	.601296	6.11	.398704	14
47	.569488	5.27	.967826	.84	.601663	6.11	.398337	13
48	.569804	5.26	.967775	.84	.602029	6.10	.397971	12
49	.570120	5.26	.967725	.84	.602395	6.10	.397605	11
50	9.570435	5.25	9.967674	.84	9.602761	6.10	0.397239	10
51	.570751	5.25	.967624	.84	.603127	6.09	.396873	9
52	.571066	5.24	.967573	.85	.603493	6.09	.396507	8
53	.571380	5.24	.967522	.85	.603858	6.09	.396142	7
54	.571695	5.24	.967471	.85	.604223	6.08	.395777	6
55	.572009	5.23	.967421	.85	.604588	6.08	.395412	5
56	.572323	5.23	.967370	.85	.604953	6.07	.395047	4
57	.572636	5.22	.967319	.85	.605317	6.07	.394683	3
58	572950	5.22	.967268	.85	.605682	6.07	.394318	2
59	.573263	5.21	.967217	.85	.606046	6.06	.393954	1
60	.573575		.967166		.606410		.393590	0
M.	Cosine.	D. 1".	Sine.	D. 1".	Cotang.	D. 1".	Tang.	M.

32° | 157°

M.	Sine.	D. 1″.	Cosine.	D. 1″.	Tang.	D. 1″.	Cotang.	M.
0	9.573575	5.21	9.967166	.85	9.606410	6.06	0.393590	60
1	.573888	5.20	.967115	.85	.606773	6.06	.393227	59
2	.574200	5.20	.967064	.85	.607137	6.05	.392863	58
3	.574512	5.20	.967013	.85	.607500	6.05	.392500	57
4	.574824	5.19	.966961	.85	.607863	6.05	.392137	56
5	.575136	5.19	.966910	.85	.608225	6.04	.391775	55
6	.575447	5.18	.966859	.85	.608588	6.04	.391412	54
7	.575758	5.18	.966808	.86	.608950	6.03	.391050	53
8	.576069	5.17	.966756	.86	.609312	6.03	.390688	52
9	.576379	5.17	.966705	.86	.609674	6.03	.390326	51
10	9.576689	5.17	9.966653	.86	9.610036	6.02	0.389964	50
11	.576999	5.16	.966602	.86	.610397	6.02	.389603	49
12	.577309	5.16	.966550	.86	.610759	6.02	.389241	48
13	.577618	5.15	.966499	.86	.611120	6.01	.388880	47
14	.577927	5.15	.966447	.86	.611480	6.01	.388520	46
15	.578236	5.14	.966395	.86	.611841	6.01	.388159	45
16	.578545	5.14	.966344	.86	.612201	6.00	.387799	44
17	.578853	5.14	.966292	.86	.612561	6.00	.387439	43
18	.579162	5.13	.966240	.86	.612921	6.00	.387079	42
19	.579470	5.13	.966188	.86	.613281	5.99	.386719	41
20	9.579777	5.12	9.966136	.87	9.613641	5.99	0.386359	40
21	.580085	5.12	.966085	.87	.614000	5.98	.386000	39
22	.580392	5.11	.966033	.87	.614359	5.98	.385641	38
23	.580699	5.11	.965981	.87	.614718	5.98	.385282	37
24	.581005	5.11	.965929	.87	.615077	5.97	.384923	36
25	.581312	5.10	.965876	.87	.615435	5.97	.384565	35
26	.581618	5.10	.965824	.87	.615793	5.97	.384207	34
27	.581924	5.09	.965772	.87	.616151	5.96	.383849	33
28	.582229	5.09	.965720	.87	.616509	5.96	.383491	32
29	.582535	5.09	.965668	.87	.616867	5.96	.383133	31
30	9.582840	5.08	9.965615	.87	9.617224	5.95	0.382776	30
31	.583145	5.08	.965563	.87	.617582	5.95	.382418	29
32	.583449	5.07	.965511	.87	.617939	5.95	.382061	28
33	.583754	5.07	.965458	.87	.618295	5.94	.381705	27
34	.584058	5.06	.965406	.88	.618652	5.94	.381348	26
35	.584361	5.06	.965353	.88	.619008	5.94	.380992	25
36	.584665	5.06	.965301	.88	.619364	5.93	.380636	24
37	.584968	5.05	.965248	.88	.619720	5.93	.380280	23
38	.585272	5.05	.965195	.88	.620076	5.93	.379924	22
39	.585574	5.04	.965143	.88	.620432	5.92	.379568	21
40	9.585877	5.04	9.965090	.88	9.620787	5.92	0.379213	20
41	.586179	5.04	.965037	.88	.621142	5.92	.378858	19
42	.586482	5.03	.964984	.88	.621497	5.91	.378503	18
43	.586783	5.03	.964931	.88	.621852	5.91	.378148	17
44	.587085	5.02	.964879	.88	.622207	5.91	.377793	16
45	.587386	5.02	.964826	.88	.622561	5.90	.377439	15
46	.587688	5.01	.964773	.88	.622915	5.90	.377085	14
47	.587989	5.01	.964720	.88	.623269	5.90	.376731	13
48	.588289	5.01	.964666	.89	.623623	5.89	.376377	12
49	.588590	5.00	.964613	.89	.623976	5.89	.376024	11
50	9.588890	5.00	9.964560	.89	9.624330	5.89	0.375670	10
51	.589190	4.99	.964507	.89	.624683	5.88	.375317	9
52	.589489	4.99	.964454	.89	.625036	5.88	.374964	8
53	.589789	4.99	.964400	.89	.625388	5.88	.374612	7
54	.590088	4.98	.964347	.89	.625741	5.87	.374259	6
55	.590387	4.98	.964294	.89	.626093	5.87	.373907	5
56	.590686	4.97	.964240	.89	.626445	5.87	.373555	4
57	.590984	4.97	.964187	.89	.626797	5.86	.373203	3
58	.591282	4.97	.964133	.89	.627149	5.86	.372851	2
59	.591580	4.96	.964080	.89	.627501	5.86	.372499	1
60	.591878		.964026		.627852		.372148	0
M.	Cosine.	D. 1″.	Sine.	D. 1″.	Cotang.	D. 1″.	Tang.	M.

112° | 67°

M.	Sine.	D. 1″.	Cosine.	D. 1″.	Tang.	D. 1″.	Cotang.	M.
0	9.591878	4.96	9.964026	.89	9.627852	5.85	0.372148	60
1	.592176	4.95	.963972	.89	.628203	5.85	.371797	59
2	.592473	4.95	.963919	.90	.628554	5.85	.371446	58
3	.592770	4.95	.963865	.90	.628905	5.84	.371095	57
4	.593067	4.94	.963811	.90	.629255	5.84	.370745	56
5	.593363	4.94	.963757	.90	.629606	5.84	.370394	55
6	.593659	4.93	.963704	.90	.629956	5.83	.370044	54
7	.593955	4.93	.963650	.90	.630306	5.83	.369694	53
8	.594251	4.93	.963596	.90	.630656	5.83	.369344	52
9	.594547	4.92	.963542	.90	.631005	5.82	.368995	51
10	9.594842	4.92	9.963488	.90	9.631355	5.82	0.368645	50
11	.595137	4.91	.963434	.90	.631704	5.82	.368296	49
12	.595432	4.91	.963379	.90	.632053	5.81	.367947	48
13	.595727	4.91	.963325	.90	.632402	5.81	.367598	47
14	.596021	4.90	.963271	.90	.632750	5.81	.367250	46
15	.596315	4.90	.963217	.90	.633099	5.80	.366901	45
16	.596609	4.89	.963163	.91	.633447	5.80	.366553	44
17	.596903	4.89	.963108	.91	.633795	5.80	.366205	43
18	.597196	4.89	.963054	.91	.634143	5.79	.365857	42
19	.597490	4.88	.962999	.91	.634490	5.79	.365510	41
20	9.597783	4.88	9.962945	.91	9.634838	5.79	0.365162	40
21	.598075	4.88	.962890	.91	.635185	5.78	.364815	39
22	.598368	4.87	.962836	.91	.635532	5.78	.364468	38
23	.598660	4.87	.962781	.91	.635879	5.78	.364121	37
24	.598952	4.86	.962727	.91	.636226	5.78	.363774	36
25	.599244	4.86	.962672	.91	.636572	5.77	.363428	35
26	.599536	4.86	.962617	.91	.636919	5.77	.363081	34
27	.599827	4.85	.962562	.91	.637265	5.77	.362735	33
28	.600118	4.85	.962508	.91	.637611	5.76	.362389	32
29	.600409	4.84	.962453	.92	.637956	5.76	.362044	31
30	9.600700	4.84	9.962398	.92	9.638302	5.76	0.361698	30
31	.600990	4.84	.962343	.92	.638647	5.75	.361353	29
32	.601280	4.83	.962288	.92	.638992	5.75	.361008	28
33	.601570	4.83	.962233	.92	.639337	5.75	.360663	27
34	.601860	4.83	.962178	.92	.639682	5.74	.360318	26
35	.602150	4.82	.962123	.92	.640027	5.74	.359973	25
36	.602439	4.82	.962067	.92	.640371	5.74	.359629	24
37	.602728	4.81	.962012	.92	.640716	5.73	.359284	23
38	.603017	4.81	.961957	.92	.641060	5.73	.358940	22
39	.603305	4.81	.961902	.92	.641404	5.73	.358596	21
40	9.603594	4.80	9.961846	.92	9.641747	5.73	0.358253	20
41	.603882	4.80	.961791	.92	.642091	5.72	.357909	19
42	.604170	4.79	.961735	.92	.642434	5.72	.357566	18
43	.604457	4.79	.961680	.93	.642777	5.72	.357223	17
44	.604745	4.79	.961624	.93	.643120	5.71	.356880	16
45	.605032	4.78	.961569	.93	.643463	5.71	.356537	15
46	.605319	4.78	.961513	.93	.643806	5.71	.356194	14
47	.605606	4.78	.961458	.93	.644148	5.70	.355852	13
48	.605892	4.77	.961402	.93	.644490	5.70	.355510	12
49	.606179	4.77	.961346	.93	.644832	5.70	.355168	11
50	9.606465	4.76	9.961290	.93	9.645174	5.69	0.354826	10
51	.606751	4.76	.961235	.93	.645516	5.69	.354484	9
52	.607036	4.76	.961179	.93	.645857	5.69	.354143	8
53	.607322	4.75	.961123	.93	.646199	5.69	.353801	7
54	.607607	4.75	.961067	.93	.646540	5.68	.353460	6
55	.607892	4.74	.961011	.93	.646881	5.68	.353119	5
56	.608177	4.74	.960955	.93	.647222	5.68	.352778	4
57	.608461	4.74	.960899	.94	.647562	5.67	.352438	3
58	.608745	4.73	.960843	.94	.647903	5.67	.352097	2
59	.609029	4.73	.960786	.94	.648243	5.67	.351757	1
60	.609313		.960730		.648583		.351417	0
M.	Cosine.	D. 1″.	Sine.	D. 1″.	Cotang.	D. 1″.	Tang.	M.

M.	Sine.	D. 1″.	Cosine.	D. 1″.	Tang.	D. 1″.	Cotang.	M.
0	9.609313	4.73	9.960730	.94	9.648583	5.67	0.351417	60
1	.609597	4.72	.960674	.94	.648923	5.66	.351077	59
2	.609880	4.72	.960618	.94	.649263	5.66	.350737	58
3	.610164	4.72	.960561	.94	.649602	5.66	.350398	57
4	.610447	4.71	.960505	.94	.649942	5.65	.350058	56
5	.610729	4.71	.960448	.94	.650281	5.65	.349719	55
6	.611012	4.71	.960392	.94	.650620	5.65	.349380	54
7	.611294	4.70	.960335	.94	.650959	5.64	.349041	53
8	.611576	4.70	.960279	.94	.651297	5.64	.348703	52
9	.611858	4.69	.960222	.94	.651636	5.64	.348364	51
10	9.612140	4.69	9.960165	.95	9.651974	5.64	0.348026	50
11	.612421	4.69	.960109	.95	.652312	5.63	.347688	49
12	.612702	4.68	.960052	.95	.652650	5.63	.347350	48
13	.612983	4.68	.959995	.95	.652988	5.63	.347012	47
14	.613264	4.68	.959938	.95	.653326	5.62	.346674	46
15	.613545	4.67	.959882	.95	.653663	5.62	.346337	45
16	.613825	4.67	.959825	.95	.654000	5.62	.346000	44
17	.614105	4.67	.959768	.95	.654337	5.62	.345663	43
18	.614385	4.66	.959711	.95	.654674	5.61	.345326	42
19	.614665	4.66	.959654	.95	.655011	5.61	.344989	41
20	9.614944	4.65	9.959596	.95	9.655348	5.61	0.344652	40
21	.615223	4.65	.959539	.95	.655684	5.61	.344316	39
22	.615502	4.65	.959482	.95	.656020	5.60	.343980	38
23	.615781	4.64	.959425	.95	.656356	5.60	.343644	37
24	.616060	4.64	.959368	.95	.656692	5.60	.343308	36
25	.616338	4.64	.959310	.96	.657028	5.59	.342972	35
26	.616616	4.63	.959253	.96	.657364	5.59	.342636	34
27	.616894	4.63	.959195	.96	.657699	5.59	.342301	33
28	.617172	4.63	.959138	.96	.658034	5.58	.341966	32
29	.617450	4.62	.959080	.96	.658369	5.58	.341631	31
30	9.617727	4.62	9.959023	.96	9.658704	5.58	0.341296	30
31	.618004	4.61	.958965	.96	.659039	5.58	.340961	29
32	.618281	4.61	.958908	.96	.659373	5.57	.340627	28
33	.618558	4.61	.958850	.96	.659708	5.57	.340292	27
34	.618834	4.60	.958792	.96	.660042	5.57	.339958	26
35	.619110	4.60	.958734	.96	.660376	5.56	.339624	25
36	.619386	4.60	.958677	.96	.660710	5.56	.339290	24
37	.619662	4.59	.958619	.97	.661043	5.56	.338957	23
38	.619938	4.59	.958561	.97	.661377	5.56	.338623	22
39	.620213	4.59	.958503	.97	.661710	5.55	.338290	21
40	9.620488	4.58	9.958445	.97	9.662043	5.55	0.337957	20
41	.620763	4.58	.958387	.97	.662376	5.55	.337624	19
42	.621038	4.58	.958329	.97	.662709	5.54	.337291	18
43	.621313	4.57	.958271	.97	.663042	5.54	.336958	17
44	.621587	4.57	.958213	.97	.663375	5.54	.336625	16
45	.621861	4.57	.958154	.97	.663707	5.54	.336293	15
46	.622135	4.56	.958096	.97	.664039	5.53	.335961	14
47	.622409	4.56	.958038	.97	.664371	5.53	.335629	13
48	.622682	4.56	.957979	.97	.664703	5.53	.335297	12
49	.622956	4.55	.957921	.97	.665035	5.53	.334965	11
50	9.623229	4.55	9.957863	.97	9.665366	5.52	0.334634	10
51	.623502	4.54	.957804	.98	.665698	5.52	.334302	9
52	.623774	4.54	.957746	.98	.666029	5.52	.333971	8
53	.624047	4.54	.957687	.98	.666360	5.51	.333640	7
54	.624319	4.53	.957628	.98	.666691	5.51	.333309	6
55	.624591	4.53	.957570	.98	.667021	5.51	.332979	5
56	.624863	4.53	.957511	.98	.667352	5.51	.332648	4
57	.625135	4.52	.957452	.98	.667682	5.50	.332318	3
58	.625406	4.52	.957393	.98	.668013	5.50	.331987	2
59	.625677	4.52	.957335	.98	.668343	5.50	.331657	1
60	.625948		.957276		.668673		.331327	0
M.	Cosine.	D. 1″.	Sine.	D. 1″.	Cotang.	D. 1″.	Tang.	M.

25° 154°

M.	Sine.	D. 1".	Cosine.	D. 1".	Tang.	D. 1".	Cotang.	M.
0	9.625948	4.51	9.957276	.98	9.668673	5.50	0.331327	60
1	.626219	4.51	.957217	.98	.669002	5.49	.330998	59
2	.626490	4.51	.957158	.98	.669332	5.49	.330668	58
3	.626760	4.50	.957099	.98	.669661	5.49	.330339	57
4	.627030	4.50	.957040	.99	.669991	5.49	.330009	56
5	.627300	4.50	.956981	.99	.670320	5.48	.329680	55
6	.627570	4.49	.956921	.99	.670649	5.48	.329351	54
7	.627840	4.49	.956862	.99	.670977	5.48	.329023	53
8	.628109	4.49	.956803	.99	.671306	5.47	.328694	52
9	.628378	4.48	.956744	.99	.671635	5.47	.328365	51
10	9.628647	4.48	9.956684	.99	9.671963	5.47	0.328037	50
11	.628916	4.48	.956625	.99	.672291	5.47	.327709	49
12	.629185	4.47	.956566	.99	.672619	5.46	.327381	48
13	.629453	4.47	.956506	.99	.672947	5.46	.327053	47
14	.629721	4.47	.956447	.99	.673274	5.46	.326726	46
15	.629989	4.46	.956387	.99	.673602	5.46	.326398	45
16	.630257	4.46	.956327	.99	.673929	5.45	.326071	44
17	.630524	4.46	.956268	.99	.674257	5.45	.325743	43
18	.630792	4.45	.956208	1.00	.674584	5.45	.325416	42
19	.631059	4.45	.956148	1.00	.674911	5.45	.325089	41
20	9.631326	4.45	9.956089	1.00	9.675237	5.44	0.324763	40
21	.631593	4.44	.956029	1.00	.675564	5.44	.324436	39
22	.631859	4.44	.955969	1.00	.675890	5.44	.324110	38
23	.632125	4.44	.955909	1.00	.676217	5.44	.323783	37
24	.632392	4.43	.955849	1.00	.676543	5.43	.323457	36
25	.632658	4.43	.955789	1.00	.676869	5.43	.323131	35
26	.632923	4.43	.955729	1.00	.677194	5.43	.322806	34
27	.633189	4.42	.955669	1.00	.677520	5.42	.322480	33
28	.633454	4.42	.955609	1.00	.677846	5.42	.322154	32
29	.633719	4.42	.955548	1.00	.678171	5.42	321829	31
30	9.633984	4.41	9.955488	1.00	9.678496	5.42	0.321504	30
31	.634249	4.41	.955428	1.01	.678821	5.41	.321179	29
32	.634514	4.41	.955368	1.01	.679146	5.41	.320854	28
33	.634778	4.40	.955307	1.01	.679471	5.41	.320529	27
34	.635042	4.40	.955247	1.01	.679795	5.41	.320205	26
35	.635306	4.40	.955186	1.01	.680120	5.40	.319880	25
36	.635570	4.39	.955126	1.01	.680444	5.40	.319556	24
37	.635834	4.39	.955065	1.01	.680768	5.40	.319232	23
38	.636097	4.39	.955005	1.01	.681092	5.40	.318908	22
39	.636360	4.38	.954944	1.01	.681416	5.39	.318584	21
40	9.636623	4.38	9.954883	1.01	9.681740	5.39	0.318260	20
41	.636886	4.38	.954823	1.01	.682063	5.39	.317937	19
42	.637148	4.37	.954762	1.01	.682387	5.39	.317613	18
43	.637411	4.37	.954701	1.01	.682710	5.38	.317290	17
44	.637673	4.37	.954640	1.02	.683033	5.38	.316967	16
45	.637935	4.36	.954579	1.02	.683356	5.38	.316644	15
46	.638197	4.36	.954518	1.02	.683679	5.38	.316321	14
47	.638458	4.36	.954457	1.02	.684001	5.37	.315999	13
48	.638720	4.35	.954396	1.02	.684324	5.37	.315676	12
49	.638981	4.35	.954335	1.02	.684646	5.37	.315354	11
50	9.639242	4.35	9.954274	1.02	9.684968	5.37	0.315032	10
51	.639503	4.34	.954213	1.02	.685290	5.36	.314710	9
52	.639764	4.34	.954152	1.02	.685612	5.36	.314388	8
53	.640024	4.34	.954090	1.02	.685934	5.36	.314066	7
54	.640284	4.33	.954029	1.02	.686255	5.36	.313745	6
55	.640544	4.33	.953968	1.02	.686577	5.35	.313423	5
56	.640804	4.33	.953906	1.02	.686898	5.35	.313102	4
57	.641064	4.32	.953845	1.03	.687219	5.35	.312781	3
58	.641324	4.32	.953783	1.03	687540	5.35	.312460	2
59	.641583	4.32	.953722	1.03	687861	5.35	.312139	1
60	.641842		.953660		.688182		.311818	0
M.	Cosine.	D. 1".	Sine.	D. 1".	Cotang.	D. 1'.	Tang.	M.

115° 64°

M.	Sine.	D. 1″.	Cosine.	D. 1″.	Tang.	D. 1″.	Cotang.	M.
0	9.641842	4.32	9.953660	1.03	9.688182	5.34	0.311818	60
1	.642101	4.31	.953599	1.03	.688502	5.34	.311498	59
2	.642360	4.31	.953537	1.03	.688823	5.34	.311177	58
3	.642618	4.31	.953475	1.03	.689143	5.34	.310857	57
4	.642877	4.30	.953413	1.03	.689463	5.33	.310537	56
5	.643135	4.30	.953352	1.03	.689783	5.33	.310217	55
6	.643393	4.30	.953290	1.03	.690103	5.33	.309897	54
7	.643650	4.29	.953228	1.03	.690423	5.33	.309577	53
8	.643908	4.29	.953166	1.03	.690742	5.32	.309258	52
9	.644165	4.29	.953104	1.03	.691062	5.32	.308938	51
10	9.644423	4.28	9.953042	1.03	9.691381	5.32	0.308619	50
11	.644680	4.28	.952980	1.04	.691700	5.32	.308300	49
12	.644936	4.28	.952918	1.04	.692019	5.31	.307981	48
13	.645193	4.27	.952855	1.04	.692338	5.31	.307662	47
14	.645450	4.27	.952793	1.04	.692656	5.31	.307344	46
15	.645706	4.27	.952731	1.04	.692975	5.31	.307025	45
16	.645962	4.26	.952669	1.04	.693293	5.30	.306707	44
17	.646218	4.26	.952606	1.04	.693612	5.30	.306388	43
18	.646474	4.26	.952544	1.04	.693930	5.30	.306070	42
19	.646729	4.26	.952481	1.04	.694248	5.30	.305752	41
20	9.646984	4.25	9.952419	1.04	9.694566	5.29	0.305434	40
21	.647240	4.25	.952356	1.04	.694883	5.29	.305117	39
22	.647494	4.25	.952294	1.04	.695201	5.29	.304799	38
23	.647749	4.24	.952231	1.04	.695518	5.29	.304482	37
24	.648004	4.24	.952168	1.05	.695836	5.29	.304164	36
25	.648258	4.24	.952106	1.05	.696153	5.28	.303847	35
26	.648512	4.23	.952043	1.05	.696470	5.28	.303530	34
27	.648766	4.23	.951980	1.05	.696787	5.28	.303213	33
28	.649020	4.23	.951917	1.05	.697103	5.28	.302897	32
29	.649274	4.22	.951854	1.05	.697420	5.27	.302580	31
30	9.649527	4.22	9.951791	1.05	9.697736	5.27	0.302264	30
31	.649781	4.22	.951728	1.05	.698053	5.27	.301947	29
32	.650034	4.22	.951665	1.05	.698369	5.27	.301631	28
33	.650287	4.21	.951602	1.05	.698685	5.26	.301315	27
34	.650539	4.21	.951539	1.05	.699001	5.26	.300999	26
35	.650792	4.21	.951476	1.05	.699316	5.26	.300684	25
36	.651044	4.20	.951412	1.05	.699632	5.26	.300368	24
37	.651297	4.20	.951349	1.06	.699947	5.26	.300053	23
38	.651549	4.20	.951286	1.06	.700263	5.25	.299737	22
39	.651800	4.19	.951222	1.06	.700578	5.25	.299422	21
40	9.652052	4.19	9.951159	1.06	9.700893	5.25	0.299107	20
41	.652304	4.19	.951096	1.06	.701208	5.25	.298792	19
42	.652555	4.18	.951032	1.06	.701523	5.24	.298477	18
43	.652806	4.18	.950968	1.06	.701837	5.24	.298163	17
44	.653057	4.18	.950905	1.06	.702152	5.24	.297848	16
45	.653308	4.18	.950841	1.06	.702466	5.24	.297534	15
46	.653558	4.17	.950778	1.06	.702781	5.24	.297219	14
47	.653808	4.17	.950714	1.06	.703095	5.23	.296905	13
48	.654059	4.17	.950650	1.06	.703409	5.23	.296591	12
49	.654309	4.16	.950586	1.06	.703722	5.23	.296278	11
50	9.654558	4.16	9.950522	1.07	9.704036	5.23	0.295964	10
51	.654808	4.16	.950458	1.07	.704350	5.22	.295650	9
52	.655058	4.15	.950394	1.07	.704663	5.22	.295337	8
53	.655307	4.15	.950330	1.07	.704976	5.22	.295024	7
54	.655556	4.15	.950266	1.07	.705290	5.22	.294710	6
55	.655805	4.15	.950202	1.07	.705603	5.22	.294397	5
56	.656054	4.14	.950138	1.07	.705916	5.21	.294084	4
57	.656302	4.14	.950074	1.07	.706228	5.21	.293772	3
58	.656551	4.14	.950010	1.07	.706541	5.21	.293459	2
59	.656799	4.13	.949945	1.07	.706854	5.21	.293146	1
60	.657047		.949881		.707166		.292834	0
M.	Cosine.	D. 1″.	Sine.	D. 1″.	Cotang.	D. 1″.	Tang.	M.

M.	Sine.	D. 1″.	Cosine.	D. 1″	Tang.	D. 1″.	Cotang.	M.
0	9.657047	4.13	9.949881	1.07	9.707166	5.20	0.292834	60
1	.657295	4.13	.919816	1.07	.707478	5.20	.292522	59
2	.657542	4.12	.949752	1.07	.707790	5.20	.292210	58
3	.657790	4.12	.949688	1.08	.708102	5.20	.291898	57
4	.658037	4.12	.949623	1.08	.708414	5.20	.291586	56
5	.658284	4.12	.949558	1.08	.708726	5.19	.291274	55
6	.658531	4.11	.949494	1.08	.709037	5.19	.290963	54
7	.658778	4.11	.949429	1.08	.709349	5.19	.290651	53
8	.659025	4.11	.949364	1.08	.709660	5.19	.290340	52
9	.659271	4.10	.949300	1.08	.709971	5.18	.290029	51
10	9.659517	4.10	9.949235	1.08	9.710282	5.18	0.289718	50
11	.659763	4.10	.949170	1.08	.710593	5.18	.289407	49
12	.660009	4.10	.949105	1.08	.710904	5.18	.289096	48
13	.660255	4.09	.949040	1.08	.711215	5.18	.288785	47
14	.660501	4.09	.948975	1.08	.711525	5.17	.288475	46
15	.660746	4.09	.948910	1.08	.711836	5.17	.288164	45
16	.660991	4.08	.948845	1.09	.712146	5.17	.287854	44
17	.661236	4.08	.948780	1.09	.712456	5.17	.287544	43
18	.661481	4.08	.948715	1.09	.712766	5.17	.287234	42
19	.661726	4.08	.948650	1.09	.713076	5.16	.286924	41
20	9.661970	4.07	9.948584	1.09	9.713386	5.16	0.286614	40
21	.662214	4.07	.948519	1.09	.713696	5.16	.286304	39
22	.662459	4.07	.948454	1.09	.714005	5.16	.285995	38
23	.662703	4.06	.948388	1.09	.714314	5.15	.285686	37
24	.662946	4.06	.948323	1.09	.714624	5.15	.285376	36
25	.663190	4.06	.948257	1.09	.714933	5.15	.285067	35
26	.663433	4.05	.948192	1.09	.715242	5.15	.284758	34
27	.663677	4.05	.948126	1.09	.715551	5.15	.284449	33
28	.663920	4.05	.948060	1.09	.715860	5.14	.284140	32
29	.664163	4.05	.947995	1.10	.716168	5.14	.283832	31
30	9.664406	4.04	9.947929	1.10	9.716477	5.14	0.283523	30
31	.664648	4.04	.947863	1.10	.716785	5.14	.283215	29
32	.664891	4.04	.947797	1.10	.717093	5.14	.282907	28
33	.665133	4.03	.947731	1.10	.717401	5.13	.282599	27
34	.665375	4.03	.947665	1.10	.717709	5.13	.282291	26
35	.665617	4.03	.947600	1.10	.718017	5.13	.281983	25
36	.665859	4.03	.947533	1.10	.718325	5.13	.281675	24
37	.666100	4.02	.947467	1.10	.718633	5.13	.281367	23
38	.666342	4.02	.947401	1.10	.718940	5.12	.281060	22
39	.666583	4.02	.947335	1.10	.719248	5.12	.280752	21
40	9.666824	4.01	9.947269	1.10	9.719555	5.12	0.280445	20
41	.667065	4.01	.947203	1.11	.719862	5.12	.280138	19
42	.667305	4.01	.917136	1.11	.720169	5.11	.279831	18
43	.667546	4.01	.947070	1.11	.720476	5.11	.279524	17
44	.667786	4.00	.947004	1.11	.720783	5.11	.279217	16
45	.668027	4.00	.946937	1.11	.721089	5.11	.278911	15
46	.668267	4.00	.946871	1.11	.721396	5.11	.278604	14
47	.668506	3.99	.946804	1.11	.721702	5.10	.278298	13
48	.668746	3.99	.946738	1.11	.722009	5.10	.277991	12
49	.668986	3 99	.946671	1.11	.722315	5.10	.277685	11
50	9.669225	3.99	9.916604	1.11	9.722621	5.10	0.277379	10
51	.669464	3.98	.946538	1.11	.722927	5.10	.277073	9
52	.669703	3.98	.946471	1.11	.723232	5.09	.276768	8
53	.669942	3.98	.946404	1.11	.723538	5.09	.276462	7
54	.670181	3.98	.946337	1.12	.723844	5.09	.276156	6
55	.670419	3.97	.946270	1.12	.724149	5.09	.275851	5
56	.670658	3.97	.946203	1.12	.724454	5.09	.275546	4
57	.670896	3.97	.946136	1.12	.724760	5.08	.275240	3
58	.671134	3.96	.946069	1.12	.725065	5.08	.274935	2
59	.671372	3.96	.946002	1 12	.725370	5.08	.274630	1
60	.671609		.945935		.725674		.274326	0
M.	Cosine.	D. 1″.	Sine.	D. 1″.	Cotang.	D. 1″.	Tang.	M.

M.	Sine.	D. 1″.	Cosine.	D. 1″.	Tang.	D 1″.	Cotang.	M.
0	9.671609	3.96	9.945935	1.12	9.725674	5.08	0.274326	60
1	.671847	3.96	.945868	1.12	.725979	5.08	.274021	59
2	.672084	3.95	.945800	1.12	.726284	5.07	.273716	58
3	.672321	3.95	.945733	1.12	.726588	5.07	.273412	57
4	.672558	3.95	.945666	1.12	.726892	5.07	.273108	56
5	.672795	3.94	.945598	1.12	.727197	5.07	.272803	55
6	.673032	3.94	.945531	1.12	.727501	5.07	.272499	54
7	.673268	3.94	.945464	1.13	.727805	5.06	.272195	53
8	.673505	3.94	.945396	1.13	.728109	5.06	.271891	52
9	.673741	3.93	.945328	1.13	.728412	5.06	.271588	51
10	9.673977	3.93	9.945261	1.13	9.728716	5.06	0.271284	50
11	.674213	3.93	.945193	1.13	.729020	5.06	.270980	49
12	.674448	3.93	.945125	1.13	.729323	5.05	.270677	48
13	.674684	3.92	.945058	1.13	.729626	5.05	.270374	47
14	.674919	3.92	.944990	1.13	.729929	5.05	.270071	46
15	.675155	3.92	.944922	1.13	.730233	5.05	.269767	45
16	.675390	3.91	.944854	1.13	.730535	5.05	.269465	44
17	.675624	3.91	.944786	1.13	.730838	5.05	.269162	43
18	.675859	3.91	.944718	1.13	.731141	5.04	.268859	42
19	.676094	3.91	.944650	1.13	.731444	5.04	.268556	41
20	9.676328	3.90	9.944582	1.14	9.731746	5.04	0.268254	40
21	.676562	3.90	.944514	1.14	.732048	5.04	.267952	39
22	.676796	3.90	.944446	1.14	.732351	5.04	.267649	38
23	.677030	3.90	.944377	1.14	.732653	5.03	.267347	37
24	.677264	3.89	.944309	1.14	.732955	5.03	.267045	36
25	.677498	3.89	.944241	1.14	.733257	5.03	.266743	35
26	.677731	3.89	.944172	1.14	.733558	5.03	.266442	34
27	.677964	3.88	.944104	1.14	.733860	5.03	.266140	33
28	.678197	3.88	.944036	1.14	.734162	5.02	.265838	32
29	.678430	3.88	.943967	1.14	.734463	5.02	.265537	31
30	9.678663	3.88	9.943899	1.14	9.734764	5.02	0.265236	30
31	.678895	3.87	.943830	1.14	.735066	5.02	.264934	29
32	.679128	3.87	.943761	1.15	.735367	5.02	.261633	28
33	.679360	3.87	.943693	1.15	.735668	5.01	.264332	27
34	.679592	3.87	.943624	1.15	.735969	5.01	.264031	26
35	.679824	3.86	.943555	1.15	.736269	5.01	.263731	25
36	.680056	3.86	.943486	1.15	.736570	5.01	.263430	24
37	.680288	3.86	.943417	1.15	.736870	5.01	.263130	23
38	.680519	3.86	.943348	1.15	.737171	5.01	.262829	22
39	.680750	3.85	.943279	1.15	.737471	5.00	.262529	21
40	9.680982	3.85	9.943210	1.15	9.737771	5.00	0.262229	20
41	.681213	3.85	.943141	1.15	.738071	5.00	.261929	19
42	.681443	3.84	.943072	1.15	.738371	5.00	.261629	18
43	.681674	3.84	.943003	1.15	.733671	5.00	.261329	17
44	.681905	3.84	.942934	1.15	.738971	4.99	.261029	16
45	.682135	3.84	.942864	1.16	.739271	4.99	.260729	15
46	.682365	3.83	.942795	1.16	.739570	4.99	.260430	14
47	.682595	3.83	.942726	1.16	.739870	4.99	.260130	13
48	.682825	3.83	.942656	1.16	.740169	4.99	.259831	12
49	.683055	3.83	.942587	1.16	.740468	4.98	.259532	11
50	9.683284	3.82	9.942517	1.16	9.740767	4.98	0.259233	10
51	.683514	3.82	.942448	1.16	.741066	4.98	.258934	9
52	.683743	3.82	.942378	1.16	.741365	4.98	.258635	8
53	.683972	3.82	.942308	1.16	.741664	4.98	.258336	7
54	.684201	3.81	.942239	1.16	.741962	4.98	.258038	6
55	.684430	3.81	.942169	1.16	.742261	4.97	.257739	5
56	.684658	3.81	.942099	1.16	.742559	4.97	.257441	4
57	.684887	3.80	.942029	1.17	.742858	4.97	.257142	3
58	.685115	3.80	.941959	1.17	.743156	4.97	.256844	2
59	.685343	3.80	.941889	1.17	.743454	4.97	.256546	1
60	.685571		.941819		.743752		.256248	0
M.	Cosine.	D. 1″.	Sine.	D. 1″.	Cotang.	D. 1″.	Tang.	M.

29° 150°

M	Sine.	D. 1″.	Cosine.	D. 1″.	Tang.	D. 1″.	Cotang.	M.
0	9.685571	3.80	9.941819	1.17	9.743752	4.96	0.256248	60
1	.685799	3.79	.941749	1.17	.744050	4.96	.255950	59
2	.686027	3.79	.941679	1.17	.744348	4.96	.255652	58
3	.686254	3.79	.941609	1.17	.744645	4.96	.255355	57
4	.686482	3.79	.941539	1.17	.744943	4.96	.255057	56
5	.686709	3.78	.941469	1.17	.745240	4.96	.254760	55
6	.686936	3.78	.941398	1.17	.745538	4.95	.254462	54
7	.687163	3.78	.941328	1.17	.745835	4.95	.254165	53
8	.687389	3.78	.941258	1.17	.746132	4.95	.253868	52
9	.687616	3.77	.941187	1.17	.746429	4.95	.253571	51
10	9.687843	3.77	9.941117	1.18	9.746726	4.95	0.253274	50
11	.688069	3.77	.941046	1.18	.747023	4.95	.252977	49
12	.688295	3.77	.940975	1.18	.747319	4.94	.252681	48
13	.688521	3.76	.940905	1.18	.747616	4.94	.252384	47
14	.688747	3.76	.940834	1.18	.747913	4.94	.252087	46
15	.688972	3.76	.940763	1.18	.748209	4.94	.251791	45
16	.689198	3.76	.940693	1.18	.748505	4.94	.251495	44
17	.689423	3.75	.940622	1.18	.748801	4.93	.251199	43
18	.689648	3.75	.940551	1.18	.749097	4.93	.250903	42
19	.689873	3.75	.940480	1.18	.749393	4.93	.250607	41
20	9.690098	3.75	9.940409	1.18	9.749689	4.93	0.250311	40
21	.690323	3.74	.940338	1.18	.749985	4.93	.250015	39
22	.690548	3.74	.940267	1.19	.750281	4.93	.249719	38
23	.690772	3.74	.940196	1.19	.750576	4.92	.249424	37
24	.690996	3.74	.940125	1.19	.750872	4.92	.249128	36
25	.691220	3.73	.940054	1.19	.751167	4.92	.248833	35
26	.691444	3.73	.939982	1.19	.751462	4.92	.248538	34
27	.691668	3.73	.939911	1.19	.751757	4.92	.248243	33
28	.691892	3.73	.939840	1.19	.752052	4.92	.247948	32
29	.692115	3.72	.939768	1.19	.752347	4.91	.247653	31
30	9.692339	3.72	9.939697	1.19	9.752642	4.91	0.247358	30
31	.692562	3.72	.939625	1.19	.752937	4.91	.247063	29
32	.692785	3.72	.939554	1.19	.753231	4.91	.246769	28
33	.693008	3.71	.939482	1.19	.753526	4.91	.246474	27
34	.693231	3.71	939410	1.19	.753820	4.91	.246180	26
35	.693453	3.71	.939339	1.20	.754115	4.90	.245885	25
36	.693676	3.71	.939267	1.20	.754409	4.90	.245591	24
37	.693898	3.70	.939195	1.20	.754703	4.90	.245297	23
38	.694120	3.70	.939123	1.20	.754997	4.90	.245003	22
39	.694342	3.70	.939052	1.20	.755291	4.90	.244709	21
40	9.694564	3.70	9.938980	1.20	9.755585	4.89	0.244415	20
41	.694786	3.69	.938908	1.20	.755878	4.89	.244122	19
42	.695007	3.69	.938836	1.20	.756172	4.89	.243828	18
43	.695229	3.69	.938763	1.20	.756465	4.89	.243535	17
44	.695450	3.69	.938691	1.20	.756759	4.89	.243241	16
45	.695671	3.68	.938619	1.20	.757052	4.89	.242948	15
46	.695892	3.68	.938547	1.20	.757345	4.88	.242655	14
47	.696113	3.68	.938475	1.21	.757638	4.88	.242362	13
48	.696334	3.68	.938402	1.21	.757931	4.88	.242069	12
49	.696554	3.67	.938330	1.21	.758224	4.88	.241776	11
50	9.696775	3.67	9.938258	1.21	9.758517	4.88	0.241483	10
51	.696995	3.67	.938185	1.21	.758810	4.88	.241190	9
52	.697215	3.67	.938113	1.21	.759102	4.87	.240898	8
53	.697435	3.66	.938040	1.21	.759395	4.87	.240605	7
54	.697654	3.66	.937967	1.21	.759687	4.87	.240313	6
55	.697874	3.66	.937895	1.21	.759979	4.87	.240021	5
56	.698094	3.66	.937822	1.21	.760272	4.87	.239728	4
57	.698313	3.65	.937749	1.21	.760564	4.87	.239436	3
58	.698532	3.65	.937676	1.21	.760856	4.86	.239144	2
59	.698751	3.65	.937604	1.22	.761148	4.86	.238852	1
60	.698970		.937531		.761439		.238561	0
M.	Cosine.	D. 1″.	Sine.	D. 1″.	Cotang.	D. 1″.	Tang.	M.

119° 60°

30° | 149°

M.	Sine.	D. 1".	Cosine.	D. 1".	Tang.	D. 1".	Cotang.	M.
0	9.698970	3.65	9.937531	1.22	9.761439	4.86	0.238561	60
1	.699189	3.64	.937458	1.22	.761731	4.86	.238269	59
2	.699407	3.64	.937385	1.22	.762023	4.86	.237977	58
3	.699626	3.64	.937312	1.22	.762314	4.86	.237686	57
4	.699844	3.64	.937238	1.22	.762606	4.86	.237394	56
5	.700062	3.63	.937165	1.22	.762897	4.85	.237103	55
6	.700280	3.63	.937092	1.22	.763188	4.85	.236812	54
7	.700498	3.63	.937019	1.22	.763479	4.85	.236521	53
8	.700716	3.63	.936946	1.22	.763770	4.85	.236230	52
9	.700933	3.62	.936872	1.22	.764061	4.85	.235939	51
10	9.701151	3.62	9.936799	1.22	9.764352	4.85	0.235648	50
11	.701368	3.62	.936725	1.23	.764643	4.84	.235357	49
12	.701585	3.62	.936652	1.23	.764933	4.84	.235067	48
13	.701802	3.61	.936578	1.23	.765224	4.84	.234776	47
14	.702019	3.61	.936505	1.23	.765514	4.84	.234486	46
15	.702236	3.61	.936431	1.23	.765805	4.84	.234195	45
16	.702452	3.61	.936357	1.23	.766095	4.84	.233905	44
17	.702669	3.60	.936284	1.23	.766385	4.83	.233615	43
18	.702885	3.60	.936210	1.23	.766675	4.83	.233325	42
19	.703101	3.60	.936136	1.23	.766965	4.83	.233035	41
20	9.703317	3.60	9.936062	1.23	9.767255	4.83	0.232745	40
21	.703533	3.59	.935988	1.23	.767545	4.83	.232455	39
22	.703749	3.59	.935914	1.23	.767834	4.83	.232166	38
23	.703964	3.59	.935840	1.23	.768124	4.82	.231876	37
24	.704179	3.59	.935766	1.24	.768414	4.82	.231586	36
25	.704395	3.59	.935692	1.24	.768703	4.82	.231297	35
26	.704610	3.58	.935618	1.24	.768992	4.82	.231008	34
27	.704825	3.58	.935543	1.24	.769281	4.82	.230719	33
28	.705040	3.58	.935469	1.24	.769571	4.82	.230429	32
29	.705254	3.58	.935395	1.24	.769860	4.82	.230140	31
30	9.705469	3.57	9.935320	1.24	9.770148	4.81	0.229852	30
31	.705683	3.57	.935246	1.24	.770437	4.81	.229563	29
32	.705898	3.57	.935171	1.24	.770726	4.81	.229274	28
33	.706112	3.57	.935097	1.24	.771015	4.81	.228985	27
34	.706326	3.56	.935022	1.24	.771303	4.81	.228697	26
35	.706539	3.56	.934948	1.24	.771592	4.81	.228408	25
36	.706753	3.56	.934873	1.25	.771880	4.80	.228120	24
37	.706967	3.56	.934798	1.25	.772168	4.80	.227832	23
38	.707180	3.55	.934723	1.25	.772457	4.80	.227543	22
39	.707393	3.55	.934649	1.25	.772745	4.80	.227255	21
40	9.707606	3.55	9.934574	1.25	9.773033	4.80	0.226967	20
41	.707819	3.55	.934499	1.25	.773321	4.80	.226679	19
42	.708032	3.54	.934424	1.25	.773608	4.80	.226392	18
43	.708245	3.54	.934349	1.25	.773896	4.79	.226104	17
44	.708458	3.54	.934274	1.25	.774184	4.79	.225816	16
45	.708670	3.54	.934199	1.25	.774471	4.79	.225529	15
46	.708882	3.54	.934123	1.25	.774759	4.79	.225241	14
47	.709094	3.53	.934048	1.25	.775046	4.79	.224954	13
48	.709306	3.53	.933973	1.26	.775333	4.79	.224667	12
49	.709518	3.53	.933898	1.26	.775621	4.78	.224379	11
50	9.709730	3.53	9.933822	1.26	9.775908	4.78	0.224092	10
51	.709941	3.52	.933747	1.26	.776195	4.78	.223805	9
52	.710153	3.52	.933671	1.26	.776482	4.78	.223518	8
53	.710364	3.52	.933596	1.26	.776768	4.78	.223232	7
54	.710575	3.52	933520	1.26	.777055	4.78	.222945	6
55	.710786	3.51	933445	1.26	.777342	4.77	.222658	5
56	.710997	3.51	933369	1.26	.777628	4.77	.222372	4
57	.711208	3.51	933293	1.26	.777915	4.77	.222085	3
58	.711419	3.51	933217	1.26	.778201	4.77	.221799	2
59	.711629	3.51	.933141	1.26	.778488	4.77	.221512	1
60	.711839		.933066		.778774		.221226	0
M.	Cosine.	D. 1".	Sine.	D. 1".	Cotang.	D. 1".	Tang.	M.

120° | 59°

31° 148°

M.	Sine.	D. 1″.	Cosine.	D. 1″.	Tang.	D. 1″.	Cotang	M.
0	9.711839	3.50	9.933066	1.27	9.778774	4.77	0.221226	60
1	.712050	3.50	.932990	1.27	.779060	4.77	.220940	59
2	.712260	3.50	.932914	1.27	.779346	4.77	.220654	58
3	.712469	3.50	.932838	1.27	.779632	4.76	.220368	57
4	.712679	3.49	.932762	1.27	.779918	4.76	.220082	56
5	.712889	3.49	.932685	1.27	.780203	4.76	.219797	55
6	.713098	3.49	.932609	1.27	.780489	4.76	.219511	54
7	.713308	3.49	.932533	1.27	.780775	4.76	.219225	53
8	.713517	3.48	.932457	1.27	.781060	4.76	.218940	52
9	.713726	3.48	.932380	1.27	.781346	4.76	.218654	51
10	9.713935	3.48	9.932304	1.27	9.781631	4.75	0.218369	50
11	.714144	3.48	.932228	1.27	.781916	4.75	.218084	49
12	.714352	3.48	.932151	1.28	.782201	4.75	.217799	48
13	.714561	3.47	.932075	1.28	.782486	4.75	.217514	47
14	.714769	3.47	.931998	1.28	.782771	4.75	.217229	46
15	.714978	3.47	.931921	1.28	.783056	4.75	.216944	45
16	.715186	3.47	.931845	1.28	.783341	4.75	.216659	44
17	.715394	3.46	.931768	1.28	.783626	4.74	.216374	43
18	.715602	3.46	.931691	1.28	.783910	4.74	.216090	42
19	.715809	3.46	.931614	1.28	.784195	4.74	.215805	41
20	9.716017	3.46	9.931537	1.28	9.784479	4.74	0.215521	40
21	.716224	3.46	.931460	1.28	.784764	4.74	.215236	39
22	.716432	3.45	.931383	1.28	.785048	4.74	.214952	38
23	.716639	3.45	.931306	1.28	.785332	4.74	.214668	37
24	.716846	3.45	.931229	1.29	.785616	4.73	.214384	36
25	.717053	3.45	.931152	1.29	.785900	4.73	.214100	35
26	.717259	3.44	.931075	1.29	.786184	4.73	.213816	34
27	.717466	3.44	.930998	1.29	.786468	4.73	.213532	33
28	.717673	3.44	.930921	1.29	.786752	4.73	.213248	32
29	.717879	3.44	.930843	1.29	.787036	4.73	.212964	31
30	9.718085	3.43	9.930766	1.29	9.787319	4.73	0.212681	30
31	.718291	3.43	.930688	1.29	.787603	4.72	.212397	29
32	.718497	3.43	.930611	1.29	.787886	4.72	.212114	28
33	.718703	3.43	.930533	1.29	.788170	4.72	.211830	27
34	.718909	3.43	.930456	1.29	.788453	4.72	.211547	26
35	.719114	3.42	.930378	1.29	.788736	4.72	211264	25
36	.719320	3.42	.930300	1.30	.789019	4.72	.210981	24
37	.719525	3.42	.930223	1.30	.789302	4.72	.210698	23
38	.719730	3 42	.930145	1.30	.789585	4.71	.210415	22
39	.719935	3.41	.930067	1.30	.789868	4.71	.210132	21
40	9.720140	3.41	9.929989	1.30	9.790151	4.71	0.209849	20
41	.720345	3.41	.929911	1.30	.790434	4.71	.209566	19
42	.720549	3.41	.929833	1.30	.790716	4.71	.209284	18
43	.720754	3.41	.929755	1.30	.790999	4.71	.209001	17
44	.720958	3.40	.929677	1.30	.791281	4.71	.208719	16
45	.721162	3.40	.929599	1.30	.791563	4.70	.208437	15
46	.721366	3.40	.929521	1.30	.791846	4.70	.208154	14
47	.721570	3.40	.929442	1.31	.792128	4.70	.207872	13
48	.721774	3.39	.929364	1.31	.792410	4.70	.207590	12
49	.721978	3.39	.929286	1.31	.792692	4.70	.207308	11
50	9.722181	3.39	9.929207	1.31	9.792974	4.70	0.207026	10
51	.722385	3.39	.929129	1.31	.793256	4.70	.206744	9
52	.722588	3.39	.929050	1.31	.793538	4.70	.206462	8
53	.722791	3.38	.928972	1.31	.793819	4.69	.206181	7
54	.722994	3.38	.928893	1.31	.794101	4.69	.205899	6
55	.723197	3.38	.928815	1.31	.794383	4.69	.205617	5
56	.723400	3.38	.928736	1.31	.794664	4.69	.205336	4
57	.723603	3.37	.928657	1.31	.794946	4.69	.205054	3
58	.723805	3.37	.928578	1.31	.795227	4.69	.204773	2
59	.724007	3.37	.928499	1.32	.795508	4.69	.204492	1
60	.724210		.928420		.795789		.204211	0
M.	Cosine.	D. 1″.	Sine.	D. 1″.	Cotang.	D. 1″.	Tang.	M

121° 58°

32° 147°

M.	Sine.	D. 1″.	Cosine.	D. 1″.	Tang.	D. 1″.	Cotang.	M.
0	9.724210	3.37	9.928420	1.32	9.795789	4.68	0.204211	60
1	.724412	3.37	.928342	1.32	.796070	4.68	.203930	59
2	.724614	3.36	.928263	1.32	.796351	4.68	.203649	58
3	.724816	3.36	.928183	1.32	.796632	4.68	.203368	57
4	.725017	3.36	.928104	1.32	.796913	4.68	.203087	56
5	.725219	3.36	.928025	1.32	.797194	4.68	.202806	55
6	.725420	3.36	.927946	1.32	.797474	4.68	.202526	54
7	.725622	3.35	.927867	1.32	.797755	4.68	.202245	53
8	.725823	3.35	.927787	1.32	.798036	4.67	.201964	52
9	.726024	3.35	.927708	1.32	.798316	4.67	.201684	51
10	9.726225	3.35	9.927629	1.32	9.798596	4.67	0.201404	50
11	.726426	3.34	.927549	1.33	.798877	4.67	.201123	49
12	.726626	3.34	.927470	1.33	.799157	4.67	.200843	48
13	.726827	3.34	.927390	1.33	.799437	4.67	.200563	47
14	.727027	3.34	.927310	1.33	.799717	4.67	.200283	46
15	.727228	3.34	.927231	1.33	.799997	4.66	.200003	45
16	.727428	3.33	.927151	1.33	.800277	4.66	.199723	44
17	.727628	3.33	.927071	1.33	.800557	4.66	.199443	43
18	.727828	3.33	.926991	1.33	.800836	4.66	.199164	42
19	.728027	3.33	.926911	1.33	.801116	4.66	.198884	41
20	9.728227	3.33	9.926831	1.33	9.801396	4.66	0.198604	40
21	.728427	3.32	.926751	1.33	.801675	4.66	.198325	39
22	.728626	3.32	.926671	1.33	.801955	4.66	.198045	38
23	.728825	3.32	.926591	1.34	.802234	4.65	.197766	37
24	.729024	3.32	.926511	1.34	.802513	4.65	.197487	36
25	.729223	3.31	.926431	1.34	.802792	4.65	.197208	35
26	.729422	3.31	.926351	1.34	.803072	4.65	.196928	34
27	.729621	3.31	.926270	1.34	.803351	4.65	.196649	33
28	.729820	3.31	.926190	1.34	.803630	4.65	.196370	32
29	.730018	3.31	.926110	1.34	.803909	4.65	.196091	31
30	9.730217	3.30	9.926029	1.34	9.804187	4.65	0.195813	30
31	.730415	3.30	.925949	1.34	.804466	4.64	.195534	29
32	.730613	3.30	.925868	1.34	.804745	4.64	.195255	28
33	.730811	3.30	.925788	1.34	.805023	4.64	.194977	27
34	.731009	3.30	.925707	1.35	.805302	4.64	.194698	26
35	.731206	3.29	.925626	1.35	.805580	4.64	.194420	25
36	.731404	3.29	.925545	1.35	.805859	4.64	.194141	24
37	.731602	3.29	.925465	1.35	.806137	4.64	.193863	23
38	.731799	3.29	.925384	1.35	.806415	4.64	.193585	22
39	.731996	3.28	.925303	1.35	.806693	4.63	.193307	21
40	9.732193	3.28	9.925222	1.35	9.806971	4.63	0.193029	20
41	.732390	3.28	.925141	1.35	.807249	4.63	.192751	19
42	.732587	3.28	.925060	1.35	.807527	4.63	.192473	18
43	.732784	3.28	.924979	1.35	.807805	4.63	.192195	17
44	.732980	3.27	.924897	1.35	.808083	4.63	.191917	16
45	.733177	3.27	.924816	1.35	.808361	4.63	.191639	15
46	.733373	3.27	.924735	1.36	.808638	4.63	.191362	14
47	.733569	3.27	.924654	1.36	.808916	4.62	.191084	13
48	.733765	3.27	.924572	1.36	.809193	4.62	.190807	12
49	.733961	3.26	.924491	1.36	.809471	4.62	.190529	11
50	9.734157	3.26	9.924409	1.36	9.809748	4.62	0.190252	10
51	.734353	3.26	.924328	1.36	.810025	4.62	.189975	9
52	.734549	3.26	.924246	1.36	.810302	4.62	.189698	8
53	.734744	3.26	.924164	1.36	.810580	4.62	.189420	7
54	.734939	3.25	.924083	1.36	.810857	4.62	.189143	6
55	.735135	3.25	.924001	1.36	.811134	4.61	.188866	5
56	.735330	3.25	.923919	1.36	.811410	4.61	.188590	4
57	.735525	3.25	.923837	1.37	.811687	4.61	.188313	3
58	.735719	3.25	.923755	1.37	.811964	4.61	.188036	2
59	.735914	3.24	.923673	1.37	.812241	4.61	.187759	1
60	.736109		.923591		.812517		.187483	0
M.	Cosine.	D. 1″.	Sine.	D. 1″.	Cotang	D. 1″	Tang	M.

122° 57°

M.	Sine.	D. 1″.	Cosine.	D. 1″.	Tang.	D. 1″.	Cotang.	M.
0	9.736109	3.24	9.923591	1.37	9.812517	4.61	0.187483	60
1	.736303	3.24	.923509	1.37	.812794	4.61	.187206	59
2	.736498	3.24	.923427	1.37	.813070	4.61	.186930	58
3	.736692	3.23	.923345	1.37	.813347	4.61	.186653	57
4	.736886	3.23	.923263	1.37	.813623	4.60	.186377	56
5	.737080	3.23	.923181	1.37	.813899	4.60	.186101	55
6	.737274	3.23	.923098	1.37	.814176	4.60	.185824	54
7	.737467	3.23	.923016	1.37	.814452	4.60	.185548	53
8	.737661	3.22	.922933	1.37	.814728	4.60	.185272	52
9	.737855	3.22	.922851	1.38	.815004	4.60	.184996	51
10	9.738048	3.22	9.922768	1.38	9.815280	4.60	0.184720	50
11	.738241	3.22	.922686	1.38	.815555	4.60	.184445	49
12	.738434	3.22	.922603	1.38	.815-31	4.59	.184169	48
13	.738627	3.21	.922520	1.38	.816107	4.59	.183893	47
14	.738820	3.21	.922438	1.38	.816382	4.59	.183618	46
15	.739013	3.21	.922355	1.38	.816658	4.59	.183342	45
16	.739206	3.21	.922272	1.38	.816933	4.59	.183067	44
17	.739398	3.21	.922189	1.38	.817209	4.59	.182791	43
18	.739590	3.20	.922106	1.38	.817484	4.59	.182516	42
19	.739783	3.20	.922023	1.38	.817759	4.59	.182241	41
20	9.739975	3.20	9.921940	1.39	9.818035	4.59	0.181965	40
21	.740167	3.20	.921857	1.39	.818310	4.58	.181690	39
22	.740359	3.20	.921774	1.39	.818585	4.58	.181415	38
23	.740550	3.19	.921691	1.39	.818860	4.58	.181140	37
24	.740742	3.19	.921607	1.39	.819135	4.58	.180865	36
25	.740934	3.19	.921524	1.39	.819410	4.58	.180590	35
26	.741125	3.19	.921441	1.39	.819684	4.58	.180316	34
27	.741316	3.19	.921357	1.39	.819959	4.58	.180041	33
28	.741508	3.18	.921274	1.39	.820234	4.58	.179766	32
29	.741699	3.18	.921190	1.39	.820508	4.58	.179492	31
30	9.741889	3.18	9.921107	1.39	9.820783	4.57	0.179217	30
31	.742080	3.18	.921023	1.39	.821057	4.57	.178943	29
32	.742271	3.18	.920939	1.40	.821332	4.57	.178668	28
33	.742462	3.17	.920856	1.40	.821606	4.57	.178394	27
34	.742652	3.17	.920772	1.40	.821880	4.57	.178120	26
35	.742842	3.17	.920688	1.40	.822154	4.57	.177846	25
36	.743033	3.17	.920604	1.40	.822429	4.57	.177571	24
37	.743233	3.17	.920520	1.40	.822703	4.57	.177297	23
38	.743413	3.16	.920436	1.40	.822977	4.57	.177023	22
39	.743602	3.16	.920352	1.40	.823251	4.56	.176749	21
40	9.743792	3.16	9.920268	1.40	9.823524	4.56	0.176476	20
41	.743982	3.16	.920184	1.40	.823798	4.56	.176202	19
42	.744171	3.16	.920099	1.40	.824072	4.56	.175928	18
43	.744361	3.15	.920015	1.41	.824345	4.56	.175655	17
44	.744550	3.15	.919931	1.41	.824619	4.56	.175381	16
45	.744739	3.15	.919846	1.41	.824893	4.56	.175107	15
46	.744928	3.15	.919762	1.41	.825166	4.56	.174834	14
47	.745117	3.15	.919677	1.41	.825439	4.56	.174561	13
48	.745306	3.14	.919593	1.41	.825713	4.55	.174287	12
49	.745594	3.14	.919508	1.41	.825986	4.55	.174014	11
50	9.745683	3.14	9.919424	1.41	9.826259	4.55	0.173741	10
51	.745871	3.14	.919339	1.41	.826532	4.55	.173468	9
52	.746060	3.14	.919254	1.41	.826805	4.55	.173195	8
53	.746248	3.13	.919169	1.41	.827078	4.55	.172922	7
54	.746436	3.13	.919085	1.42	.827351	4.55	.172649	6
55	.746624	3.13	.919000	1.42	.827624	4.55	.172376	5
56	.746812	3.13	.918915	1.42	.827897	4.55	.172103	4
57	.746999	3.13	.918830	1.42	.828170	4.54	.171830	3
58	.747187	3.12	.918745	1.42	.828442	4.54	.171558	2
59	.747374	3.12	.918659	1.42	.828715	4.54	.171285	1
60	.747562		.918574		.828987		.171013	0
M.	Cosine.	D. 1″.	Sine.	D. 1″.	Cotang.	D. 1″.	Tang.	M.

34° | 145°

M.	Sine.	D. 1″.	Cosine.	D. 1″.	Tang.	D. 1″.	Cotang.	M.
0	9.747562	3.12	9.918574	1.42	9.828987	4.54	0.171013	60
1	.747749	3.12	.918489	1.42	.829260	4.54	.170740	59
2	.747936	3.12	.918404	1.42	.829532	4.54	.170468	58
3	.748123	3.11	.918318	1.42	.829805	4.54	.170195	57
4	.748310	3.11	.918233	1.42	.830077	4.54	.169923	56
5	.748497	3.11	.918147	1.43	.830349	4.54	.169651	55
6	.748683	3.11	.918062	1.43	.830621	4.53	.169379	54
7	.748870	3.11	.917976	1.43	.830893	4.53	.169107	53
8	.749056	3.10	.917891	1.43	.831165	4.53	.168835	52
9	.749243	3.10	.917805	1.43	.831437	4.53	.168563	51
10	9.749429	3.10	9.917719	1.43	9.831709	4.53	0.168291	50
11	.749615	3.10	.917634	1.43	.831981	4.53	.168019	49
12	.749801	3.10	.917548	1.43	.832253	4.53	.167747	48
13	.749987	3.10	.917462	1.43	.832525	4.53	.167475	47
14	.750172	3.09	.917376	1.43	.832796	4.53	.167204	46
15	.750358	3.09	.917290	1.43	.833068	4.53	.166932	45
16	.750543	3.09	.917204	1.43	.833339	4.53	.166661	44
17	.750729	3.09	.917118	1.44	.833611	4.52	.166389	43
18	.750914	3.09	.917032	1.44	.833882	4.52	.166118	42
19	.751099	3.08	.916946	1.44	.834154	4.52	.165846	41
20	9.751284	3.08	9.916859	1.44	9.834425	4.52	0.165575	40
21	.751469	3.08	.916773	1.44	.834696	4.52	.165304	39
22	.751654	3.08	.916687	1.44	.834967	4.52	.165033	38
23	.751839	3.08	.916600	1.44	.835238	4.52	.164762	37
24	.752023	3.07	.916514	1.44	.835509	4.52	.164491	36
25	.752208	3.07	.916427	1.44	.835780	4.52	.164220	35
26	.752392	3.07	.916341	1.44	.836051	4.51	.163949	34
27	.752576	3.07	.916254	1.44	.836322	4.51	.163678	33
28	.752760	3.07	.916167	1.45	.836593	4.51	.163407	32
29	.752944	3.06	.916081	1.45	.836864	4.51	.163136	31
30	9.753128	3.06	9.915994	1.45	9.837134	4.51	0.162866	30
31	.753312	3.06	.915907	1.45	.837405	4.51	.162595	29
32	.753495	3.06	.915820	1.45	.837675	4.51	.162325	28
33	.753679	3.06	.915733	1.45	.837946	4.51	.162054	27
34	.753862	3.05	.915646	1.45	.838216	4.51	.161784	26
35	.754046	3.05	.915559	1.45	.838487	4.51	.161513	25
36	.754229	3.05	.915472	1.45	.838757	4.50	.161243	24
37	.754412	3.05	.915385	1.45	.839027	4.50	.160973	23
38	.754595	3.05	.915297	1.45	.839297	4.50	.160703	22
39	.754778	3.05	.915210	1.46	.839568	4.50	.160432	21
40	9.754960	3.04	9.915123	1.46	9.839838	4.50	0.160162	20
41	.755143	3.04	.915035	1.46	.840108	4.50	.159892	19
42	.755326	3.04	.914948	1.46	.840378	4.50	.159622	18
43	.755508	3.04	.914860	1.46	.840648	4.50	.159352	17
44	.755690	3.04	.914773	1.46	.840917	4.50	.159083	16
45	.755872	3.03	.914685	1.46	.841187	4.49	.158813	15
46	.756054	3.03	.914598	1.46	.841457	4.49	.158543	14
47	.756236	3.03	.914510	1.46	.841727	4.49	.158273	13
48	.756418	3.03	.914422	1.46	.841996	4.49	.158004	12
49	.756600	3.03	.914334	1.46	.842266	4.49	.157734	11
50	9.756782	3.02	9.914246	1.47	9.842535	4.49	0.157465	10
51	.756963	3.02	.914158	1.47	.842805	4.49	.157195	9
52	.757144	3.02	.914070	1.47	.843074	4.49	.156926	8
53	.757326	3.02	.913982	1.47	.843343	4.49	.156657	7
54	.757507	3.02	.913894	1.47	.843612	4.49	.156388	6
55	.757688	3.02	.913866	1.47	.843882	4.49	.156118	5
56	.757869	3.01	.913718	1.47	.844151	4.48	.155849	4
57	.758050	3.01	913630	1.47	.844420	4.48	.155580	3
58	.758230	3.01	.913541	1.47	.844689	4.48	.1553[illegible]	2
59	.758411	3.01	.913453	1.47	.844958	4.48	.15[illegible]	[illegible]
60	.758591		.913365		.845227		.1547[illegible]	[illegible]
M.	Cosine.	D. 1″.	Sine.	D. 1″.	Cotang.	D. 1″.	Tang.	M.

M.	Sine.	D. 1".	Cosine.	D. 1".	Tang.	D. 1".	Cotang.	M.
0	9.758591	3.01	9.913365	1.47	9.845227	4.48	0.154773	60
1	.758772	3.00	.913276	1.48	.845496	4.48	.154504	59
2	.758952	3.00	.913187	1.48	.845764	4.48	.154236	58
3	.759132	3.00	.913099	1.48	.846033	4.48	.153967	57
4	.759312	3.00	.913010	1.48	.846302	4.48	.153698	56
5	.759492	3.00	.912922	1.48	.846570	4.48	.153430	55
6	.759672	2.99	.912833	1.48	.846839	4.48	.153161	54
7	.759852	2.99	.912744	1.48	.847108	4.47	.152892	53
8	.760031	2.99	.912655	1.48	.847376	4.47	.152624	52
9	.760211	2.99	.912566	1.48	.847644	4.47	.152356	51
10	9.760390	2.99	9.912477	1.48	9.847913	4.47	0.152087	50
11	.760569	2.99	.912388	1.48	.848181	4.47	.151819	49
12	.760748	2.98	.912299	1.49	.848449	4.47	.151551	48
13	.760927	2.98	.912210	1.49	.848717	4.47	.151283	47
14	.761106	2.98	.912121	1.49	.848986	4.47	.151014	46
15	.761285	2.98	.912031	1.49	.849254	4.47	.150746	45
16	.761464	2.98	.911942	1.49	.849522	4.47	.150478	44
17	.761642	2.97	.911853	1.49	.849790	4.46	.150210	43
18	.761821	2.97	.911763	1.49	.850057	4.46	.149943	42
19	.761999	2.97	.911674	1.49	.850325	4.46	.149675	41
20	9.762177	2.97	9.911584	1.49	9.850593	4.46	0.149407	40
21	.762356	2.97	.911495	1.49	.850861	4.46	.149139	39
22	.762534	2.97	.911405	1.49	.851129	4.46	.148871	38
23	.762712	2.96	.911315	1.50	.851396	4.46	.148604	37
24	.762889	2.96	.911226	1.50	.851664	4.46	.148336	36
25	.763067	2.96	.911136	1.50	.851931	4.46	.148069	35
26	.763245	2.96	.911046	1.50	.852199	4.46	.147801	34
27	.763422	2.96	.910956	1.50	.852466	4.46	.147534	33
28	.763600	2.95	.910866	1.50	.852733	4.46	.147267	32
29	.763777	2.95	.910776	1.50	.853001	4.45	.146999	31
30	9.763954	2.95	9.910686	1.50	9.853268	4.45	0.146732	30
31	.764131	2.95	.910596	1.50	.853535	4.45	.146465	29
32	.764308	2.95	.910506	1.50	.853802	4.45	.146198	28
33	.764485	2.95	.910415	1.51	.854069	4.45	.145931	27
34	.764662	2.94	.910325	1.51	.854336	4.45	.145664	26
35	.764838	2.94	.910235	1.51	.854603	4.45	.145397	25
36	.765015	2.94	.910144	1.51	.854870	4.45	.145130	24
37	.765191	2.94	.910054	1.51	.855137	4.45	.144863	23
38	.765367	2.94	.909963	1.51	.855404	4.45	.144596	22
39	.765544	2.93	.909873	1.51	.855671	4.44	.144329	21
40	9.765720	2.93	9.909782	1.51	9.855938	4.44	0.144062	20
41	.765896	2.93	.909691	1.51	.856204	4.44	.143796	19
42	.766072	2.93	.909601	1.51	.856471	4.44	.143529	18
43	.766247	2.93	.909510	1.51	.856737	4.44	.143263	17
44	.766423	2.93	.909419	1.52	.857004	4.44	.142996	16
45	.766598	2.92	.909328	1.52	.857270	4.44	.142730	15
46	.766774	2.92	.909237	1.52	.857537	4.44	.142463	14
47	.766949	2.92	.909146	1.52	.857803	4.44	.142197	13
48	.767124	2.92	.909055	1.52	.858069	4.44	.141931	12
49	.767300	2.92	.908964	1.52	.858336	4.44	.141664	11
50	9.767475	2.91	9.908873	1.52	9.858602	4.44	0.141398	10
51	.767649	2.91	.908781	1.52	.858868	4.43	.141132	9
52	.767824	2.91	.908690	1.52	.859134	4.43	.140866	8
53	.767999	2.91	.908599	1.52	.859400	4.43	.140600	7
54	.768173	2.91	.908507	1.52	.859666	4.43	.140334	6
55	.768348	2.91	.908416	1.53	.859932	4.43	.140068	5
56	.768522	2.90	.908324	1.53	.860198	4.43	.139802	4
57	.768697	2.90	.908233	1.53	.860464	4.43	.139536	3
58	.768871	2.90	.908141	1.53	.860730	4.43	.139270	2
59	.769045	2.90	.908049	1.53	.860995	4.43	.139005	1
60	.769219		.907958		.861261		.138739	0
M.	Cosine.	D. 1".	Sine.	D. 1".	Cotang.	D. 1".	Tang.	M.

36° | 143°

M.	Sine.	D. 1″.	Cosine.	D. 1″.	Tang.	D. 1″.	Cotang.	M.
0	9.769219	2.90	9.907958	1.53	9.861261	4.43	0.138739	60
1	.769393	2.90	.907866	1.53	.861527	4.43	.138473	59
2	.769566	2.89	.907774	1.53	.861792	4.43	.138208	58
3	.769740	2.89	.907682	1.53	.862058	4.42	.137942	57
4	.769913	2.89	.907590	1.53	.862323	4.42	.137677	56
5	.770087	2.89	.907498	1.53	.862589	4.42	.137411	55
6	.770260	2.89	.907406	1.54	.862854	4.42	.137146	54
7	.770433	2.88	.907314	1.54	.863119	4.42	.136881	53
8	.770606	2.88	.907222	1.54	.863385	4.42	.136615	52
9	.770779	2.88	.907129	1.54	.863650	4.42	.136350	51
10	9.770952	2.88	9.907037	1.54	9.863915	4.42	0.136085	50
11	.771125	2.88	.906945	1.54	.864180	4.42	.135820	49
12	.771298	2.88	.906852	1.54	.864445	4.42	.135555	48
13	.771470	2.87	.906760	1.54	.864710	4.42	.135290	47
14	.771643	2.87	.906667	1.54	.864975	4.42	.135025	46
15	.771815	2.87	.906575	1.54	.865240	4.42	.134760	45
16	.771987	2.87	.906482	1.55	.865505	4.41	.134495	44
17	.772159	2.87	.906389	1.55	.865770	4.41	.134230	43
18	.772331	2.87	.906296	1.55	.866035	4.41	.133965	42
19	.772503	2.86	.906204	1.55	.866300	4.41	.133700	41
20	9.772675	2.86	9.906111	1.55	9.866564	4.41	0.133436	40
21	.772847	2.86	.906018	1.55	.866829	4.41	.133171	39
22	.773018	2.86	.905925	1.55	.867094	4.41	.132906	38
23	.773190	2.86	.905832	1.55	.867358	4.41	.132642	37
24	.773361	2.85	.905739	1.55	.867623	4.41	.132377	36
25	.773533	2.85	.905645	1.55	.867887	4.41	.132113	35
26	.773704	2.85	.905552	1.55	.868152	4.41	.131848	34
27	.773875	2.85	.905459	1.56	.868416	4.41	.131584	33
28	.774046	2.85	.905366	1.56	.868680	4.40	.131320	32
29	.774217	2.85	.905272	1.56	.868945	4.40	.131055	31
30	9.774388	2.84	9.905179	1.56	9.869209	4.40	0.130791	30
31	.774558	2.84	.905085	1.56	.869473	4.40	.130527	29
32	.774729	2.84	.904992	1.56	.869737	4.40	.130263	28
33	.774899	2.84	.904898	1.56	.870001	4.40	.129999	27
34	.775070	2.84	.904804	1.56	.870265	4.40	.129735	26
35	.775240	2.84	.904711	1.56	.870529	4.40	.129471	25
36	.775410	2.83	.904617	1.56	.870793	4.40	.129207	24
37	.775580	2.83	.904523	1.57	.871057	4.40	.128943	23
38	.775750	2.83	.904429	1.57	.871321	4.40	.128679	22
39	.775920	2.83	.904335	1.57	.871585	4.40	.128415	21
40	9.776090	2.83	9.904241	1.57	9.871849	4.40	0.128151	20
41	.776259	2.83	.904147	1.57	.872112	4.39	.127888	19
42	.776429	2.82	.904053	1.57	.872376	4.39	.127624	18
43	.776598	2.82	.903959	1.57	.872640	4.39	.127360	17
44	.776768	2.82	.903864	1.57	.872903	4.39	.127097	16
45	.776937	2.82	.903770	1.57	.873167	4.39	.126833	15
46	.777106	2.82	.903676	1.57	.873430	4.39	.126570	14
47	.777275	2.82	.903581	1.57	.873694	4.39	.126306	13
48	.777444	2.81	.903487	1.58	.873957	4.39	.126043	12
49	.777613	2.81	.903392	1.58	.874220	4.39	.125780	11
50	9.777781	2.81	9.903298	1.58	9.874484	4.39	0.125516	10
51	.777950	2.81	.903203	1.58	.874747	4.39	.125253	9
52	.778119	2.81	.903108	1.58	.875010	4.39	.124990	8
53	.778287	2.81	.903014	1.58	.875273	4.39	.124727	7
54	.778455	2.80	.902919	1.58	.875537	4.38	.124463	6
55	.778624	2.80	.902824	1.58	.875800	4.38	.124200	5
56	.778792	2.80	.902729	1.58	.876063	4.38	.123937	4
57	.778960	2.80	.902634	1.58	.876326	4.38	.123674	3
58	.779128	2.80	.902539	1.59	.876589	4.38	.123411	2
59	.779295	2.79	.902444	1.59	.876852	4.38	.123148	1
60	.779463		.902349		.877114		.122886	0
M.	Cosine.	D. 1″.	Sine.	D. 1″.	Cotang.	D. 1″.	Tang.	M.

126° | 53°

37° | 142°

M.	Sine.	D. 1″.	Cosine.	D. 1″.	Tang.	D. 1″.	Cotang.	M.
0	9.779463	2.79	9.902349	1.59	9.877114	4.38	0.122886	60
1	.779631	2.79	.902253	1.59	.877377	4.38	.122623	59
2	.779798	2.79	.902158	1.59	.877640	4.38	.122360	58
3	.779966	2.79	.902063	1.59	.877903	4.38	.122097	57
4	.780133	2.79	.901967	1.59	.878165	4.38	.121835	56
5	.780300	2.78	.901872	1.59	.878428	4.38	.121572	55
6	.780467	2.78	.901776	1.59	.878691	4.38	.121309	54
7	.780634	2.78	.901681	1.59	.878953	4.38	.121047	53
8	.780801	2.78	.901585	1.59	.879216	4.37	.120784	52
9	.780968	2.78	.901490	1.60	.879478	4.37	.120522	51
10	9.781134	2.78	9.901394	1.60	9.879741	4.37	0.120259	50
11	.781301	2.77	.901298	1.60	.880003	4.37	.119997	49
12	.781468	2.77	.901202	1.60	.880265	4.37	.119735	48
13	.781634	2.77	.901106	1.60	.880528	4.37	.119472	47
14	.781800	2.77	.901010	1.60	.880790	4.37	.119210	46
15	.781966	2.77	.900914	1.60	.881052	4.37	.118948	45
16	.782132	2.77	.900818	1.60	.881314	4.37	.118686	44
17	.782298	2.76	.900722	1.60	.881577	4.37	.118423	43
18	.782464	2.76	.900626	1.60	.881839	4.37	.118161	42
19	.782630	2.76	.900529	1.61	.882101	4.37	.117899	41
20	9.782796	2.76	9.900433	1.61	9.882363	4.37	0.117637	40
21	.782961	2.76	.900337	1.61	.882625	4.37	.117375	39
22	.783127	2.76	.900240	1.61	.882887	4.36	.117113	38
23	.783292	2.75	.900144	1.61	.883148	4.36	.116852	37
24	.783458	2.75	.900047	1.61	.883410	4.36	.116590	36
25	.783623	2.75	.899951	1.61	.883672	4.36	.116328	35
26	.783788	2.75	.899854	1.61	.883934	4.36	.116066	34
27	.783953	2.75	.899757	1.61	.884196	4.36	.115804	33
28	.784118	2.75	.899660	1.61	.884457	4.36	.115543	32
29	.784282	2.74	.899564	1.62	.884719	4.36	.115281	31
30	9.784447	2.74	9.899467	1.62	9.884980	4.36	0.115020	30
31	.784612	2.74	.899370	1.62	.885242	4.36	.114758	29
32	.784776	2.74	.899273	1.62	.885504	4.36	.114496	28
33	.784941	2.74	.899176	1.62	.885765	4.36	.114235	27
34	.785105	2.74	.899078	1.62	.886026	4.36	.113974	26
35	.785269	2.73	.898981	1.62	.886288	4.36	.113712	25
36	.785433	2.73	.898884	1.62	.886549	4.36	.113451	24
37	.785597	2.73	.898787	1.62	.886811	4.35	.113189	23
38	.785761	2.73	.898689	1.62	.887072	4.35	.112928	22
39	.785925	2.73	.898592	1.62	.887333	4.35	.112667	21
40	9.786089	2.73	9.898494	1.63	9.887594	4.35	0.112406	20
41	.786252	2.73	.898397	1.63	.887855	4.35	.112145	19
42	.786416	2.72	.898299	1.63	.888116	4.35	.111884	18
43	.786579	2.72	.898202	1.63	.888378	4.35	.111622	17
44	.786742	2.72	.898104	1.63	.888639	4.35	.111361	16
45	.786906	2.72	.898006	1.63	.888900	4.35	.111100	15
46	.787069	2.72	.897908	1.63	.889161	4.35	.110839	14
47	.787232	2.72	.897810	1.63	.889421	4.35	.110579	13
48	.787395	2.71	.897712	1.63	.889682	4.35	.110318	12
49	.787557	2.71	.897614	1.63	.889943	4.35	.110057	11
50	9.787720	2.71	9.897516	1.64	9.890204	4.35	0.109796	10
51	.787883	2.71	.897418	1.64	.890465	4.35	.109535	9
52	.788045	2.71	.897320	1.64	.890725	4.34	.109275	8
53	.788208	2.71	.897222	1.64	.890986	4.34	.109014	7
54	.788370	2.70	.897123	1.64	.891247	4.34	.108753	6
55	.788532	2.70	.897025	1.64	.891507	4.34	.108493	5
56	.788694	2.70	.896926	1.64	.891768	4.34	.108232	4
57	.788856	2.70	.896828	1.64	.892028	4.34	.107972	3
58	.789018	2.70	.896729	1.64	.892289	4.34	.107711	2
59	.789180	2.70	.896631	1.64	.892549	4.34	.107451	1
60	.789342		.896532		.892810		.107190	0
M.	Cosine.	D. 1″.	Sine.	D. 1″.	Cotang.	D. 1″.	Tang.	M.

127° | 52°

M.	Sine.	D. 1″.	Cosine.	D. 1″.	Tang.	D. 1″.	Cotang.	M
0	9.789342	2.69	9.896532	1.65	9.892810	4.34	0.107190	60
1	.789504	2.69	.896433	1.65	.893070	4.34	.106930	59
2	.789665	2.69	.896335	1.65	.893331	4.34	.106669	58
3	.789827	2.69	.896236	1.65	.893591	4.34	.106409	57
4	.789988	2.69	.896137	1.65	.893851	4.34	.106149	56
5	.790149	2.69	.896038	1.65	.894111	4.34	.105889	55
6	.790310	2.68	.895939	1.65	.894372	4.34	.105628	54
7	.790471	2.68	.895840	1.65	.894632	4.34	.105368	53
8	.790632	2.68	.895741	1.65	.894892	4.33	.105108	52
9	.790793	2.68	.895641	1.65	.895152	4.33	.104848	51
10	9.790954	2.68	9.895542	1.66	9.895412	4.33	0.104588	50
11	.791115	2.68	.895443	1.66	.895672	4.33	.104328	49
12	.791275	2.67	.895343	1.66	.895932	4.33	.104068	48
13	.791436	2.67	.895244	1.66	.896192	4.33	.103808	47
14	.791596	2.67	.895145	1.66	.896452	4.33	.103548	46
15	.791757	2.67	.895045	1.66	.896712	4.33	.103288	45
16	.791917	2.67	.894945	1.66	.896971	4.33	.103029	44
17	.792077	2.67	.894846	1.66	.897231	4.33	.102769	43
18	.792237	2.67	.894746	1.66	.897491	4.33	.102509	42
19	.792397	2.66	.894646	1.66	.897751	4.33	.102249	41
20	9.792557	2.66	9.894546	1.67	9.898010	4.33	0.101990	40
21	.792716	2.66	.894446	1.67	.898270	4.33	.101730	39
22	.792876	2.66	.894346	1.67	.898530	4.33	.101470	38
23	.793035	2.66	.894246	1.67	.898789	4.33	.101211	37
24	.793195	2.66	.894146	1.67	.899049	4.33	.100951	36
25	.793354	2.65	.894046	1.67	.899308	4.32	.100692	35
26	.793514	2.65	.893946	1.67	.899568	4.32	.100432	34
27	.793673	2.65	.893846	1.67	.899827	4.32	.100173	33
28	.793832	2.65	.893745	1.67	.900087	4.32	.099913	32
29	.793991	2.65	.893645	1.67	.900346	4.32	.099654	31
30	9.794150	2.65	9.893544	1.68	9.900605	4.32	0.099395	30
31	.794308	2.64	.893444	1.68	.900864	4.32	.099136	29
32	.794467	2.64	.893343	1.68	.901124	4.32	.098876	28
33	.794626	2.64	.893243	1.68	.901383	4.32	.098617	27
34	.794784	2.64	.893142	1.68	.901642	4.32	.098358	26
35	.794942	2.64	.893041	1.68	.901901	4.32	.098099	25
36	.795101	2.64	.892940	1.68	.902160	4.32	.097840	24
37	.795259	2.64	.892839	1.68	.902420	4.32	.097580	23
38	.795417	2.63	.892739	1.68	.902679	4.32	.097321	22
39	.795575	2.63	.892638	1.68	.902938	4.32	.097062	21
40	9.795733	2.63	9.892536	1.69	9.903197	4.32	0.096803	20
41	.795891	2.63	.892435	1.69	.903456	4.32	.096544	19
42	.796049	2.63	.892334	1.69	.903714	4.31	.096286	18
43	.796206	2.63	.892233	1.69	.903973	4.31	.096027	17
44	.796364	2.62	.892132	1.69	.904232	4.31	.095768	16
45	.796521	2.62	.892030	1.69	.904491	4.31	.095509	15
46	.796679	2.62	.891929	1.69	.904750	4.31	.095250	14
47	.796836	2.62	.891827	1.69	.905008	4.31	.094992	13
48	.796993	2.62	.891726	1.69	.905267	4.31	.094733	12
49	.797150	2.61	.891624	1.69	.905526	4.31	.094474	11
50	9.797307	2.61	9.891523	1.70	9.905785	4.31	0.094215	10
51	.797464	2.61	.891421	1.70	.906043	4.31	.093957	9
52	.797621	2.61	.891319	1.70	.906302	4.31	.093698	8
53	.797777	2.61	.891217	1.70	.906560	4.31	.093440	7
54	.797934	2.61	.891115	1.70	.906819	4.31	.093181	6
55	.798091	2.61	.891013	1.70	.907077	4.31	.092923	5
56	.798247	2.61	.890911	1.70	.907336	4.31	.092664	4
57	.798403	2.60	.890809	1.70	.907594	4.31	.092406	3
58	.798560	2.60	.890707	1.70	.907853	4.31	.092147	2
59	.798716	2.60	.890605	1.70	.908111	4.31	.091889	1
60	.798872		.890503		.908369		.091631	0
M.	Cosine.	D. 1″.	Sine.	D. 1″.	Cotang.	D. 1″.	Tang.	M.

M.	Sine.	D. 1″.	Cosine.	D. 1″.	Tang.	D. 1″.	Cotang.	M.
0	9.798872	2.60	9.890503	1.71	9.908369	4.30	0.091631	60
1	.799028	2.60	.890400	1.71	.908628	4.30	.091372	59
2	.799184	2.60	.890298	1.71	.908886	4.30	.091114	58
3	.799339	2.59	.890195	1.71	.909144	4.30	.090856	57
4	.799495	2.59	.890093	1.71	.909402	4.30	.090598	56
5	.799651	2.59	.889990	1.71	.909660	4.30	.090340	55
6	.799806	2.59	.889888	1.71	.909918	4.30	.090082	54
7	.799962	2.59	.889785	1.71	.910177	4.30	.089823	53
8	.800117	2.59	.889682	1.71	.910435	4.30	.089565	52
9	.800272	2.59	.889579	1.71	.910693	4.30	.089307	51
10	9.800427	2.58	9.889477	1.72	9.910951	4.30	0.089049	50
11	.800582	2.58	.889374	1.72	.911209	4.30	.088791	49
12	.800737	2.58	.889271	1.72	.911467	4.30	.088533	48
13	.800892	2.58	.889168	1.72	.911725	4.30	.088275	47
14	.801047	2.58	.889064	1.72	.911982	4.30	.088018	46
15	.801201	2.58	.888961	1.72	.912240	4.30	.087760	45
16	.801356	2.57	.888858	1.72	.912498	4.30	.087502	44
17	.801511	2.57	.888755	1.72	.912756	4.30	.087244	43
18	.801665	2.57	.888651	1.72	.913014	4.30	.086986	42
19	.801819	2.57	.888548	1.72	.913271	4.30	.086729	41
20	9.801973	2.57	9.888444	1.73	9.913529	4.29	0.086471	40
21	.802128	2.57	.888341	1.73	.913787	4.29	.086213	39
22	.802282	2.57	.888237	1.73	.914044	4.29	.085956	38
23	.802436	2.56	.888134	1.73	.914302	4.29	.085698	37
24	.802589	2.56	.888030	1.73	.914560	4.29	.085440	36
25	.802743	2.56	.887926	1.73	.914817	4.29	.085183	35
26	.802897	2.56	.887822	1.73	.915075	4.29	.084925	34
27	.803050	2.56	.887718	1.73	.915332	4.29	.084668	33
28	.803204	2.56	.887614	1.73	.915590	4.29	.084410	32
29	.803357	2.55	.887510	1.74	.915847	4.29	.084153	31
30	9.803511	2.55	9.887406	1.74	9.916104	4.29	0.083896	30
31	.803664	2.55	.887302	1.74	.916362	4.29	.083638	29
32	.803817	2.55	.887198	1.74	.916619	4.29	.083381	28
33	.803970	2.55	.887093	1.74	.916877	4.29	.083123	27
34	.804123	2.55	.886989	1.74	.917134	4.29	.082866	26
35	.804276	2.55	.886885	1.74	.917391	4.29	.082609	25
36	.804428	2.54	.886780	1.74	.917648	4.29	.082352	24
37	.804581	2.54	.886676	1.74	.917906	4.29	.082094	23
38	.804734	2.54	.886571	1.74	.918163	4.29	.081837	22
39	.804886	2.54	.886466	1.75	.918420	4.29	.081580	21
40	9.805039	2.54	9.886362	1.75	9.918677	4.28	0.081323	20
41	.805191	2.54	.886257	1.75	.918934	4.28	.081066	19
42	.805343	2.54	.886152	1.75	.919191	4.28	.080809	18
43	.805495	2.53	.886047	1.75	.919448	4.28	.080552	17
44	.805647	2.53	.885942	1.75	.919705	4.28	.080295	16
45	.805799	2.53	.885837	1.75	.919962	4.28	.080038	15
46	.805951	2.53	.885732	1.75	.920219	4.28	.079781	14
47	.806103	2.53	.885627	1.75	.920476	4.28	.079524	13
48	.806254	2.53	.885522	1.75	.920733	4.28	.079267	12
49	.806406	2.52	.885416	1.76	.920990	4.28	.079010	11
50	9.806557	2.52	9.885311	1.76	9.921247	4.28	0.078753	10
51	.806709	2.52	.885205	1.76	.921503	4.28	.078497	9
52	.806860	2.52	.885100	1.76	.921760	4.28	.078240	8
53	.807011	2.52	.884994	1.76	.922017	4.28	.077983	7
54	.807163	2.52	.884889	1.76	.922274	4.28	.077726	6
55	.807314	2.52	.884783	1.76	.922530	4.28	.077470	5
56	.807465	2.51	.884677	1.76	.922787	4.28	.077213	4
57	.807615	2.51	.884572	1.76	.923044	4.28	.076956	3
58	.807766	2.51	.884466	1.77	.923300	4.28	.076700	2
59	.807917	2.51	.884360	1.77	.923557	4.28	.076443	1
60	.808067		.884254		.923814		.076186	0
M.	Cosine.	D. 1″.	Sine.	D. 1″.	Cotang.	D. 1″.	Tang.	M.

40° | 139°

M.	Sine.	D. 1″.	Cosine.	D. 1″.	Tang.	D. 1″.	Cotang.	M.
0	9.808067	2.51	9.884254	1.77	9.923814	4.28	0.076186	60
1	.808218	2.51	.884148	1.77	.924070	4.28	.075930	59
2	.808368	2.51	.884042	1.77	.924327	4.27	.075673	58
3	.808519	2.50	.883936	1.77	.924583	4.27	.075417	57
4	.808669	2.50	.883829	1.77	.924840	4.27	.075160	56
5	.808819	2.50	.883723	1.77	.925096	4.27	.074904	55
6	.808969	2.50	.883617	1.77	.925352	4.27	.074648	54
7	.809119	2.50	.883510	1.77	.925609	4.27	.074391	53
8	.809269	2.50	.883404	1.78	.925865	4.27	.074135	52
9	.809419	2.50	.883297	1.78	.926122	4.27	.073878	51
10	9.809569	2.49	9.883191	1.78	9.926378	4.27	0.073622	50
11	.809718	2.49	.883084	1.78	.926634	4.27	.073366	49
12	.809868	2.49	.882977	1.78	.926890	4.27	.073110	48
13	.810017	2.49	.882871	1.78	.927147	4.27	.072853	47
14	.810167	2.49	.882764	1.78	.927403	4.27	.072597	46
15	.810316	2.49	.882657	1.78	.927659	4.27	.072341	45
16	.810465	2.48	.882550	1.78	.927915	4.27	.072085	44
17	.810614	2.48	.882443	1.79	.928171	4.27	.071829	43
18	.810763	2.48	.882336	1.79	.928427	4.27	.071573	42
19	.810912	2.48	.882229	1.79	.928684	4.27	.071316	41
20	9.811061	2.48	9.882121	1.79	9.928940	4.27	0.071060	40
21	.811210	2.48	.882014	1.79	.929196	4.27	.070804	39
22	.811358	2.48	.881907	1.79	.929452	4.27	.070548	38
23	.811507	2.47	.881799	1.79	.929708	4.27	.070292	37
24	.811655	2.47	.881692	1.79	.929964	4.27	.070036	36
25	.811804	2.47	.881584	1.79	.930220	4.27	.069780	35
26	.811952	2.47	.881477	1.79	.930475	4.26	.069525	34
27	.812100	2.47	.881369	1.80	.930731	4.26	.069269	33
28	.812248	2.47	.881261	1.80	.930987	4.26	.069013	32
29	.812396	2.47	.881153	1.80	.931243	4.26	.068757	31
30	9.812544	2.46	9.881046	1.80	9.931499	4.26	0.068501	30
31	.812692	2.46	.880938	1.80	.931755	4.26	.068245	29
32	.812840	2.46	.880830	1.80	.932010	4.26	.067990	28
33	.812988	2.46	.880722	1.80	.932266	4.26	.067734	27
34	.813135	2.46	.880613	1.80	.932522	4.26	.067478	26
35	.813283	2.46	.880505	1.80	.932778	4.26	.067222	25
36	.813430	2.46	.880397	1.81	.933033	4.26	.066967	24
37	.813578	2.45	.880289	1.81	.933289	4.26	.066711	23
38	.813725	2.45	.880180	1.81	.933545	4.26	.066455	22
39	.813872	2.45	.880072	1.81	.933800	4.26	.066200	21
40	9.814019	2.45	9.879963	1.81	9.934056	4.26	0.065944	20
41	.814166	2.45	.879855	1.81	.934311	4.26	.065689	19
42	.814313	2.45	.879746	1.81	.934567	4.26	.065433	18
43	.814460	2.45	.879637	1.81	.934822	4.26	.065178	17
44	.814607	2.44	.879529	1.81	.935078	4.26	.064922	16
45	.814753	2.44	.879420	1.81	.935333	4.26	.064667	15
46	.814900	2.44	.879311	1.82	.935589	4.26	.064411	14
47	.815046	2.44	.879202	1.82	.935844	4.26	.064156	13
48	.815193	2.44	.879093	1.82	.936100	4.26	.063900	12
49	.815339	2.44	.878984	1.82	.936355	4.26	.063645	11
50	9.815485	2.44	9.878875	1.82	9.936611	4.26	0.063389	10
51	.815632	2.43	.878766	1.82	.936866	4.26	.063134	9
52	.815778	2.43	.878656	1.82	.937121	4.26	.062879	8
53	.815924	2.43	.878547	1.82	.937377	4.25	.062623	7
54	.816069	2.43	.878438	1.82	.937632	4.25	.062368	6
55	.816215	2.43	.878328	1.83	.937887	4.25	.062113	5
56	.816361	2.43	.878219	1.83	.938142	4.25	.061858	4
57	.816507	2.43	.878109	1.83	.938398	4.25	.061602	3
58	.816652	2.42	.877999	1.83	.938653	4.25	.061347	2
59	.816798	2.42	.877890	1.83	.938908	4.25	.061092	1
60	.816943		.877780		.939163		.060837	0
M.	Cosine.	D. 1″.	Sine.	D. 1″.	Cotang.	D. 1″.	Tang.	M.

130° | 49°

41° 138°

M.	Sine.	D. 1″.	Cosine.	D. 1″.	Tang.	D. 1″.	Cotang.	M.
0	9.816943	2.42	9.877780	1.83	9.939163	4.25	0.060837	60
1	.817088	2.42	.877670	1.83	.939418	4.25	.060582	59
2	.817233	2.42	.877560	1.83	.939673	4.25	.060327	58
3	.817379	2.42	.877450	1.83	.939928	4.25	.060072	57
4	.817524	2.42	.877340	1.84	.940183	4.25	.059817	56
5	.817668	2.41	.877230	1.84	.940439	4.25	.059561	55
6	.817813	2.41	.877120	1.84	.940694	4.25	.059306	54
7	.817958	2.41	.877010	1.84	.940949	4.25	.059051	53
8	.818103	2.41	.876899	1.84	.941204	4.25	.058796	52
9	.818247	2.41	.876789	1.84	.941459	4.25	.058541	51
10	9.818392	2.41	9.876678	1.84	9.941713	4.25	0.058287	50
11	.818536	2.41	.876568	1.84	.941968	4.25	.058032	49
12	.818681	2.40	.876457	1.84	.942223	4.25	.057777	48
13	.818825	2.40	.876347	1.84	.942478	4.25	.057522	47
14	.818969	2.40	.876236	1.85	.942733	4.25	.057267	46
15	.819113	2.40	.876125	1.85	.942988	4.25	.057012	45
16	.819257	2.40	.876014	1.85	.943243	4.25	.056757	44
17	.819401	2.40	.875904	1.85	.943498	4.25	.056502	43
18	.819545	2.40	.875793	1.85	.943752	4.25	.056248	42
19	.819689	2.39	.875682	1.85	.944007	4.25	.055993	41
20	9.819832	2.39	9.875571	1.85	9.944262	4.25	0.055738	40
21	.819976	2.39	.875459	1.85	.944517	4.25	.055483	39
22	.820120	2.39	.875348	1.85	.944771	4.24	.055229	38
23	.820263	2.39	.875237	1.86	.945026	4.24	.054974	37
24	.820406	2.39	.875126	1.86	.945281	4.24	.054719	36
25	.820550	2.39	.875014	1.86	.945535	4.24	.054465	35
26	.820693	2.38	.874903	1.86	.945790	4.24	.054210	34
27	.820836	2.38	.874791	1.86	.946045	4.24	.053955	33
28	.820979	2.38	.874680	1.86	.946299	4.24	.053701	32
29	.821122	2.38	.874568	1.86	.946554	4.24	.053446	31
30	9.821265	2.38	9.874456	1.86	9.946808	4.24	0.053192	30
31	.821407	2.38	.874344	1.86	.947063	4.24	.052937	29
32	.821550	2.38	.874232	1.87	.947318	4.24	.052682	28
33	.821693	2.37	.874121	1.87	.947572	4.24	.052428	27
34	.821835	2.37	.874009	1.87	.947827	4.24	.052173	26
35	.821977	2.37	.873896	1.87	.948081	4.24	.051919	25
36	.822120	2.37	.873784	1.87	.948335	4.24	.051665	24
37	.822262	2.37	.873672	1.87	.948590	4.24	.051410	23
38	.822404	2.37	.873560	1.87	.948844	4.24	.051156	22
39	.822546	2.37	.873448	1.87	.949099	4.24	.050901	21
40	9.822688	2.37	9.873335	1.87	9.949353	4.24	0.050647	20
41	.822830	2.36	.873223	1.88	.949608	4.24	.050392	19
42	.822972	2.36	.873110	1.88	.949862	4.24	.050138	18
43	.823114	2.36	.872998	1.88	.950116	4.24	.049884	17
44	.823255	2.36	.872885	1.88	.950371	4.24	.049629	16
45	.823397	2.36	.872772	1.88	.950625	4.24	.049375	15
46	.823539	2.36	.872659	1.88	.950879	4.24	.049121	14
47	.823680	2.36	.872547	1.88	.951133	4.24	.048867	13
48	.823821	2.35	.872434	1.88	.951388	4.24	.048612	12
49	.823963	2.35	.872321	1.88	.951642	4.24	.048358	11
50	9.824104	2.35	9.872208	1.89	9.951896	4.24	0.048104	10
51	.824245	2.35	.872095	1.89	.952150	4.24	.047850	9
52	.824386	2.35	.871981	1.89	.952405	4.24	.047595	8
53	.824527	2.35	.871868	1.89	.952659	4.24	.047341	7
54	.824668	2.35	.871755	1.89	.952913	4.24	.047087	6
55	.824808	2.34	.871641	1.89	.953167	4.24	.046833	5
56	.824949	2.34	.871528	1.89	.953421	4.24	.046579	4
57	.825090	2.34	.871414	1.89	.953675	4.23	.046325	3
58	.825230	2.34	.871301	1.89	.953929	4.23	.046071	2
59	.825371	2.34	.871187	1.90	.954183	4.23	.045817	1
60	.825511		.871073		.954437		.045563	0
M.	Cosine.	D. 1″.	Sine.	D. 1″.	Cotang.	D. 1″.	Tang.	M.

131° 48°

42° **137°**

M.	Sine.	D. 1″.	Cosine.	D. 1″.	Tang.	D. 1″.	Cotang.	M.
0	9.825511	2.34	9.871073	1.90	9.954437	4.23	0.045563	60
1	.825651	2.34	.870960	1.90	.954691	4.23	.045309	59
2	.825791	2.33	.870846	1.90	.954946	4.23	.045054	58
3	.825931	2.33	.870732	1.90	.955200	4.23	.044800	57
4	.826071	2.33	.870618	1.90	.955454	4.23	.044546	56
5	.826211	2.33	.870504	1.90	.955708	4.23	.044292	55
6	.826351	2.33	.870390	1.90	.955961	4.23	.044039	54
7	.826491	2.33	.870276	1.90	.956215	4.23	.043785	53
8	.826631	2.33	.870161	1.91	.956469	4.23	.043531	52
9	.826770	2.33	.870047	1.91	.956723	4.23	.043277	51
10	9.826910	2.32	9.869933	1.91	9.956977	4.23	0.043023	50
11	.827049	2.32	.869818	1.91	.957231	4.23	.042769	49
12	.827189	2.32	.869704	1.91	.957485	4.23	.042515	48
13	.827328	2.32	.869589	1.91	.957739	4.23	.042261	47
14	.827467	2.32	.869474	1.91	.957993	4.23	.042007	46
15	.827606	2.32	.869360	1.91	.958247	4.23	.041753	45
16	.827745	2.32	.869245	1.91	.958500	4.23	.041500	44
17	.827884	2.31	.869130	1.92	.958754	4.23	.041246	43
18	.828023	2.31	.869015	1.92	.959008	4.23	.040992	42
19	.828162	2.31	.868900	1.92	.959262	4.23	.040738	41
20	9.828301	2.31	9.868785	1.92	9.959516	4.23	0.040484	40
21	.828439	2.31	.868670	1.92	.959769	4.23	.040231	39
22	.828578	2.31	.868555	1.92	.960023	4.23	.039977	38
23	.828716	2.31	.868440	1.92	.960277	4.23	.039723	37
24	.828855	2.31	.868324	1.92	.960530	4.23	.039470	36
25	.828993	2.30	.868209	1.92	.960784	4.23	.039216	35
26	.829131	2.30	.868093	1.93	.961038	4.23	.038962	34
27	.829269	2.30	.867978	1.93	.961292	4.23	.038708	33
28	.829407	2.30	.867862	1.93	.961545	4.23	.038455	32
29	.829545	2.30	.867747	1.93	.961799	4.23	.038201	31
30	9.829683	2.30	9.867631	1.93	9.962052	4.23	0.037948	30
31	.829821	2.30	.867515	1.93	.962306	4.23	.037694	29
32	.829959	2.29	.867399	1.93	.962560	4.23	.037440	28
33	.830097	2.29	.867283	1.93	.962813	4.23	.037187	27
34	.830234	2.29	.867167	1.93	.963067	4.23	.036933	26
35	.830372	2.29	.867051	1.94	.963320	4.23	.036680	25
36	.830509	2.29	.866935	1.94	.963574	4.23	.036426	24
37	.830646	2.29	.866819	1.94	.963828	4.23	.036172	23
38	.830784	2.29	.866703	1.94	.964081	4.23	.035919	22
39	.830921	2.29	.866586	1.94	.964335	4.23	.035665	21
40	9.831058	2.28	9.866470	1.94	9.964588	4.22	0.035412	20
41	.831195	2.28	.866353	1.94	.964842	4.22	.035158	19
42	.831332	2.28	.866237	1.94	.965095	4.22	.034905	18
43	.831469	2.28	.866120	1.94	.965349	4.22	.034651	17
44	.831606	2.28	.866004	1.95	.965602	4.22	.034398	16
45	.831742	2.28	.865887	1.95	.965855	4.22	.034145	15
46	.831879	2.28	.865770	1.95	.966109	4.22	.033891	14
47	.832015	2.27	.865653	1.95	.966362	4.22	.033638	13
48	.832152	2.27	.865536	1.95	.966616	4.22	.033384	12
49	.832288	2.27	.865419	1.95	.966869	4.22	.033131	11
50	9.832425	2.27	9.865302	1.95	9.967123	4.22	0.032877	10
51	.832561	2.27	.865185	1.95	.967376	4.22	.032624	9
52	.832697	2.27	.865068	1.95	.967629	4.22	.032371	8
53	.832833	2.27	.864950	1.96	.967883	4.22	.032117	7
54	.832969	2.27	.864833	1.96	.968136	4.22	.031864	6
55	.833105	2.26	.864716	1.96	.968389	4.22	.031611	5
56	.833241	2.26	.864598	1.96	.968643	4.22	.031357	4
57	.833377	2.26	.864481	1.96	.968896	4.22	.031104	3
58	.833512	2.26	.864363	1.96	.969149	4.22	.030851	2
59	.833648	2.26	.864245	1.96	.969403	4.22	.030597	1
60	.833783		.864127		.969656		.030344	0
M.	Cosine.	D. 1″.	Sine.	D. 1″.	Cotang.	D. 1″.	Tang.	M.

132° **47°**

M.	Sine.	D. 1''.	Cosine.	D. 1''.	Tang.	D. 1''.	Cotang.	M.
0	9.833783	2.26	9.864127	1.96	9.969656	4.22	0.030344	60
1	.833919	2.26	.864010	1.97	.969909	4.22	.030091	59
2	.834054	2.25	.863892	1.97	.970162	4.22	.029838	58
3	.834189	2.25	.863774	1.97	.970416	4.22	.029584	57
4	.834325	2.25	.863656	1.97	.970669	4.22	.029331	56
5	.834460	2.25	.863538	1.97	.970922	4.22	.029078	55
6	.834595	2.25	.863419	1.97	.971175	4.22	.028825	54
7	.834730	2.25	.863301	1.97	.971429	4.22	.028571	53
8	.834865	2.25	.863183	1.97	.971682	4.22	.028318	52
9	.834999	2.25	.863064	1.97	.971935	4.22	.028065	51
10	9.835134	2.24	9.862946	1.98	9.972188	4.22	0.027812	50
11	.835269	2.24	.862827	1.98	.972441	4.22	.027559	49
12	.835403	2.24	.862709	1.98	.972695	4.22	.027305	48
13	.835538	2.24	.862590	1.98	.972948	4.22	.027052	47
14	.835672	2.24	.862471	1.98	.973201	4.22	.026799	46
15	.835807	2.24	.862353	1.98	.973454	4.22	.026546	45
16	.835941	2.24	.862234	1.98	.973707	4.22	.026293	44
17	.836075	2.23	.862115	1.98	.973960	4.22	.026040	43
18	.836209	2.23	.861996	1.98	.974213	4.22	.025787	42
19	.836343	2.23	.861877	1.99	.974466	4.22	.025534	41
20	9.836477	2.23	9.861758	1.99	9.974720	4.22	0.025280	40
21	.836611	2.23	.861638	1.99	.974973	4.22	.025027	39
22	.836745	2.23	.861519	1.99	.975226	4.22	.024774	38
23	.836878	2.23	.861400	1.99	.975479	4.22	.024521	37
24	.837012	2.23	.861280	1.99	.975732	4.22	.024268	36
25	.837146	2.22	.861161	1.99	.975985	4.22	.024015	35
26	.837279	2.22	.861041	1.99	.976238	4.22	.023762	34
27	.837412	2.22	.860922	2.00	.976491	4.22	.023509	33
28	.837546	2.22	.860802	2.00	.976744	4.22	.023256	32
29	.837679	2.22	.860682	2.00	.976997	4.22	.023003	31
30	9.837812	2.22	9.860562	2.00	9.977250	4.22	0.022750	30
31	.837945	2.22	.860442	2.00	.977503	4.22	.022497	29
32	.838078	2.22	.860322	2.00	.977756	4.22	.022244	28
33	.838211	2.21	.860202	2.00	.978009	4.22	.021991	27
34	.838344	2.21	.860082	2.00	.978262	4.22	.021738	26
35	.838477	2.21	.859962	2.00	.978515	4.22	.021485	25
36	.838610	2.21	.859842	2.01	.978768	4.22	.021232	24
37	.838742	2.21	.859721	2.01	.979021	4.22	.020979	23
38	.838875	2.21	.859601	2.01	.979274	4.22	.020726	22
39	.839007	2.21	.859480	2.01	.979527	4.22	.020473	21
40	9.839140	2.21	9.859360	2.01	9.979780	4.22	0.020220	20
41	.839272	2.20	.859239	2.01	.980033	4.22	.019967	19
42	.839404	2.20	.859119	2.01	.980286	4.22	.019714	18
43	.839536	2.20	.858998	2.01	.980538	4.22	.019462	17
44	.839668	2.20	.858877	2.02	.980791	4.22	.019209	16
45	.839800	2.20	.858756	2.02	.981044	4.21	.018956	15
46	.839932	2.20	.858635	2.02	.981297	4.21	.018703	14
47	.840064	2.20	.858514	2.02	.981550	4.21	.018450	13
48	.840196	2.19	.858393	2.02	.981803	4.21	.018197	12
49	.840328	2.19	.858272	2.02	.982056	4.21	.017944	11
50	9.840459	2.19	9.858151	2.02	9.982309	4.21	0.017691	10
51	.840591	2.19	.858029	2.02	.982562	4.21	.017438	9
52	.840722	2.19	.857908	2.02	.982814	4.21	.017186	8
53	.840854	2.19	.857786	2.03	.983067	4.21	.016933	7
54	.840985	2.19	.857665	2.03	.983320	4.21	.016680	6
55	.841116	2.19	.857543	2.03	.983573	4.21	.016427	5
56	.841247	2.18	.857422	2.03	.983826	4.21	.016174	4
57	.841378	2.18	.857300	2.03	.984079	4.21	.015921	3
58	.841509	2.18	.857178	2.03	.984332	4.21	.015668	2
59	.841640	2.18	.857056	2.03	.984584	4.21	.015416	1
60	.841771		.856934		.984837		.015163	0
M.	Cosine.	D. 1''.	Sine.	D. 1''.	Cotang.	D. 1''.	Tang.	M.

44° | 135°

M.	Sine.	D 1″.	Cosine.	D. 1″.	Tang.	D. 1″.	Cotang.	M.
0	9.841771	2.18	9.856934	2.03	9.984837	4.21	0.015163	60
1	.841902	2.18	.856812	2.04	.985090	4.21	.014910	59
2	.842033	2.18	.856690	2.04	.985343	4.21	.014657	58
3	.842163	2.18	.856568	2.04	.985596	4.21	.014404	57
4	.842294	2.17	.856446	2.04	.985848	4.21	.014152	56
5	.842424	2.17	.856323	2.04	.986101	4.21	.013899	55
6	.842555	2.17	.856201	2.04	.986354	4.21	.013646	54
7	.842685	2.17	.856078	2.04	.986607	4.21	.013393	53
8	.842815	2.17	.855956	2.04	.986860	4.21	.013140	52
9	.842946	2.17	.855833	2.04	.987112	4.21	.012888	51
10	9.843076	2.17	9.855711	2.05	9.987365	4.21	0.012635	50
11	.843206	2.17	.855588	2.05	.987618	4.21	.012382	49
12	.843336	2.16	.855465	2.05	.987871	4.21	.012129	48
13	.843466	2.16	.855342	2.05	.988123	4.21	.011877	47
14	.843595	2.16	.855219	2.05	.988376	4.21	.011624	46
15	.843725	2.16	.855096	2.05	.988629	4.21	.011371	45
16	.843855	2.16	.854973	2.05	.988882	4.21	.011118	44
17	.843984	2.16	.854850	2.05	.989134	4.21	.010866	43
18	.844114	2.16	.854727	2.06	.989387	4.21	.010613	42
19	.844243	2.16	.854603	2.06	.989640	4.21	.010360	41
20	9.844372	2.15	9.854480	2.06	9.989893	4.21	0.010107	40
21	.844502	2.15	.854356	2.06	.990145	4.21	.009855	39
22	.844631	2.15	.854233	2.06	.990398	4.21	.009602	38
23	.844760	2.15	.854109	2.06	.990651	4.21	.009349	37
24	.844889	2.15	.853986	2.06	.990903	4.21	.009097	36
25	.845018	2.15	.853862	2.06	.991156	4.21	.008844	35
26	.845147	2.15	.853738	2.06	.991409	4.21	.008591	34
27	.845276	2.15	.853614	2.07	.991662	4.21	.008338	33
28	.845405	2.14	.853490	2.07	.991914	4.21	.008086	32
29	.845533	2.14	.853366	2.07	.992167	4.21	.007833	31
30	9.845662	2.14	9.853242	2.07	9.992420	4.21	0.007580	30
31	.845790	2.14	.853118	2.07	.992672	4.21	.007328	29
32	.845919	2.14	.852994	2.07	.992925	4.21	.007075	28
33	.846047	2.14	.852869	2.07	.993178	4.21	.006822	27
34	.846175	2.14	.852745	2.07	.993431	4.21	.006569	26
35	.846304	2.14	.852620	2.08	.993683	4.21	.006317	25
36	.846432	2.13	.852496	2.08	.993936	4.21	.006064	24
37	.846560	2.13	.852371	2.08	.994189	4.21	.005811	23
38	.846688	2.13	.852247	2.08	.994441	4.21	.005559	22
39	.846816	2.13	.852122	2.08	.994694	4.21	.005306	21
40	9.846944	2.13	9.851997	2.08	9.994947	4.21	0.005053	20
41	.847071	2.13	.851872	2.08	.995199	4.21	.004801	19
42	.847199	2.13	.851747	2.08	.995452	4.21	.004548	18
43	.847327	2.13	.851622	2.09	.995705	4.21	.004295	17
44	.847454	2.12	.851497	2.09	.995957	4.21	.004043	16
45	.847582	2.12	.851372	2.09	.996210	4.21	.003790	15
46	.847709	2.12	.851246	2.09	.996463	4.21	.003537	14
47	.847836	2.12	.851121	2.09	.996715	4.21	.003285	13
48	.847964	2.12	.850996	2.09	.996968	4.21	.003032	12
49	.848091	2.12	.850870	2.09	.997221	4.21	.002779	11
50	9.848218	2.12	9.850745	2.09	9.997473	4.21	0.002527	10
51	.848345	2.12	.850619	2.10	.997726	4.21	.002274	9
52	.848472	2.11	.850493	2.10	.997979	4.21	.002021	8
53	.848599	2.11	.850368	2.10	.998231	4.21	.001769	7
54	848726	2.11	.850242	2.10	.998484	4.21	.001516	6
55	.848852	2.11	.850116	2.10	.998737	4.21	.001263	5
56	.848979	2.11	.849990	2.10	.998989	4.21	.001011	4
57	.849106	2.11	.849864	2.10	.999242	4.21	.000758	3
58	.849232	2.11	.849738	2.10	.999495	4.21	.000505	2
59	.849359	2.11	.849611	2.11	.999747	4.21	.000253	1
60	.849485		.849485		0.000000		.000000	0
M.	Cosine.	D. 1″.	Sine.	D. 1″.	Cotang.	D. 1″.	Tang.	M.

134° | 45°

TABLE XIV.

NATURAL SINES AND COSINES

	0°		1°		2°		3°		4°		
M.	Sine.	Cosin.	Sine.	Cosin.	Sine.	Cosin.	Sine.	Cosin.	Sine.	Cosin.	M.
0	.00000	One.	.01745	.99985	.03490	.99939	.05234	.99863	.06976	.99756	60
1	.00029	One.	.01774	.99984	.03519	.99938	.05263	.99861	.07005	.99754	59
2	.00058	One.	.01803	.99984	.03548	.99937	.05292	.99860	.07034	.99752	58
3	.00087	One.	.01832	.99983	.03577	.99936	.05321	.99858	.07063	.99750	57
4	.00116	One.	.01862	.99983	.03606	.99935	.05350	.99857	.07092	.99748	56
5	.00145	One.	.01891	.99982	.03635	.99934	.05379	.99855	.07121	.99746	55
6	.00175	One.	.01920	.99982	.03664	.99933	.05408	.99854	.07150	.99744	54
7	.00204	One.	.01949	.99981	.03693	.99932	.05437	.99852	.07179	.99742	53
8	.00233	One.	.01978	.99980	.03723	.99931	.05466	.99851	.07208	.99740	52
9	.00262	One.	.02007	.99980	.03752	.99930	.05495	.99849	.07237	.99738	51
10	.00291	One.	.02036	.99979	.03781	.99929	.05524	.99847	.07266	.99736	50
11	.00320	.99999	.02065	.99979	.03810	.99927	.05553	.99846	.07295	.99734	49
12	.00349	.99999	.02094	.99978	.03839	.99926	.05582	.99844	.07324	.99731	48
13	.00378	.99999	.02123	.99977	.03868	.99925	.05611	.99842	.07353	.99729	47
14	.00407	.99999	.02152	.99977	.03897	.99924	.05640	.99841	.07382	.99727	46
15	.00436	.99999	.02181	.99976	.03926	.99923	.05669	.99839	.07411	.99725	45
16	.00465	.99999	.02211	.99976	.03955	.99922	.05698	.99838	.07440	.99723	44
17	.00495	.99999	.02240	.99975	.03984	.99921	.05727	.99836	.07469	.99721	43
18	.00524	.99999	.02269	.99974	.04013	.99919	.05756	.99834	.07498	.99719	42
19	.00553	.99998	.02298	.99974	.04042	.99918	.05785	.99833	.07527	.99716	41
20	.00582	.99998	.02327	.99973	.04071	.99917	.05814	.99831	.07556	.99714	40
21	.00611	.99998	.02356	.99972	.04100	.99916	.05844	.99829	.07585	.99712	39
22	.00640	.99998	.02385	.99972	.04129	.99915	.05873	.99827	.07614	.99710	38
23	.00669	.99998	.02414	.99971	.04159	.99913	.05902	.99826	.07643	.99708	37
24	.00698	.99998	.02443	.99970	.04188	.99912	.05931	.99824	.07672	.99705	36
25	.00727	.99997	.02472	.99969	.04217	.99911	.05960	.99822	.07701	.99703	35
26	.00756	.99997	.02501	.99969	.04246	.99910	.05989	.99821	.07730	.99701	34
27	.00785	.99997	.02530	.99968	.04275	.99909	.06018	.99819	.07759	.99699	33
28	.00814	.99997	.02560	.99967	.04304	.99907	.06047	.99817	.07788	.99696	32
29	.00844	.99996	.02589	.99966	.04333	.99906	.06076	.99815	.07817	.99694	31
30	.00873	.99996	.02618	.99966	.04362	.99905	.06105	.99813	.07846	.99692	30
31	.00902	.99996	.02647	.99965	.04391	.99904	.06134	.99812	.07875	.99689	29
32	.00931	.99996	.02676	.99964	.04420	.99902	.06163	.99810	.07904	.99687	28
33	.00960	.99995	.02705	.99963	.04449	.99901	.06192	.99808	.07933	.99685	27
34	.00989	.99995	.02734	.99963	.04478	.99900	.06221	.99806	.07962	.99683	26
35	.01018	.99995	.02763	.99962	.04507	.99898	.06250	.99804	.07991	.99680	25
36	.01047	.99995	.02792	.99961	.04536	.99897	.06279	.99803	.08020	.99678	24
37	.01076	.99994	.02821	.99960	.04565	.99896	.06308	.99801	.08049	.99676	23
38	.01105	.99994	.02850	.99959	.04594	.99894	.06337	.99799	.08078	.99673	22
39	.01134	.99994	.02879	.99959	.04623	.99893	.06366	.99797	.08107	.99671	21
40	.01164	.99993	.02908	.99958	.04653	.99892	.06395	.99795	.08136	.99668	20
41	.01193	.99993	.02938	.99957	.04682	.99890	.06424	.99793	.08165	.99666	19
42	.01222	.99993	.02967	.99956	.04711	.99889	.06453	.99792	.08194	.99664	18
43	.01251	.99992	.02996	.99955	.04740	.99888	.06482	.99790	.08223	.99661	17
44	.01280	.99992	.03025	.99954	.04769	.99886	.06511	.99788	.08252	.99659	16
45	.01309	.99991	.03054	.99953	.04798	.99885	.06540	.99786	.08281	.99657	15
46	.01338	.99991	.03083	.99952	.04827	.99883	.06569	.99784	.08310	.99654	14
47	.01367	.99991	.03112	.99952	.04856	.99882	.06598	.99782	.08339	.99652	13
48	.01396	.99990	.03141	.99951	.04885	.99881	.06627	.99780	.08368	.99649	12
49	.01425	.99990	.03170	.99950	.04914	.99879	.06656	.99778	.08397	.99647	11
50	.01454	.99989	.03199	.99949	.04943	.99878	.06685	.99776	.08426	.99644	10
51	.01483	.99989	.03228	.99948	.04972	.99876	.06714	.99774	.08455	.99642	9
52	.01513	.99989	.03257	.99947	.05001	.99875	.06743	.99772	.08484	.99639	8
53	.01542	.99988	.03286	.99946	.05030	.99873	.06773	.99770	.08513	.99637	7
54	.01571	.99988	.03316	.99945	.05059	.99872	.06802	.99768	.08542	.99635	6
55	.01600	.99987	.03345	.99944	.05088	.99870	.06831	.99766	.08571	.99632	5
56	.01629	.99987	.03374	.99943	.05117	.99869	.06860	.99764	.08600	.99630	4
57	.01658	.99986	.03403	.99942	.05146	.99867	.06889	.99762	.08629	.99627	3
58	.01687	.99986	.03432	.99941	.05175	.99866	.06918	.99760	.08658	.99625	2
59	.01716	.99985	.03461	.99940	.05205	.99864	.06947	.99758	.08687	.99622	1
60	.01745	.99985	.03490	.99939	.05234	.99863	.06976	.99756	.08716	.99619	0
M.	Cosin.	Sine.	Cosin.	Sine.	Cosin.	Sine.	Cosin.	Sine.	Cosin.	Sine.	M.
	89°		88°		87°		86°		85°		

	5°		6°		7°		8°		9°		
M.	Sine.	Cosin.	Sine.	Cosin.	Sine.	Cosin.	Sine.	Cosin.	Sine.	Cosin.	M.
0	.08716	.99619	.10453	.99452	.12187	.99255	.13917	.99027	.15643	.98769	60
1	.08745	.99617	.10482	.99449	.12216	.99251	.13946	.99023	.15672	.98764	59
2	.08774	.99614	.10511	.99446	.12245	.99248	.13975	.99019	.15701	.98760	58
3	.08803	.99612	.10540	.99443	.12274	.99244	.14004	.99015	.15730	.98755	57
4	.08831	.99609	.10569	.99440	.12302	.99240	.14033	.99011	.15758	.98751	56
5	.08860	.99607	.10597	.99437	.12331	.99237	.14061	.99006	.15787	.98746	55
6	.08889	.99604	.10626	.99434	.12360	.99233	.14090	.99002	.15816	.98741	54
7	.08918	.99602	.10655	.99431	.12389	.99230	.14119	.98998	.15845	.98737	53
8	.08947	.99599	.10684	.99428	.12418	.99226	.14148	.98994	.15873	.98732	52
9	.08976	.99596	.10713	.99424	.12447	.99222	.14177	.98990	.15902	.98728	51
10	.09005	.99594	.10742	.99421	.12476	.99219	.14205	.98986	.15931	.98723	50
11	.09034	.99591	.10771	.99418	.12504	.99215	.14234	.98982	.15959	.98718	49
12	.09063	.99588	.10800	.99415	.12533	.99211	.14263	.98978	.15988	.98714	48
13	.09092	.99586	.10829	.99412	.12562	.99208	.14292	.98973	.16017	.98709	47
14	.09121	.99583	.10858	.99409	.12591	.99204	.14320	.98969	.16046	.98704	46
15	.09150	.99580	.10887	.99406	.12620	.99200	.14349	.98965	.16074	.98700	45
16	.09179	.99578	.10916	.99402	.12649	.99197	.14378	.98961	.16103	.98695	44
17	.09208	.99575	.10945	.99399	.12678	.99193	.14407	.98957	.16132	.98690	43
18	.09237	.99572	.10973	.99396	.12706	.99189	.14436	.98953	.16160	.98686	42
19	.09266	.99570	.11002	.99393	.12735	.99186	.14464	.98948	.16189	.98681	41
20	.09295	.99567	.11031	.99390	.12764	.99182	.14493	.98944	.16218	.98676	40
21	.09324	.99564	.11060	.99386	.12793	.99178	.14522	.98940	.16246	.98671	39
22	.09353	.99562	.11089	.99383	.12822	.99175	.14551	.98936	.16275	.98667	38
23	.09382	.99559	.11118	.99380	.12851	.99171	.14580	.98931	.16304	.98662	37
24	.09411	.99556	.11147	.99377	.12880	.99167	.14608	.98927	.16333	.98657	36
25	.09440	.99553	.11176	.99374	.12908	.99163	.14637	.98923	.16361	.98652	35
26	.09469	.99551	.11205	.99370	.12937	.99160	.14666	.98919	.16390	.98648	34
27	.09498	.99548	.11234	.99367	.12966	.99156	.14695	.98914	.16419	.98643	33
28	.09527	.99545	.11263	.99364	.12995	.99152	.14723	.98910	.16447	.98638	32
29	.09556	.99542	.11291	.99360	.13024	.99148	.14752	.98906	.16476	.98633	31
30	.09585	.99540	.11320	.99357	.13053	.99144	.14781	.98902	.16505	.98629	30
31	.09614	.99537	.11349	.99354	.13081	.99141	.14810	.98897	.16533	.98624	29
32	.09642	.99534	.11378	.99351	.13110	.99137	.14838	.98893	.16562	.98619	28
33	.09671	.99531	.11407	.99347	.13139	.99133	.14867	.98889	.16591	.98614	27
34	.09700	.99528	.11436	.99344	.13168	.99129	.14896	.98884	.16620	.98609	26
35	.09729	.99526	.11465	.99341	.13197	.99125	.14925	.98880	.16648	.98604	25
36	.09758	.99523	.11494	.99337	.13226	.99122	.14954	.98876	.16677	.98600	24
37	.09787	.99520	.11523	.99334	.13254	.99118	.14982	.98871	.16706	.98595	23
38	.09816	.99517	.11552	.99331	.13283	.99114	.15011	.98867	.16734	.98590	22
39	.09845	.99514	.11580	.99327	.13312	.99110	.15040	.98863	.16763	.98585	21
40	.09874	.99511	.11609	.99324	.13341	.99106	.15069	.98858	.16792	.98580	20
41	.09903	.99508	.11638	.99320	.13370	.99102	.15097	.98854	.16820	.98575	19
42	.09932	.99506	.11667	.99317	.13399	.99098	.15126	.98849	.16849	.98570	18
43	.09961	.99503	.11696	.99314	.13427	.99094	.15155	.98845	.16878	.98565	17
44	.09990	.99500	.11725	.99310	.13456	.99091	.15184	.98841	.16906	.98561	16
45	.10019	.99497	.11754	.99307	.13485	.99087	.15212	.98836	.16935	.98556	15
46	.10048	.99494	.11783	.99303	.13514	.99083	.15241	.98832	.16964	.98551	14
47	.10077	.99491	.11812	.99300	.13543	.99079	.15270	.98827	.16992	.98546	13
48	.10106	.99488	.11840	.99297	.13572	.99075	.15299	.98823	.17021	.98541	12
49	.10135	.99485	.11869	.99293	.13600	.99071	.15327	.98818	.17050	.98536	11
50	.10164	.99482	.11898	.99290	.13629	.99067	.15356	.98814	.17078	.98531	10
51	.10192	.99179	.11927	.99286	.13658	.99063	.15385	.98809	.17107	.98526	9
52	.10221	.99476	.11956	.99283	.13687	.99059	.15414	.98805	.17136	.98521	8
53	.10250	.99473	.11985	.99279	.13716	.99055	.15442	.98800	.17164	.98516	7
54	.10279	.99470	.12014	.99276	.13744	.99051	.15471	.98796	.17193	.98511	6
55	.10308	.99467	.12043	.99272	.13773	.99047	15500	.98791	.17222	.98506	5
56	.10337	.99464	.12071	.99269	.13802	.99043	.15529	.98787	.17250	.98501	4
57	.10366	.99461	.12100	.99265	.13331	.99039	.15557	.98782	.17279	.98496	3
58	.10395	.99458	.12129	.99262	.13860	.99035	.15586	.98778	.17308	.98491	2
59	.10424	.99455	.12158	.99258	.13889	.99031	.15615	.98773	.17336	.98486	1
60	.10453	.99452	.12187	.99255	.13917	.99027	.15643	.98769	.17365	.98481	0
M.	Cosin.	Sine.	Cosin.	Sine.	Cosin.	Sine.	Cosin.	Sine.	Cosin.	Sine.	M.
	84°		83°		82°		81°		80°		

	10°		11°		12°		13°		14°		
M.	Sine.	Cosin.	Sine.	Cosin.	Sine.	Cosin.	Sine.	Cosin.	Sine.	Cosin.	M.
0	.17365	.98481	.19081	.98163	.20791	.97815	.22495	.97437	.24192	.97030	60
1	.17393	.98476	.19109	.98157	.20820	.97809	.22523	.97430	.24220	.97023	59
2	.17422	.98471	.19138	.98152	.20848	.97803	.22552	.97424	.24249	.97015	58
3	.17451	.98466	.19167	.98146	.20877	.97797	.22580	.97417	.24277	.97008	57
4	.17479	.98461	.19195	.98140	.20905	.97791	.22608	.97411	.24305	.97001	56
5	.17508	.98455	.19224	.98135	.20933	.97784	.22637	.97404	.24333	.96994	55
6	.17537	.98450	.19252	.98129	.20962	.97778	.22665	.97398	.24362	.96987	54
7	.17565	.98445	.19281	.98124	.20990	.97772	.22693	.97391	.24390	.96980	53
8	.17594	.98440	.19309	.98118	.21019	.97766	.22722	.97384	.24418	.96973	52
9	.17623	.98435	.19338	.98112	.21047	.97760	.22750	.97378	.24446	.96966	51
10	.17651	.98430	.19366	.98107	.21076	.97754	.22778	.97371	.24474	.96959	50
11	.17680	.98425	.19395	.98101	.21104	.97748	.22807	.97365	.24503	.96952	49
12	.17708	.98420	.19423	.98096	.21132	.97742	.22835	.97358	.24531	.96945	48
13	.17737	.98414	.19452	.98090	.21161	.97735	.22863	.97351	.24559	.96937	47
14	.17766	.98409	.19481	.98084	.21189	.97729	.22892	.97345	.24587	.96930	46
15	.17794	.98404	.19509	.98079	.21218	.97723	.22920	.97338	.24615	.96923	45
16	.17823	.98399	.19538	.98073	.21246	.97717	.22948	.97331	.24644	.96916	44
17	.17852	.98394	.19566	.98067	.21275	.97711	.22977	.97325	.24672	.96909	43
18	.17880	.98389	.19595	.98061	.21303	.97705	.23005	.97318	.24700	.96902	42
19	.17909	.98383	.19623	.98056	.21331	.97698	.23033	.97311	.24728	.96894	41
20	.17937	.98378	.19652	.98050	.21360	.97692	.23062	.97304	.24756	.96887	40
21	.17966	.98373	.19680	.98044	.21388	.97686	.23090	.97298	.24784	.96880	39
22	.17995	.98368	.19709	.98039	.21417	.97680	.23118	.97291	.24813	.96873	38
23	.18023	.98362	.19737	.98033	.21445	.97673	.23146	.97284	.24841	.96866	37
24	.18052	.98357	.19766	.98027	.21474	.97667	.23175	.97278	.24869	.96858	36
25	.18081	.98352	.19794	.98021	.21502	.97661	.23203	.97271	.24897	.96851	35
26	.18109	.98347	.19823	.98016	.21530	.97655	.23231	.97264	.24925	.96844	34
27	.18138	.98341	.19851	.98010	.21559	.97648	.23260	.97257	.24954	.96837	33
28	.18166	.98336	.19880	.98004	.21587	.97642	.23288	.97251	.24982	.96829	32
29	.18195	.98331	.19908	.97998	.21616	.97636	.23316	.97244	.25010	.96822	31
30	.18224	.98325	.19937	.97992	.21644	.97630	.23345	.97237	.25038	.96815	30
31	.18252	.98320	.19965	.97987	.21672	.97623	.23373	.97230	.25066	.96807	29
32	.18281	.98315	.19994	.97981	.21701	.97617	.23401	.97223	.25094	.96800	28
33	.18309	.98310	.20022	.97975	.21729	.97611	.23429	.97217	.25122	.96793	27
34	.18338	.98304	.20051	.97969	.21758	.97604	.23458	.97210	.25151	.96786	26
35	.18367	.98299	.20079	.97963	.21786	.97598	.23486	.97203	.25179	.96778	25
36	.18395	.98294	.20108	.97958	.21814	.97592	.23514	.97196	.25207	.96771	24
37	.18424	.98288	.20136	.97952	.21843	.97585	.23542	.97189	.25235	.96764	23
38	.18452	.98283	.20165	.97946	.21871	.97579	.23571	.97182	.25263	.96756	22
39	.18481	.98277	.20193	.97940	.21899	.97573	.23599	.97176	.25291	.96749	21
40	.18509	.98272	.20222	.97934	.21928	.97566	.23627	.97169	.25320	.96742	20
41	.18538	.98267	.20250	.97928	.21956	.97560	.23656	.97162	.25348	.96734	19
42	.18567	.98261	.20279	.97922	.21985	.97553	.23684	.97155	.25376	.96727	18
43	.18595	.98256	.20307	.97916	.22013	.97547	.23712	.97148	.25404	.96719	17
44	.18624	.98250	.20336	.97910	.22041	.97541	.23740	.97141	.25432	.96712	16
45	.18652	.98245	.20364	.97905	.22070	.97534	.23769	.97134	.25460	.96705	15
46	.18681	.98240	.20393	.97899	.22098	.97528	.23797	.97127	.25488	.96697	14
47	.18710	.98234	.20421	.97893	.22126	.97521	.23825	.97120	.25516	.96690	13
48	.18738	.98229	.20450	.97887	.22155	.97515	.23853	.97113	.25545	.96682	12
49	.18767	.98223	.20478	.97881	.22183	.97508	.23882	.97106	.25573	.96675	11
50	.18795	.98218	.20507	.97875	.22212	.97502	.23910	.97100	.25601	.96667	10
51	.18824	.98212	.20535	.97869	.22240	.97496	.23938	.97093	.25629	.96660	9
52	.18852	.98207	.20563	.97863	.22268	.97489	.23966	.97086	.25657	.96653	8
53	.18881	.98201	.20592	.97857	.22297	.97483	.23995	.97079	.25685	.96645	7
54	.18910	.98196	.20620	.97851	.22325	.97476	.24023	.97072	.25713	.96638	6
55	.18938	.98190	.20649	.97845	.22353	.97470	.24051	.97065	.25741	.96630	5
56	.18967	.98185	.20677	.97839	.22382	.97463	.24079	.97058	.25769	.96623	4
57	.18995	.98179	.20706	.97833	.22410	.97457	.24108	.97051	.25798	.96615	3
58	.19024	.98174	.20734	.97827	.22438	.97450	.24136	.97044	.25826	.96608	2
59	.19052	.98168	.20763	.97821	.22467	.97444	.24164	.97037	.25854	.96600	1
60	.19081	.98163	.20791	.97815	.22495	.97437	.24192	.97030	.25882	.96593	0
M.	Cosin.	Sine.	Cosin.	Sine.	Cosin.	Sine.	Cosin.	Sine.	Cosin.	Sine.	M
	79°		78°		77°		76°		75°		

	15°		16°		17°		18°		19°		
M.	Sine.	Cosin.	Sine.	Cosin.	Sine.	Cosin.	Sine.	Cosin.	Sine.	Cosin.	M.
0	.25882	.96593	.27564	.96126	.29237	.95630	.30902	.95106	.32557	.94552	60
1	.25910	.96585	.27592	.96118	.29265	.95622	.30929	.95097	.32584	.94542	59
2	.25938	.96578	.27620	.96110	.29293	.95613	.30957	.95088	.32612	.94533	58
3	.25966	.96570	.27648	.96102	.29321	.95605	.30985	.95079	.32639	.94523	57
4	.25994	.96562	.27676	.96094	.29348	.95596	.31012	.95070	.32667	.94514	56
5	.26022	.96555	.27704	.96086	.29376	.95588	.31040	.95061	.32694	.94504	55
6	.26050	.96547	.27731	.96078	.29404	.95579	.31068	.95052	.32722	.94495	54
7	.26079	.96540	.27759	.96070	.29432	.95571	.31095	.95043	.32749	.94485	53
8	.26107	.96532	.27787	.96062	.29460	.95562	.31123	.95033	.32777	.94476	52
9	.26135	.96524	.27815	.96054	.29487	.95554	.31151	.95024	.32804	.94466	51
10	.26163	.96517	.27843	.96046	.29515	.95545	.31178	.95015	.32832	.94457	50
11	.26191	.96509	.27871	.96037	.29543	.95536	.31206	.95006	.32859	.94447	49
12	.26219	.96502	.27899	.96029	.29571	.95528	.31233	.94997	.32887	.94438	48
13	.26247	.96494	.27927	.96021	.29599	.95519	.31261	.94988	.32914	.94428	47
14	.26275	.96486	.27955	.96013	.29626	.95511	.31289	.94979	.32942	.94418	46
15	.26303	.96479	.27983	.96005	.29654	.95502	.31316	.94970	.32969	.94409	45
16	.26331	.96471	.28011	.95997	.29682	.95493	.31344	.94961	.32997	.94399	44
17	.26359	.96463	.28039	.95989	.29710	.95485	.31372	.94952	.33024	.94390	43
18	.26387	.96456	.28067	.95981	.29737	.95476	.31399	.94943	.33051	.94380	42
19	.26415	.96448	.28095	.95972	.29765	.95467	.31427	.94933	.33079	.94370	41
20	.26443	.96440	.28123	.95964	.29793	.95459	.31454	.94924	.33106	.94361	40
21	.26471	.96433	.28150	.95956	.29821	.95450	.31482	.94915	.33134	.94351	39
22	.26500	.96425	.28178	.95948	.29849	.95441	.31510	.94906	.33161	.94342	38
23	.26528	.96417	.28206	.95940	.29876	.95433	.31537	.94897	.33189	.94332	37
24	.26556	.96410	.28234	.95931	.29904	95424	.31565	.94888	.33216	.94322	36
25	.26584	.96402	.28262	.95923	.29932	.95415	.31593	.94878	.33244	.94313	35
26	.26612	.96394	.28290	.95915	.29960	.95407	.31620	.94869	.33271	.94303	34
27	.26640	.96386	.28318	.95907	.29987	.95398	.31648	.94860	.33298	.94293	33
28	.26668	.96379	.28346	.95898	.30015	.95389	.31675	.94851	.33326	.94284	32
29	.26696	.96371	.28374	.95890	.30043	.95380	.31703	.94842	.33353	.94274	31
30	.26724	.96363	.28402	.95882	.30071	.95372	.31730	.94832	.33381	.94264	30
31	.26752	.96355	.28429	.95874	.30098	.95363	.31758	.94823	.33408	.94254	29
32	.26780	.96347	.28457	.95865	.30126	.95354	.31786	.94814	.33436	.94245	28
33	.26808	.96340	.28485	.95857	.30154	.95345	.31813	.94805	.33463	.94235	27
34	.26836	.96332	.28513	.95849	.30182	.95337	.31841	.94795	.33490	.94225	26
35	.26864	.96324	.28541	.95841	.30209	.95328	.31868	.94786	.33518	.94215	25
36	.26892	.96316	.28569	.95832	.30237	.95319	.31896	.94777	.33545	.94206	24
37	.26920	.96308	.28597	.95824	.30265	.95310	.31923	.94768	.33573	.94196	23
38	.26948	.96301	.28625	.95816	.30292	.95301	.31951	.94758	.33600	.94186	22
39	.26976	.96293	.28652	.95807	.30320	.95293	.31979	.94749	.33627	.94176	21
40	.27004	.96285	.28680	.95799	.30348	.95284	.32006	.94740	.33655	.94167	20
41	.27032	.96277	.28708	.95791	.30376	.95275	.32034	.94730	.33682	.94157	19
42	.27060	.96269	.28736	.95782	.30403	.95266	.32061	.94721	.33710	.94147	18
43	.27088	.96261	.28764	.95774	.30431	.95257	.32089	.94712	.33737	.94137	17
44	.27116	.96253	.28792	.95766	.30459	.95248	.32116	.94702	.33764	.94127	16
45	.27144	.96246	.28820	.95757	.30486	.95240	.32144	.94693	.33792	.94118	15
46	.27172	.96238	.28847	.95749	.30514	.95231	.32171	.94684	.33819	.94108	14
47	.27200	.96230	.28875	.95740	.30542	.95222	.32199	.94674	.33846	.94098	13
48	.27228	.96222	.28903	.95732	.30570	.95213	.32227	.94665	.33874	.94088	12
49	.27256	.96214	.28931	.95724	.30597	.95204	32254	.94656	.33901	.94078	11
50	.27284	.96206	.28959	.95715	.30625	.95195	.32282	.94646	.33929	.94068	10
51	.27312	.96198	.28987	.95707	.30653	.95186	.32309	.94637	.33956	.94058	9
52	.27340	.96190	.29015	.95698	.30680	.95177	.32337	.94627	.33983	.94049	8
53	.27368	.96182	.29042	.95690	.30708	.95168	.32364	.94618	.34011	.94039	7
54	.27396	.96174	.29070	.95681	.30736	.95159	.32392	.94609	.34038	.94029	6
55	.27424	.96166	.29098	.95673	.30763	.95150	.32419	.94599	.34065	.94019	5
56	.27452	.96158	.29126	.95664	.30791	.95142	.32447	.94590	.34093	.91009	4
57	.27480	.96150	.29154	.95656	.30819	.95133	.32474	.94580	.34120	.93999	3
58	.27508	.96142	.29182	.95647	.30846	.95124	.32502	.94571	.34147	.93989	2
59	.27536	.96134	.29209	.95639	.30874	.95115	.32529	.94561	.34175	.93979	1
60	.27564	.96126	.29237	.95630	.30902	.95106	.32557	.94552	.34202	.93969	0
M.	Cosin.	Sine.	Cosin.	Sine.	Cosin.	Sine.	Cosin.	Sine.	Cosin.	Sine.	M.
	74°		73°		72°		71°		70°		

	20°		21°		22°		23°		24°		
M.	Sine.	Cosin	Sine.	Cosin.	Sine.	Cosin.	Sine.	Cosin.	Sine.	Cosin.	M.
0	.34202	.93969	.35837	.93358	.37461	.92718	.39073	.92050	.40674	.91355	60
1	.34229	.93959	.35864	.93348	.37488	.92707	.39100	.92039	.40700	.91343	59
2	.34257	.93949	.35891	.93337	.37515	.92697	.39127	.92028	.40727	.91331	58
3	.34284	.93939	.35918	.93327	.37542	.92686	.39153	.92016	.40753	.91319	57
4	.34311	.93929	.35945	.93316	.37569	.92675	.39180	.92005	.40780	.91307	56
5	.34339	.93919	.35973	.93306	.37595	.92664	.39207	.91994	.40806	.91295	55
6	.34366	.93909	.36000	.93295	.37622	.92653	.39234	.91982	.40833	.91283	54
7	.34393	.93899	.36027	.93285	.37649	.92642	.39260	.91971	.40860	.91272	53
8	.34421	.93889	.36054	.93274	.37676	.92631	.39287	.91959	.40886	.91260	52
9	.34448	.93879	.36081	.93264	.37703	.92620	.39314	.91948	.40913	.91248	51
10	.34475	.93869	.36108	.93253	.37730	.92609	.39341	.91936	.40939	.91236	50
11	.34503	.93859	.36135	.93243	.37757	.92598	.39367	.91925	.40966	.91224	49
12	.34530	.93849	.36162	.93232	.37784	.92587	.39394	.91914	.40992	.91212	48
13	.34557	.93839	.36190	.93222	.37811	.92576	.39421	.91902	.41019	.91200	47
14	.34584	.93829	.36217	.93211	.37838	.92565	.39448	.91891	.41045	.91188	46
15	.34612	.93819	.36244	.93201	.37865	.92554	.39474	.91879	.41072	.91176	45
16	.34639	.93809	.36271	.93190	.37892	.92543	.39501	.91868	.41098	.91164	44
17	.34666	.93799	.36298	.93180	.37919	.92532	.39528	.91856	.41125	.91152	43
18	.34694	.93789	.36325	.93169	.37946	.92521	.39555	.91845	.41151	.91140	42
19	.34721	.93779	.36352	.93159	.37973	.92510	.39581	.91833	.41178	.91128	41
20	.34748	.93769	.36379	.93148	.37999	.92499	.39608	.91822	.41204	.91116	40
21	.34775	.93759	.36106	.93137	.38026	.92488	.39635	.91810	.41231	.91104	39
22	.34803	.93748	.36434	.93127	.38053	.92477	.39661	.91799	.41257	.91092	38
23	.34830	.93738	.36461	.93116	.38080	.92466	.39688	.91787	.41284	.91080	37
24	.34857	.93728	.36488	.93106	.38107	.92455	.39715	.91775	.41310	.91068	36
25	.34884	.93718	.36515	.93095	.38134	.92444	.39741	.91764	.41337	.91056	35
26	.34912	.93708	.36512	.93084	.38161	.92432	.39768	.91752	.41363	.91044	34
27	.34939	.93698	.36569	.93074	.38188	.92421	.39795	.91741	.41390	.91032	33
28	.34966	.93688	.36596	.93063	.38215	.92410	.39822	.91729	.41416	.91020	32
29	.34993	.93677	.36623	.93052	.38241	.92399	.39848	.91718	.41443	.91008	31
30	.35021	.93667	.36650	.93042	.38268	.92388	.39875	.91706	.41469	.90996	30
31	.35048	.93657	.36677	.93031	.38295	.92377	.39902	.91694	.41496	.90984	29
32	.35075	.93647	.36704	.93020	.38322	.92366	.39928	.91683	.41522	.90972	28
33	.35102	.93637	.36731	.93010	.38349	.92355	.39955	.91671	.41549	.90960	27
34	.35130	.93626	.36758	.92999	.38376	.92343	.39982	.91660	.41575	.90948	26
35	.35157	.93616	.36785	.92988	.38403	.92332	.40008	.91648	.41602	.90936	25
36	.35184	.93606	.36812	.92978	.38430	.92321	.40035	.91636	.41628	.90924	24
37	.35211	.93596	.36839	.92967	.38456	.92310	.40062	.91625	.41655	.90911	23
38	.35239	.93585	.36867	.92956	.38483	.92299	.40088	.91613	.41681	.90899	22
39	.35266	.93575	.36894	.92945	.38510	.92287	.40115	.91601	.41707	.90887	21
40	.35293	.93565	.36921	.92935	.38537	.92276	.40141	.91590	.41734	.90875	20
41	.35320	.93555	.36948	.92924	.38564	.92265	.40168	.91578	.41760	.90863	19
42	.35347	.93544	.36975	.92913	.38591	.92254	.40195	.91566	.41787	.90851	18
43	.35375	.93534	.37002	.92902	.38617	.92243	.40221	.91555	.41813	.90839	17
44	.35402	.93524	.37029	.92892	.38644	.92231	.40248	.91543	.41840	.90826	16
45	.35429	.93514	.37056	.92881	.38671	.92220	.40275	.91531	.41866	.90814	15
46	.35456	.93503	.37083	.92870	.38698	.92209	.40301	.91519	.41892	.90802	14
47	.35484	.93493	.37110	.92859	.38725	.92198	.40328	.91508	.41919	.90790	13
48	.35511	.93483	.37137	.92849	.38752	.92186	.40355	.91496	.41945	.90778	12
49	.35538	.93472	.37164	.92838	.38778	.92175	.40381	.91484	.41972	.90766	11
50	.35565	.93462	.37191	.92827	.38805	.92164	.40408	.91472	.41998	.90753	10
51	.35592	.93452	.37218	.92816	.38832	.92152	.40434	.91461	.42024	.90741	9
52	.35619	.93441	.37245	.92805	.38859	.92141	.40461	.91449	.42051	.90729	8
53	.35647	.93431	.37272	.92794	.38886	.92130	.40488	.91437	.42077	.90717	7
54	.35674	.93420	.37299	.92784	.38912	.92119	.40514	.91425	.42104	.90704	6
55	.35701	.93410	.37326	.92773	.38939	.92107	.40541	.91414	.42130	.90692	5
56	.35728	.93400	.37353	.92762	.38966	.92096	.40567	.91402	.42156	.90680	4
57	.35755	.93389	.37380	.92751	.38993	.92085	.40594	.91390	.42183	.90668	3
58	.35782	.93379	.37407	.92740	.39020	.92073	.40621	.91378	.42209	.90655	2
59	.35810	.93368	.37434	.92729	.39046	.92062	.40647	.91366	.42235	.90643	1
60	.35837	.93358	.37461	.92718	.39073	.92050	.40674	.91355	.42262	.90631	0
M.	Cosin.	Sine.	Cosin.	Sine.	Cosin.	Sine.	Cosin.	Sine.	Cosin.	Sine.	M.
	69°		68°		67°		66°		65°		

M.	25° Sine.	25° Cosin.	26° Sine.	26° Cosin.	27° Sine.	27° Cosin.	28° Sine.	28° Cosin.	29° Sine.	29° Cosin.	M.
0	.42262	.90631	.43837	.89879	.45399	.89101	.46947	.88295	.48481	.87462	60
1	.42288	.90618	.43863	.89867	.45425	.89087	.46973	.88281	.48506	.87448	59
2	.42315	.90606	.43889	.89854	.45451	.89074	.46999	.88267	.48532	.87434	58
3	.42341	.90594	.43916	.89841	.45477	.89061	.47024	.88254	.48557	.87420	57
4	.42367	.90582	.43942	.89828	.45503	.89048	.47050	.88240	.48583	.87406	56
5	.42394	.90569	.43968	.89816	.45529	.89035	.47076	.88226	.48608	.87391	55
6	.42420	.90557	.43994	.89803	.45554	.89021	.47101	.88213	.48634	.87377	54
7	.42446	.90545	.44020	.89790	.45580	.89008	.47127	.88199	.48659	.87363	53
8	.42473	.90532	.44046	.89777	.45606	.88995	.47153	.88185	.48684	.87349	52
9	.42499	.90520	.44072	.89764	.45632	.88981	.47178	.88172	.48710	.87335	51
10	.42525	.90507	.44098	.89752	.45658	.88968	.47204	.88158	.48735	.87321	50
11	.42552	.90495	.44124	.89739	.45684	.88955	.47229	.88144	.48761	.87306	49
12	.42578	.90483	.44151	.89726	.45710	.88942	.47255	.88130	.48786	.87292	48
13	.42604	.90470	.44177	.89713	.45736	.88928	.47281	.88117	.48811	.87278	47
14	.42631	.90458	.44203	.89700	.45762	.88915	.47306	.88103	.48837	.87264	46
15	.42657	.90446	.44229	.89687	.45787	.88902	.47332	.88089	.48862	.87250	45
16	.42683	.90433	.44255	.89674	.45813	.88888	.47358	.88075	.48888	.87235	44
17	.42709	.90421	.44281	.89662	.45839	.88875	.47383	.88062	.48913	.87221	43
18	.42736	.90408	.44307	.89649	.45865	.88862	.47409	.88048	.48938	.87207	42
19	.42762	.90396	.44333	.89636	.45891	.88848	.47434	.88034	.48964	.87193	41
20	.42788	.90383	.44359	.89623	.45917	.88835	.47460	.88020	.48989	.87178	40
21	.42815	.90371	.44385	.89610	.45942	.88822	.47486	.88006	.49014	.87164	39
22	.42841	.90358	.44411	.89597	.45968	.88808	.47511	.87993	.49040	.87150	38
23	.42867	.90346	.44437	.89584	.45994	.88795	.47537	.87979	.49065	.87136	37
24	.42894	.90334	.44464	.89571	.46020	.88782	.47562	.87965	.49090	.87121	36
25	.42920	.90321	.44490	.89558	.46046	.88768	.47588	.87951	.49116	.87107	35
26	.42946	.90309	.44516	.89545	.46072	.88755	.47614	.87937	.49141	.87093	34
27	.42972	.90296	.44542	.89532	.46097	.88741	.47639	.87923	.49166	.87079	33
28	.42999	.90284	.44568	.89519	.46123	.88728	.47665	.87909	.49192	.87064	32
29	.43025	.90271	.44594	.89506	.46149	.88715	.47690	.87896	.49217	.87050	31
30	.43051	.90259	.44620	.89493	.46175	.88701	.47716	.87882	.49242	.87036	30
31	.43077	.90246	.44646	.89480	.46201	.88688	.47741	.87868	.49268	.87021	29
32	.43104	.90233	.44672	.89467	.46226	.88674	.47767	.87854	.49293	.87007	28
33	.43130	.90221	.44698	.89454	.46252	.88661	.47793	.87840	.49318	.86993	27
34	.43156	.90208	.44724	.89441	.46278	.88647	.47818	.87826	.49344	.86978	26
35	.43182	.90196	.44750	.89428	.46304	.88634	.47844	.87812	.49369	.86964	25
36	.43209	.90183	.44776	.89415	.46330	.88620	.47869	.87798	.49394	.86949	24
37	.43235	.90171	.44802	.89402	.46355	.88607	.47895	.87784	.49419	.86935	23
38	.43261	.90158	.44828	.89389	.46381	.88593	.47920	.87770	.49445	.86921	22
39	.43287	.90146	.44854	.89376	.46407	.88580	.47946	.87756	.49470	.86906	21
40	.43313	.90133	.44880	.89363	.46433	.88566	.47971	.87743	.49495	.86892	20
41	.43340	.90120	.44906	.89350	.46458	.88553	.47997	.87729	.49521	.86878	19
42	.43366	.90108	.44932	.89337	.46484	.88539	.48022	.87715	.49546	.86863	18
43	.43392	.90095	.44958	.89324	.46510	.88526	.48048	.87701	.49571	.86849	17
44	.43418	.90082	.44984	.89311	.46536	.88512	.48073	.87687	.49596	.86834	16
45	.43445	.90070	.45010	.89298	.46561	.88499	.48099	.87673	.49622	.86820	15
46	.43471	.90057	.45036	.89285	.46587	.88485	.48124	.87659	.49647	.86805	14
47	.43497	.90045	.45062	.89272	.46613	.88472	.48150	.87645	.49672	.86791	13
48	.43523	.90032	.45088	.89259	.46639	.88458	.48175	.87631	.49697	.86777	12
49	.43549	.90019	.45114	.89245	.46664	.88445	.48201	.87617	.49723	.86762	11
50	.43575	.90007	.45140	.89232	.46690	.88431	.48226	.87603	.49748	.86748	10
51	.43602	.89994	.45166	.89219	.46716	.88417	.48252	.87589	.49773	.86733	9
52	.43628	.89981	.45192	.89206	.46742	.88404	.48277	.87575	.49798	.86719	8
53	.43654	.89968	.45218	.89193	.46767	.88390	.48303	.87561	.49824	.86704	7
54	.43680	.89956	.45243	.89180	.46793	.88377	.48328	.87546	.49849	.86690	6
55	.43706	.89943	.45269	.89167	.46819	.88363	.48354	.87532	.49874	.86675	5
56	.43733	.89930	.45295	.89153	.46844	.88349	.48379	.87518	.49899	.86661	4
57	.43759	.89918	.45321	.89140	.46870	.88336	.48405	.87504	.49924	.86646	3
58	.43785	.89905	.45347	.89127	.46896	.88322	.48430	.87490	.49950	.86632	2
59	.43811	.89892	.45373	.89114	.46921	.88308	.48456	.87476	.49975	.86617	1
60	.43837	.89879	.45399	.89101	.46947	.88295	.48481	.87462	.50000	.86603	0
M.	Cosin.	Sine.	Cosin.	Sine.	Cosin.	Sine.	Cosin.	Sine.	Cosin.	Sine.	M.
	64°		63°		62°		61°		60°		

M.	30° Sine.	30° Cosin.	31° Sine.	31° Cosin.	32° Sine.	32° Cosin.	33° Sine.	33° Cosin.	34° Sine.	34° Cosin.	M.
0	.50000	.86603	51504	.85717	.52992	.84805	.54464	.83867	.55919	.82904	60
1	.50025	.86588	51529	.85702	.53017	.84789	.54488	.83851	.55943	.82887	59
2	.50050	.86573	.51554	.85687	.53041	.84774	.54513	.83835	.55968	.82871	58
3	.50076	.86559	.51579	.85672	.53066	.84759	.54537	.83819	.55992	.82855	57
4	50101	.86544	.51604	.85657	.53091	.84743	.54561	.83804	.56016	.82839	56
5	.50126	.86530	.51628	.85642	.53115	.84728	.54586	.83788	.56040	.82822	55
6	.50151	.86515	.51653	.85627	.53140	.84712	.54610	.83772	.56064	.82806	54
7	.50176	.86501	.51678	.85612	.53164	.84697	.54635	.83756	.56088	.82790	53
8	.50201	.86486	.51703	.85597	.53189	.84681	.54659	.83740	.56112	.82773	52
9	.50227	.86471	.51728	.85582	.53214	.84666	.54683	.83724	.56136	.82757	51
10	.50252	.86457	.51753	.85567	.53238	.84650	.54708	.83708	.56160	.82741	50
11	50277	.86442	.51778	.85551	.53263	.84635	.54732	.83692	.56184	.82724	49
12	.50302	.86427	.51803	.85536	.53288	.84619	.54756	.83676	.56208	.82708	48
13	.50327	.86413	.51828	.85521	.53312	.84604	.54781	.83660	.56232	.82692	47
14	.50352	.86398	.51852	.85506	.53337	.84588	.54805	.83645	.56256	.82675	46
15	.50377	.86384	.51877	.85491	.53361	.84573	.54829	.83629	.56280	.82659	45
16	.50403	.86369	.51902	.85476	.53386	.84557	.54854	.83613	.56305	.82643	44
17	.50428	.86354	.51927	.85461	.53411	.84542	.54878	.83597	.56329	.82626	43
18	.50453	.86340	.51952	.85446	.53435	.84526	.54902	.83581	.56353	.82610	42
19	.50478	.86325	.51977	.85431	.53460	.84511	.54927	.83565	.56377	.82593	41
20	.50503	.86310	.52002	.85416	.53484	.84495	.54951	.83549	.56401	.82577	40
21	.50528	.86295	.52026	.85401	.53509	.84480	.54975	.83533	.56425	.82561	39
22	.50553	.86281	.52051	.85385	.53534	.84464	.54999	.83517	.56449	.82544	38
23	.50578	.86266	.52076	.85370	.53558	.84448	.55024	.83501	.56473	.82528	37
24	.50603	.86251	.52101	.85355	.53583	.84433	.55048	.83485	.56497	.82511	36
25	.50628	.86237	.52126	.85340	.53607	.84417	.55072	.83469	.56521	.82495	35
26	.50654	.86222	.52151	.85325	.53632	.84402	.55097	.83453	.56545	.82478	34
27	.50679	.86207	.52175	.85310	.53656	.84386	.55121	.83437	.56569	.82462	33
28	.50704	.86192	.52200	.85294	.53681	.84370	.55145	.83421	.56593	.82446	32
29	.50729	.86178	.52225	.85279	.53705	.84355	.55169	.83405	.56617	.82429	31
30	.50754	.86163	.52250	.85264	.53730	.84339	.55194	.83389	.56641	.82413	30
31	.50779	.86148	.52275	.85249	.53754	.84324	.55218	.83373	.56665	.82396	29
32	.50804	.86133	.52299	.85234	.53779	.84308	.55242	.83356	.56689	.82380	28
33	.50829	.86119	.52324	.85218	.53804	.84292	.55266	.83340	.56713	.82363	27
34	.50854	.86101	.52349	.85203	.53828	.84277	.55291	.83324	.56736	.82347	26
35	.50879	.86089	.52374	.85188	.53853	.84261	.55315	.83308	.56760	.82330	25
36	.50904	.86074	.52399	.85173	.53877	.84245	.55339	.83292	.56784	.82314	24
37	.50929	.86059	.52423	.85157	.53902	.84230	.55363	.83276	.56808	.82297	23
38	.50954	.86045	.52448	.85142	.53926	.84214	.55388	.83260	.56832	.82281	22
39	.50979	.86030	.52473	.85127	.53951	.84198	.55412	.83244	.56856	.82264	21
40	.51004	.86015	.52498	.85112	.53975	.84182	.55436	.83228	.56880	.82248	20
41	.51029	.86000	.52522	.85096	.54000	.84167	.55460	.83212	.56904	.82231	19
42	.51054	.85985	.52547	.85081	.54024	.84151	.55484	.83195	.56928	.82214	18
43	.51079	.85970	.52572	.85066	.54049	.84135	.55509	.83179	.56952	.82198	17
44	.51104	.85956	.52597	.85051	.54073	.84120	.55533	.83163	.56976	.82181	16
45	.51129	.85941	.52621	.85035	.54097	.84104	.55557	.83147	.57000	.82165	15
46	.51154	.85926	.52646	.85020	.54122	.84088	.55581	.83131	.57024	.82148	14
47	.51179	.85911	.52671	.85005	.54146	.84072	.55605	.83115	.57047	.82132	13
48	.51204	.85896	.52696	.84989	.54171	.84057	.55630	.83098	.57071	.82115	12
49	.51229	.85881	.52720	.84974	.54195	.84041	.55654	.83082	.57095	.82098	11
50	.51254	.85866	.52745	.84959	.54220	.84025	.55678	.83066	.57119	.82082	10
51	.51279	.85851	.52770	.84943	.54244	.84009	.55702	.83050	.57143	.82065	9
52	.51304	.85836	.52794	.84928	.54269	.83994	.55726	.83034	.57167	.82048	8
53	.51329	.85821	.52819	.84913	.54293	.83978	.55750	.83017	.57191	.82032	7
54	.51354	.85806	.52844	.84897	.54317	.83962	.55775	.83001	.57215	.82015	6
55	.51379	.85792	.52869	.84882	.54342	.83946	.55799	.82985	.57238	.81999	5
56	.51404	.85777	.52893	.84866	.54366	.83930	.55823	.82969	.57262	.81982	4
57	.51429	.85762	.52918	.84851	.54391	.83915	.55847	.82953	.57286	.81965	3
58	.51454	.85747	.52943	.84836	.54415	.83899	.55871	.82936	.57310	.81949	2
59	.51479	.85732	.52967	.84820	.54440	.83883	.55895	.82920	.57334	.81932	1
60	.51504	.85717	.52992	.84805	54464	.83867	.55919	.82904	.57358	.81915	0
M.	Cosin.	Sine.	Cosin.	Sine.	Cosin.	Sine.	Cosin.	Sine.	Cosin.	Sine.	M.
	59°		58°		57°		56°		55°		

	35°		36°		37°		38°		39°		
M.	Sine.	Cosin.	Sine.	Cosin.	Sine.	Cosin.	Sine.	Cosin.	Sine.	Cosin.	M.
0	.57358	.81915	.58779	.80902	.60182	.79864	.61566	.78801	.62932	.77715	60
1	.57381	.81899	.58802	.80885	.60205	.79846	.61589	.78783	.62955	.77696	59
2	.57405	.81882	.58826	.80867	.60228	.79829	.61612	.78765	.62977	.77678	58
3	.57429	.81865	.58849	.80850	.60251	.79811	.61635	.78747	.63000	.77660	57
4	.57453	.81848	.58873	.80833	.60274	.79793	.61958	.78729	.63022	.77641	56
5	.57477	.81832	.58896	.80816	.60298	.79776	.61681	.78711	.63045	.77623	55
6	.57501	.81815	.58920	.80799	.60321	.79758	.61704	.78694	.63068	.77605	54
7	.57524	.81798	.58943	.80782	.60344	.79741	.61726	.78676	.63090	.77586	53
8	.57548	.81782	.58967	.80765	.60367	.79723	.61749	.78658	.63113	.77568	52
9	.57572	.81765	.58990	.80748	.60390	.79706	.61772	.78640	.63135	.77550	51
10	.57596	.81748	.59014	.80730	.60414	.79688	.61795	.78622	.63158	.77531	50
11	.57619	.81731	.59037	.80713	.60437	.79671	.61818	.78604	.63180	.77513	49
12	.57643	.81714	.59061	.80696	.60460	.79653	.61841	.78586	.63203	.77494	48
13	.57667	.81698	.59084	.80679	.60483	.79635	.61864	.78568	.63225	.77476	47
14	.57691	.81681	.59108	.80662	.60506	.79618	.61887	.78550	.63248	.77458	46
15	.57715	.81664	.59131	.80644	.60529	.79600	.61909	.78532	.63271	.77439	45
16	.57738	.81647	.59154	.80627	.60553	.79583	.61932	.78514	.63293	.77421	44
17	.57762	.81631	.59178	.80610	.60576	.79565	.61955	.78496	.63316	.77402	43
18	.57786	.81614	.59201	.80593	.60599	.79547	.61978	.78478	.63338	.77384	42
19	.57810	.81597	.59225	.80576	.60622	.79530	.62001	.78460	.63361	.77366	41
20	.57833	.81580	.59248	.80558	.60645	.79512	.62024	.78442	.63383	.77347	40
21	.57857	.81563	.59272	.80541	.60668	.79494	.62046	.78424	.63406	.77329	39
22	.57881	.81546	.59295	.80524	.60691	.79477	.62069	.78405	.63428	.77310	38
23	.57904	.81530	.59318	.80507	.60714	.79459	.62092	.78387	.63451	.77292	37
24	.57928	.81513	.59342	.80489	.60738	.79441	.62115	.78369	.63473	.77273	36
25	.57952	.81496	.59365	.80472	.60761	.79424	.62138	.78351	.63496	.77255	35
26	.57976	.81479	.59389	.80455	.60784	.79406	.62160	.78333	.63518	.77236	34
27	.57999	.81462	.59412	.80438	.60807	.79388	.62183	.78315	.63540	.77218	33
28	.58023	.81445	.59436	.80420	.60830	.79371	.62206	.78297	.63563	.77199	32
29	.58047	.81428	.59459	.80403	.60853	.79353	.62229	.78279	.63585	.77181	31
30	.58070	.81412	.59482	.80386	.60876	.79335	.62251	.78261	.63608	.77162	30
31	.58094	.81395	.59506	.80368	.60899	.79318	.62274	.78243	.63630	.77144	29
32	.58118	.81378	.59529	.80351	.60922	.79300	.62297	.78225	.63653	.77125	28
33	.58141	.81361	.59552	.80334	.60945	.79282	.62320	.78206	.63675	.77107	27
34	.58165	.81344	.59576	.80316	.60968	.79264	.62342	.78188	.63698	.77088	26
35	.58189	.81327	.59599	.80299	.60991	.79247	.62365	.78170	.63720	.77070	25
36	.58212	.81310	.59622	.80282	.61015	.79229	.62388	.78152	.63742	.77051	24
37	.58236	.81293	.59646	.80264	.61038	.79211	.62411	.78134	.63765	.77033	23
38	.58260	.81276	.59669	.80247	.61061	.79193	.62433	.78116	.63787	.77014	22
39	.58283	.81259	.59693	.80230	.61084	.79176	.62456	.78098	.63810	.76996	21
40	.58307	.81242	.59716	.80212	.61107	.79158	.62479	.78079	.63832	.76977	20
41	.58330	.81225	.59739	.80195	.61130	.79140	.62502	.78061	.63854	.76959	19
42	.58354	.81208	.59763	.80178	.61153	.79122	.62524	.78043	.63877	.76940	18
43	.58378	.81191	.59786	.80160	.61176	.79105	.62547	.78025	.63899	.76921	17
44	.58401	.81174	.59809	.80143	.61199	.79087	.62570	.78007	.63922	.76903	16
45	.58425	.81157	.59832	.80125	.61222	.79069	.62592	.77988	.63944	.76884	15
46	.58449	.81140	.59856	.80108	.61245	.79051	.62615	.77970	.63966	.76866	14
47	.58472	.81123	.59879	.80091	.61268	.79033	.62638	.77952	.63989	.76847	13
48	.58496	.81106	.59902	.80073	.61291	.79016	.62660	.77934	.64011	.76828	12
49	.58519	.81089	.59926	.80056	.61314	.78998	.62683	.77916	.64033	.76810	11
50	.58543	.81072	.59949	.80038	.61337	.78980	.62706	.77897	.64056	.76791	10
51	.58567	.81055	.59972	.80021	.61360	.78962	.62728	.77879	.64078	.76772	9
52	.58590	.81038	.59995	.80003	.61383	.78944	.62751	.77861	.64100	.76754	8
53	.58614	.81021	.60019	.79986	.61406	.78926	.62774	.77843	.64123	.76735	7
54	.58637	.81004	.60042	.79968	.61429	.78908	.62796	.77824	.64145	.76717	6
55	.58661	.80987	.60065	.79951	.61451	.78891	.62819	.77806	.64167	.76698	5
56	.58684	.80970	.60089	.79934	.61474	.78873	.62842	.77788	.64190	.76679	4
57	.58708	.80953	.60112	.79916	.61497	.78855	.62864	.77769	.64212	.76661	3
58	.58731	.80936	.60135	.79899	.61520	.78837	.62887	.77751	.64234	.76642	2
59	.58755	.80919	.60158	.79881	.61543	.78819	.62909	.77733	.64256	.76623	1
60	.58779	.80902	.60182	.79864	.61566	.78801	.62932	.77715	.64279	.76604	0
M.	Cosin.	Sine.	Cosin.	Sine.	Cosin.	Sine.	Cosin.	Sine.	Cosin.	Sine.	M.
	54°		53°		52°		51°		50°		

	40°		41°		42°		43°		44°		
M.	Sine.	Cosin.	Sine.	Cosin.	Sine.	Cosin.	Sine.	Cosin.	Sine.	Cosin.	M.
0	.64279	.76604	.65606	.75471	.66913	.74314	.68200	.73135	.69466	.71934	60
1	.64301	.76586	.65628	.75452	.66935	.74295	.68221	.73116	.69487	.71914	59
2	.64323	.76567	.65650	.75433	.66956	.74276	.68242	.73096	.69508	.71894	58
3	.64346	.76548	.65672	.75414	.66978	.74256	.68264	.73076	.69529	.71873	57
4	.64368	.76530	.65694	.75395	.66999	.74237	.68285	.73056	.69549	.71853	56
5	.64390	.76511	.65716	.75375	.67021	.74217	.68306	.73036	.69570	.71833	55
6	.64412	.76492	.65738	.75356	.67043	.74198	.68327	.73016	.69591	.71813	54
7	.64435	.76473	.65759	.75337	.67064	.74178	.68349	.72996	.69612	.71792	53
8	.64457	.76455	.65781	.75318	.67086	.74159	.68370	.72976	.69633	.71772	52
9	.64479	.76436	.65803	.75299	.67107	.74139	.68391	.72957	.69654	.71752	51
10	.64501	.76417	.65825	.75280	.67129	.74120	.68412	.72937	.69675	.71732	50
11	.64524	.76398	.65847	.75261	.67151	.74100	.68434	.72917	.69696	.71711	49
12	.64546	.76380	.65869	.75241	.67172	.74080	.68455	.72897	.69717	.71691	48
13	.64568	.76361	.65891	.75222	.67194	.74061	.68476	.72877	.69737	.71671	47
14	.64590	.76342	.65913	.75203	.67215	.74041	.68497	.72857	.69758	.71650	46
15	.64612	.76323	.65935	.75184	.67237	.74022	.68518	.72837	.69779	.71630	45
16	.64635	.76304	.65956	.75165	.67258	.74002	.68539	.72817	.69800	.71610	44
17	.64657	.76286	.65978	.75146	.67280	.73983	.68561	.72797	.69821	.71590	43
18	.64679	.76267	.66000	.75126	.67301	.73963	.68582	.72777	.69842	.71569	42
19	.64701	.76248	.66022	.75107	.67323	.73944	.68603	.72757	.69862	.71549	41
20	.64723	.76229	.66044	.75088	.67344	.73924	.68624	.72737	.69883	.71529	40
21	.64746	.76210	.66066	.75069	.67366	.73904	.68645	.72717	.69904	.71508	39
22	.64768	.76192	.66088	.75050	.67387	.73885	.68666	.72697	.69925	.71488	38
23	.64790	.76173	.66109	.75030	.67409	.73865	.68688	.72677	.69946	.71468	37
24	.64812	.76154	.66131	.75011	.67430	.73846	.68709	.72657	.69966	.71447	36
25	.64834	.76135	.66153	.74992	.67452	.73826	.68730	.72637	.69987	.71427	35
26	.64856	.76116	.66175	.74973	.67473	.73806	.68751	.72617	.70008	.71407	34
27	.64878	.76097	.66197	.74953	.67495	.73787	.68772	.72597	.70029	.71386	33
28	.64901	.76078	.66218	.74934	.67516	.73767	.68793	.72577	.70049	.71366	32
29	.64923	.76059	.66240	.74915	.67538	.73747	.68814	.72557	.70070	.71345	31
30	.64945	.76041	.66262	.74896	.67559	.73728	.68835	.72537	.70091	.71325	30
31	.64967	.76022	.66284	.74876	.67580	.73708	.68857	.72517	.70112	.71305	29
32	.64989	.76003	.66306	.74857	.67602	.73688	.68878	.72497	.70132	.71284	28
33	.65011	.75984	.66327	.74838	.67623	.73669	.68899	.72477	.70153	.71264	27
34	.65033	.75965	.66349	.74818	.67645	.73649	.68920	.72457	.70174	.71243	26
35	.65055	.75946	.66371	.74799	.67666	.73629	.68941	.72437	.70195	.71223	25
36	.65077	.75927	.66393	.74780	.67688	.73610	.68962	.72417	.70215	.71203	24
37	.65100	.75908	.66414	.74760	.67709	.73590	.68983	.72397	.70236	.71182	23
38	.65122	.75889	.66436	.74741	.67730	.73570	.69004	.72377	.70257	.71162	22
39	.65144	.75870	.66458	.74722	.67752	.73551	.69025	.72357	.70277	.71141	21
40	.65166	.75851	.66480	.74703	.67773	.73531	.69046	.72337	.70298	.71121	20
41	.65188	.75832	.66501	.74683	.67795	.73511	.69067	.72317	.70319	.71100	19
42	.65210	.75813	.66523	.74664	.67816	.73491	.69088	.72297	.70339	.71080	18
43	.65232	.75794	.66545	.74644	.67837	.73472	.69109	.72277	.70360	.71059	17
44	.65254	.75775	.66566	.74625	.67859	.73452	.69130	.72257	.70381	.71039	16
45	.65276	.75756	.66588	.74606	.67880	.73432	.69151	.72236	.70401	.71019	15
46	.65298	.75738	.66610	.74586	.67901	.73413	.69172	.72216	.70422	.70998	14
47	.65320	.75719	.66632	.74567	.67923	.73393	.69193	.72196	.70443	.70978	13
48	.65342	.75700	.66653	.74548	.67944	.73373	.69214	.72176	.70463	.70957	12
49	.65364	.75680	.66675	.74528	.67965	.73353	.69235	.72156	.70484	.70937	11
50	.65386	.75661	.66697	.74509	.67987	.73333	.69256	.72136	.70505	.70916	10
51	.65408	.75642	.66718	.74489	.68008	.73314	.69277	.72116	.70525	.70896	9
52	.65430	.75623	.66740	.74470	.68029	.73294	.69298	.72095	.70546	.70875	8
53	.65452	.75604	.66762	.74451	.68051	.73274	.69319	.72075	.70567	.70855	7
54	.65474	.75585	.66783	.74431	.68072	.73254	.69340	.72055	.70587	.70834	6
55	.65496	.75566	.66805	.74412	.68093	.73234	.69361	.72035	.70608	.70813	5
56	.65518	.75547	.66827	.74392	.68115	.73215	.69382	.72015	.70628	.70793	4
57	.65540	.75528	.66848	.74373	.68136	.73195	.69403	.71995	.70649	.70772	3
58	.65562	.75509	.66870	.74353	.68157	.73175	.69424	.71974	.70670	.70752	2
59	.65584	.75490	.66891	.74334	.68179	.73155	.69445	.71954	.70690	.70731	1
60	.65606	.75471	.66913	.74314	.68200	.73135	.69466	.71934	.70711	.70711	0
M.	Cosin.	Sine.	Cosin.	Sine.	Cosin.	Sine.	Cosin.	Sine.	Cosin.	Sine.	M.
	49°		48°		47°		46°		45°		

TABLE XV.

NATURAL TANGENTS AND COTANGENTS

	0°		1°		2°		3°		
M.	Tang.	Cotang.	Tang.	Cotang.	Tang.	Cotang.	Tang.	Cotang.	M.
0	.00000	Infinite.	.01746	57.2900	.03492	28.6363	.05241	19.0811	60
1	.00029	3437.75	.01775	56.3506	.03521	28.3994	.05270	18.9755	59
2	.00058	1718.87	.01804	55.4415	.03550	28.1664	.05299	18.8711	58
3	.00087	1145.92	.01833	54.5613	.03579	27.9372	.05328	18.7678	57
4	.00116	859.436	.01862	53.7086	.03609	27.7117	.05357	18.6656	56
5	.00145	687.549	.01891	52.8821	.03638	27.4899	.05387	18.5645	55
6	.00175	572.957	.01920	52.0807	.03667	27.2715	.05416	18.4645	54
7	.00204	491.106	.01949	51.3032	.03696	27.0566	.05445	18.3655	53
8	.00233	429.718	.01978	50.5485	.03725	26.8450	.05474	18.2677	52
9	.00262	381.971	.02007	49.8157	.03754	26.6367	.05503	18.1708	51
10	.00291	343.774	.02036	49.1039	.03783	26.4316	.05533	18.0750	50
11	.00320	312.521	.02066	48.4121	.03812	26.2296	.05562	17.9802	49
12	.00349	286.478	.02095	47.7395	.03842	26.0307	.05591	17.8863	48
13	.00378	264.441	.02124	47.0853	.03871	25.8348	.05620	17.7934	47
14	.00407	245.552	.02153	46.4489	.03900	25.6418	.05649	17.7015	46
15	.00436	229.182	.02182	45.8294	.03929	25.4517	.05678	17.6106	45
16	.00465	214.858	.02211	45.2261	.03958	25.2644	.05708	17.5205	44
17	.00495	202.219	.02240	44.6386	.03987	25.0798	.05737	17.4314	43
18	.00524	190.984	.02269	44.0661	.04016	24.8978	.05766	17.3432	42
19	.00553	180.932	.02298	43.5081	.04046	24.7185	.05795	17.2558	41
20	.00582	171.885	.02328	42.9641	.04075	24.5418	.05824	17.1693	40
21	.00611	163.700	.02357	42.4335	.04104	24.3675	.05854	17.0837	39
22	.00640	156.259	.02386	41.9158	.04133	24.1957	.05883	16.9990	38
23	.00669	149.465	.02415	41.4106	.04162	24.0263	.05912	16.9150	37
24	.00698	143.237	.02444	40.9174	.04191	23.8593	.05941	16.8319	36
25	.00727	137.507	.02473	40.4358	.04220	23.6945	.05970	16.7496	35
26	.00756	132.219	.02502	39.9655	.04250	23.5321	.05999	16.6681	34
27	.00785	127.321	.02531	39.5059	.04279	23.3718	.06029	16.5874	33
28	.00815	122.774	.02560	39.0568	.04308	23.2137	.06058	16.5075	32
29	.00844	118.540	.02589	38.6177	.04337	23.0577	.06087	16.4283	31
30	.00873	114.589	.02619	38.1885	.04366	22.9038	.06116	16.3499	30
31	.00902	110.892	.02648	37.7686	.04395	22.7519	.06145	16.2722	29
32	.00931	107.426	.02677	37.3579	.04424	22.6020	.06175	16.1952	28
33	.00960	104.171	.02706	36.9560	.04454	22.4541	.06204	16.1190	27
34	.00989	101.107	.02735	36.5627	.04483	22.3081	.06233	16.0435	26
35	.01018	98.2179	.02764	36.1776	.04512	22.1640	.06262	15.9687	25
36	.01047	95.4895	.02793	35.8006	.04541	22.0217	.06291	15.8945	24
37	.01076	92.9085	.02822	35.4313	.04570	21.8813	.06321	15.8211	23
38	.01105	90.4633	.02851	35.0695	.04599	21.7426	.06350	15.7483	22
39	.01135	88.1436	.02881	34.7151	.04628	21.6056	.06379	15.6762	21
40	.01164	85.9398	.02910	34.3678	.04658	21.4704	.06408	15.6048	20
41	.01193	83.8435	.02939	34.0273	.04687	21.3369	.06437	15.5340	19
42	.01222	81.8470	.02968	33.6935	.04716	21.2049	.06467	15.4638	18
43	.01251	79.9434	.02997	33.3662	.04745	21.0747	.06496	15.3943	17
44	.01280	78.1263	.03026	33.0452	.04774	20.9460	.06525	15.3254	16
45	.01309	76.3900	.03055	32.7303	.04803	20.8188	.06554	15 2571	15
46	.01338	74.7292	.03084	32.4213	.04833	20.6932	.06584	15.1893	14
47	.01367	73.1390	.03114	32.1181	.04862	20.5691	.06613	15.1222	13
48	01396	71.6151	.03143	31.8205	.04891	20.4465	.06642	15.0557	12
49	.01425	70.1533	.03172	31.5284	.04920	20.3253	.06671	14.9898	11
50	.01455	68.7501	.03201	31.2416	.04949	20.2056	.06700	14.9244	10
51	.01484	67.4019	.03230	30.9599	.04978	20.0872	.06730	14.8596	9
52	.01513	66.1055	.03259	30.6833	.05007	19.9702	.06759	14.7954	8
53	.01542	64.8580	.03288	30.4116	.05037	19.8546	.06788	14.7317	7
54	.01571	63.6567	.03317	30.1446	.05066	19.7403	.06817	14.6685	6
55	.01600	62.4992	.03346	29.8823	.05095	19.6273	.06847	14.6059	5
56	.01629	61.3829	.03376	29.6245	.05124	19.5156	.06876	14.5438	4
57	.01658	60.3058	.03405	29.3711	.05153	19.4051	.06905	14.4823	3
58	.01687	59.2659	.03434	29.1220	.05182	19.2959	.06934	14.4212	2
59	.01716	58.2612	.03463	28.8771	.05212	19.1879	.06963	14.3607	1
60	.01746	57.2900	.03492	28.6363	.05241	19.0811	.06993	14.3007	0
M.	Cotang.	Tang.	Cotang.	Tang.	Cotang.	Tang.	Cotang.	Tang.	M.
	89°		88°		87°		86°		

	4°		5°		6°		7°		
M.	Tang	Cotang	Tang.	Cotang.	Tang.	Cotang.	Tang.	Cotang.	M.
0	.06993	14.3007	.08749	11.4301	.10510	9.51436	.12278	8.14435	60
1	.07022	14.2411	.08778	11.3919	.10540	9.48781	.12308	8.12481	59
2	.07051	14.1821	.08807	11.3540	.10569	9.46141	.12338	8.10536	58
3	.07080	14.1235	.08837	11.3163	.10599	9.43515	.12367	8.08600	57
4	.07110	14.0655	.08866	11.2789	.10628	9.40904	.12397	8.06674	56
5	.07139	14.0079	.08895	11.2417	.10657	9.38307	.12426	8.04756	55
6	.07168	13.9507	.08925	11.2048	.10687	9.35724	.12456	8.02848	54
7	.07197	13.8940	.08954	11.1681	.10716	9.33155	.12485	8.00948	53
8	.07227	13.8378	.08983	11.1316	.10746	9.30599	.12515	7.99058	52
9	.07256	13.7821	.09013	11.0954	.10775	9.28058	.12544	7.97176	51
10	.07285	13.7267	.09042	11.0594	.10805	9.25530	.12574	7.95302	50
11	.07314	13.6719	.09071	11.0237	.10834	9.23016	.12603	7.93438	49
12	.07344	13.6174	.09101	10.9882	.10863	9.20516	.12633	7.91582	48
13	.07373	13.5634	.09130	10.9529	.10893	9.18028	.12662	7.89734	47
14	.07402	13.5098	.09159	10.9178	.10922	9.15554	.12692	7.87895	46
15	.07431	13.4566	.09189	10.8829	.10952	9.13093	.12722	7.86064	45
16	.07461	13.4039	.09218	10.8483	.10981	9.10646	.12751	7.84242	44
17	.07490	13.3515	.09247	10.8139	.11011	9.08211	.12781	7.82428	43
18	.07519	13.2996	.09277	10.7797	.11040	9.05789	.12810	7.80622	42
19	.07548	13.2480	.09306	10.7457	.11070	9.03379	.12840	7.78825	41
20	.07578	13.1969	.09335	10.7119	.11099	9.00983	.12869	7.77035	40
21	.07607	13.1461	.09365	10.6783	.11128	8.98598	.12899	7.75254	39
22	.07636	13.0958	.09394	10.6450	.11158	8.96227	.12929	7.73480	38
23	.07665	13.0458	.09423	10.6118	.11187	8.93867	.12958	7.71715	37
24	.07695	12.9962	.09453	10.5789	.11217	8.91520	.12988	7.69957	36
25	.07724	12.9469	.09482	10.5462	.11246	8.89185	.13017	7.68208	35
26	.07753	12.8981	.09511	10.5136	.11276	8.86862	.13047	7.66466	34
27	.07782	12.8496	.09541	10.4813	.11305	8.84551	.13076	7.64732	33
28	.07812	12.8014	.09570	10.4491	.11335	8.82252	.13106	7.63005	32
29	.07841	12.7536	.09600	10.4172	.11364	8.79964	.13136	7.61287	31
30	.07870	12.7062	.09629	10.3854	.11394	8.77689	.13165	7.59575	30
31	.07899	12.6591	.09658	10.3538	.11423	8.75425	.13195	7.57872	29
32	.07929	12.6124	.09688	10.3224	.11452	8.73172	.13224	7.56176	28
33	.07958	12.5660	.09717	10.2913	.11482	8.70931	.13254	7.54487	27
34	.07987	12.5199	.09746	10.2602	.11511	8.68701	.13284	7.52806	26
35	.08017	12.4742	.09776	10.2294	.11541	8.66482	.13313	7.51132	25
36	.08046	12.4288	.09805	10.1988	.11570	8.64275	.13343	7.49465	24
37	.08075	12.3838	.09834	10.1683	.11600	8.62078	.13372	7.47806	23
38	.08104	12.3390	.09864	10.1381	.11629	8.59893	.13402	7.46154	22
39	.08134	12.2946	.09893	10.1080	.11659	8.57718	.13432	7.44509	21
40	.08163	12.2505	.09923	10.0780	.11688	8.55555	.13461	7.42871	20
41	.08192	12.2067	.09952	10.0483	.11718	8.53402	.13491	7.41240	19
42	.08221	12.1632	.09981	10.0187	.11747	8.51259	.13521	7.39616	18
43	.08251	12.1201	.10011	9.98931	.11777	8.49128	.13550	7.37999	17
44	.08280	12.0772	.10040	9.96007	.11806	8.47007	.13580	7.36389	16
45	.08309	12.0346	.10069	9.93101	.11836	8.44896	.13609	7.34786	15
46	.08339	11.9923	.10099	9.90211	.11865	8.42795	.13639	7.33190	14
47	.08368	11.9504	.10128	9.87338	.11895	8.40705	.13669	7.31600	13
48	.08397	11.9087	.10158	9.84482	.11924	8.38625	.13698	7.30018	12
49	.08427	11.8673	.10187	9.81641	.11954	8.36555	.13728	7.28442	11
50	.08456	11.8262	.10216	9.78817	.11983	8.34496	.13758	7.26873	10
51	.08485	11.7853	.10246	9.76009	.12013	8.32446	.13787	7.25310	9
52	.08514	11.7448	.10275	9.73217	.12042	8.30406	.13817	7.23754	8
53	.08544	11.7045	.10305	9.70441	.12072	8.28376	.13846	7.22204	7
54	.08573	11.6645	.10334	9.67680	.12101	8.26355	.13876	7.20661	6
55	.08602	11.6248	.10363	9.64935	.12131	8.24345	.13906	7.19125	5
56	.08632	11.5853	.10393	9.62205	.12160	8.22344	.13935	7.17594	4
57	.08661	11.5461	.10422	9.59490	.12190	8.20352	.13965	7.16071	3
58	.08690	11.5072	.10452	9.56791	.12219	8.18370	.13995	7.14553	2
59	.08720	11.4685	.10481	9.54106	.12249	8.16398	.14024	7.13042	1
60	.08749	11.4301	.10510	9.51436	.12278	8.14435	.14054	7.11537	0
M.	Cotang.	Tang.	Cotang.	Tang.	Cotang.	Tang.	Cotang.	Tang.	M.
	85°		84°		83°		82°		

M.	8° Tang.	8° Cotang.	9° Tang.	9° Cotang.	10° Tang.	10° Cotang.	11° Tang.	11° Cotang.	M.
0	.14054	7.11537	.15838	6.31375	.17633	5.67128	.19438	5.14455	60
1	.14084	7.10038	15868	6.30189	.17663	5.66165	.19468	5.13658	59
2	.14113	7.08546	.15898	6.29007	.17693	5.65205	.19498	5.12862	58
3	.14143	7.07059	.15928	6.27829	.17723	5.64248	.19529	5.12069	57
4	.14173	7.05579	.15958	6.26655	.17753	5.63295	.19559	5.11279	56
5	.14202	7.04105	.15988	6.25486	.17783	5.62344	.19589	5.10490	55
6	.14232	7.02637	.16017	6.24321	.17813	5.61397	.19619	5.09704	54
7	.14262	6.91174	.16047	6.23160	.17843	5.60452	.19649	5.08921	53
8	.14291	6.99718	.16077	6.22003	.17873	5.59511	.19680	5.08139	52
9	.14321	6.98268	.16107	6.20851	.17903	5.58573	.19710	5.07360	51
10	.14351	6.96823	.16137	6.19703	.17933	5.57638	.19740	5.06584	50
11	.14381	6.95385	.16167	6.18559	.17963	5.56706	.19770	5.05809	49
12	.14410	6.93952	.16196	6.17419	.17993	5.55777	.19801	5.05037	48
13	.14440	6.92525	.16226	6.16283	.18023	5.54851	.19831	5.04267	47
14	.14470	6.91104	.16256	6.15151	.18053	5.53927	.19861	5.03499	46
15	.14499	6.89688	.16286	6.14023	.18083	5.53007	.19891	5.02734	45
16	.14529	6.88278	.16316	6.12899	.18113	5.52090	.19921	5.01971	44
17	.14559	6.86874	.16346	6.11779	.18143	5.51176	.19952	5.01210	43
18	.14588	6.85475	.16376	6.10664	.18173	5.50264	.19982	5.00451	42
19	.14618	6.84082	.16405	6.09552	.18203	5.49356	.20012	4.99695	41
20	.14648	6.82694	.16435	6.08444	.18233	5.48451	.20042	4.98940	40
21	.14678	6.81312	.16465	6.07340	.18263	5.47548	.20073	4.98188	39
22	.14707	6.79936	.16495	6.06240	.18293	5.46648	.20103	4.97438	38
23	.14737	6.78564	.16525	6.05143	.18323	5.45751	.20133	4.96690	37
24	.14767	6.77199	.16555	6.04051	.18353	5.44857	.20164	4.95945	36
25	.14796	6.75838	.16585	6.02962	.18384	5.43966	.20194	4.95201	35
26	.14826	6.74483	.16615	6.01878	.18414	5.43077	.20224	4.94460	34
27	.14856	6.73133	.16645	6.00797	.18444	5.42192	.20254	4.93721	33
28	.14886	6.71789	.16674	5.99720	.18474	5.41309	.20285	4.92984	32
29	.14915	6.70450	.16704	5.98646	.18504	5.40429	.20315	4.92249	31
30	.14945	6.69116	.16734	5.97576	.18534	5.39552	.20345	4.91516	30
31	.14975	6.67787	.16764	5.96510	.18564	5.38677	.20376	4.90785	29
32	.15005	6.66463	.16794	5.95448	.18594	5.37805	.20406	4.90056	28
33	.15034	6.65144	.16824	5.94390	.18624	5.36936	.20436	4.89330	27
34	.15064	6.63831	.16854	5.93335	.18654	5.36070	.20466	4.88605	26
35	.15094	6.62523	.16884	5.92283	.18684	5.35206	.20497	4.87882	25
36	.15124	6.61219	16914	5.91236	.18714	5.34345	.20527	4.87162	24
37	.15153	6.59921	.16944	5.90191	.18745	5.33487	.20557	4.86444	23
38	.15183	6.58627	.16974	5.89151	.18775	5.32631	.20588	4.85727	22
39	.15213	6.57339	.17004	5.88114	.18805	5.31778	.20618	4.85013	21
40	.15243	6.56055	.17033	5.87080	.18835	5.30928	.20648	4.84300	20
41	.15272	6.54777	.17063	5.86051	.18865	5.30080	.20679	4.83590	19
42	.15302	6.53503	.17093	5.85024	.18895	5.29235	.20709	4.82882	18
43	.15332	6.52234	.17123	5.84001	.18925	5.28393	.20739	4.82175	17
44	.15362	6.50970	.17153	5.82982	.18955	5.27553	.20770	4.81471	16
45	.15391	6.49710	.17183	5.81966	.18986	5.26715	.20800	4.80769	15
46	.15421	6.48456	.17213	5.80953	.19016	5.25880	.20830	4.80068	14
47	.15451	6.47206	.17243	5.79944	.19046	5.25048	.20861	4.79370	13
48	.15481	6.45961	.17273	5.78938	.19076	5.24218	.20891	4.78673	12
49	.15511	6.44720	.17303	5.77936	.19106	5.23391	.20921	4.77978	11
50	.15540	6.43484	.17333	5.76937	.19136	5.22566	.20952	4.77286	10
51	.15570	6.42253	.17363	5.75941	.19166	5.21744	.20982	4.76595	9
52	.15600	6.41026	.17393	5.74949	.19197	5.20925	.21013	4.75906	8
53	.15630	6.39804	.17423	5.73960	.19227	5.20107	.21043	4.75219	7
54	.15660	6.38587	.17453	5.72974	.19257	5.19293	.21073	4.74534	6
55	.15689	6.37374	.17483	5.71992	.19287	5.18480	.21104	4.73851	5
56	.15719	6.36165	.17513	5.71013	.19317	5.17671	.21134	4.73170	4
57	.15749	6.34961	.17543	5 70037	.19347	5.16863	.21164	4.72490	3
58	.15779	6.33761	.17573	5.69064	.19378	5.16058	.21195	4.71813	2
59	.15809	6.32566	.17603	5.68094	.19408	5.15256	.21225	4.71137	1
60	.15838	6.31375	.17633	5.67128	.19438	5.14455	.21256	4.70463	0
M.	Cotang.	Tang.	Cotang.	Tang	Cotang.	Tang.	Cotang.	Tang.	M.
	81°		80°		79°		78°		

	12°		13°		14°		15°		
M.	Tang.	Cotang.	Tang.	Cotang.	Tang.	Cotang.	Tang.	Cotang.	M.
0	.21256	4.70463	.23087	4.33148	.24933	4.01078	.26795	3.73205	60
1	.21286	4.69791	.23117	4.32573	.24964	4.00582	.26826	3.72771	59
2	.21316	4.69121	.23148	4.32001	.24995	4.00086	.26857	3.72338	58
3	.21347	4.68452	.23179	4.31430	.25026	3.99592	.26888	3.71907	57
4	.21377	4.67786	.23209	4.30860	.25056	3.99099	.26920	3.71476	56
5	.21408	4.67121	.23240	4.30291	.25087	3.98607	.26951	3.71046	55
6	.21438	4.66458	.23271	4.29724	.25118	3.98117	.26982	3.70616	54
7	.21469	4.65797	.23301	4.29159	.25149	3.97627	.27013	3.70188	53
8	.21499	4.65138	.23332	4.28595	.25180	3.97139	.27044	3.69761	52
9	.21529	4.64480	.23363	4.28032	.25211	3.96651	.27076	3.69335	51
10	.21560	4.63825	.23393	4.27471	.25242	3.96165	.27107	3.68909	50
11	.21590	4.63171	.23424	4.26911	.25273	3.95680	.27138	3.68485	49
12	.21621	4.62518	.23455	4.26352	.25304	3.95196	.27169	3.68061	48
13	.21651	4.61868	.23485	4.25795	.25335	3.94713	.27201	3.67638	47
14	.21682	4.61219	.23516	4.25239	.25366	3.94232	.27232	3.67217	46
15	.21712	4.60572	.23547	4.24685	.25397	3.93751	.27263	3.66796	45
16	.21743	4.59927	.23578	4.24132	.25428	3.93271	.27294	3.66376	44
17	.21773	4.59283	.23608	4.23580	.25459	3.92793	.27326	3.65957	43
18	.21804	4.58641	.23639	4.23030	.25490	3.92316	.27357	3.65538	42
19	.21834	4.58001	.23670	4.22481	.25521	3.91839	.27388	3.65121	41
20	.21864	4.57363	.23700	4.21933	.25552	3.91364	.27419	3.64705	40
21	.21895	4.56726	.23731	4.21387	.25583	3.90890	.27451	3.64289	39
22	.21925	4.56091	.23762	4.20842	.25614	3.90417	.27482	3.63874	38
23	.21956	4.55458	.23793	4.20298	.25645	3.89945	.27513	3.63461	37
24	.21986	4.54826	.23823	4.19756	.25676	3.89474	.27545	3.63048	36
25	.22017	4.54196	.23854	4.19215	.25707	3.89004	.27576	3.62636	35
26	.22047	4.53568	.23885	4.18675	.25738	3.88536	.27607	3.62224	34
27	.22078	4.52941	.23916	4.18137	.25769	3.88068	.27638	3.61814	33
28	.22108	4.52316	.23946	4.17600	.25800	3.87601	.27670	3.61405	32
29	.22139	4.51693	.23977	4.17064	.25831	3.87136	.27701	3.60996	31
30	.22169	4.51071	.24008	4.16530	.25862	3.86671	.27732	3.60588	30
31	.22200	4.50451	.24039	4.15997	.25893	3.86208	.27764	3.60181	29
32	.22231	4.49832	.24069	4.15465	.25924	3.85745	.27795	3.59775	28
33	.22261	4.49215	.24100	4.14934	.25955	3.85284	.27826	3.59370	27
34	.22292	4.48600	.24131	4.14405	.25986	3.84824	.27858	3.58966	26
35	.22322	4.47986	.24162	4.13877	.26017	3.84364	.27889	3.58562	25
36	.22353	4.47374	.24193	4.13350	.26048	3.83906	.27921	3.58160	24
37	.22383	4.46764	.24223	4.12825	.26079	3.83449	.27952	3.57758	23
38	.22414	4.46155	.24254	4.12301	.26110	3.82992	.27983	3.57357	22
39	.22444	4.45548	.24285	4.11778	.26141	3.82537	.28015	3.56957	21
40	.22475	4.44942	.24316	4.11256	.26172	3.82083	.28046	3.56557	20
41	.22505	4.44338	.24347	4.10736	.26203	3.81630	.28077	3.56159	19
42	.22536	4.43735	.24377	4.10216	.26235	3.81177	.28109	3.55761	18
43	.22567	4.43134	.24408	4.09699	.26266	3.80726	.28140	3.55364	17
44	.22597	4.42534	.24439	4.09182	.26297	3.80276	.28172	3.54968	16
45	.22628	4.41936	.24470	4.08666	.26328	3.79827	.28203	3.54573	15
46	.22658	4.41340	.24501	4.08152	.26359	3.79378	.28234	3.54179	14
47	.22689	4.40745	.24532	4.07639	.26390	3.78931	.28266	3.53785	13
48	.22719	4.40152	.21562	4.07127	.26421	3.78485	.28297	3.53393	12
49	.22750	4.39560	.24593	4.06616	.26452	3.78040	.28329	3.53001	11
50	.22781	4.38969	.24624	4.06107	.26483	3.77595	.28360	3.52609	10
51	.22811	4.38381	.24655	4.05599	.26515	3.77152	.28391	3.52219	9
52	.22842	4.37793	.24686	4.05092	.26546	3.76709	.28423	3.51829	8
53	.22872	4.37207	.24717	4.04586	.26577	3.76268	.28454	3.51441	7
54	.22903	4.36623	.24747	4.04081	.26608	3.75828	.28486	3.51053	6
55	.22934	4.36040	.24778	4.03578	.26639	3.75388	.28517	3.50666	5
56	.22964	4.35459	.24809	4.03076	.26670	3.74950	.28549	3.50279	4
57	.22995	4.34879	.24840	4.02574	.26701	3.74512	.28580	3.49894	3
58	.23026	4.34300	.24871	4.02074	.26733	3.74075	.28612	3.49509	2
59	.23056	4.33723	.24902	4.01576	.26764	3.73640	.28643	3.49125	1
60	.23087	4.33148	.24933	4.01078	.26795	3.73205	.28675	3.48741	0
M.	Cotang.	Tang.	Cotang.	Tang.	Cotang.	Tang.	Cotang.	Tang.	M.
	77°		76°		75°		74°		

	16°		17°		18°		19°		
M.	Tang.	Cotang.	Tang.	Cotang.	Tang.	Cotang.	Tang.	Cotang.	M.
0	.28675	3.48741	.30573	3.27085	.32492	3.07768	.34433	2.90421	60
1	.28706	3.48359	.30605	3.26745	.32524	3.07464	.34465	2.90147	59
2	.28738	3.47977	.30637	3.26406	.32556	3.07160	.34498	2.89873	58
3	.28769	3.47596	.30669	3.26067	.32588	3.06857	.34530	2.89600	57
4	.28800	3.47216	.30700	3.25729	.32621	3.06554	.34563	2.89327	56
5	.28832	3.46837	.30732	3.25392	.32653	3.06252	.34596	2.89055	55
6	.28864	3.46458	.30764	3.25055	.32685	3.05950	.34628	2.88783	54
7	.28895	3.46080	.30796	3.24719	.32717	3.05649	.34661	2.88511	53
8	.28927	3.45703	.30828	3.24383	.32749	3.05349	.34693	2.88240	52
9	.28958	3.45327	.30860	3.24049	.32782	3.05049	.34726	2.87970	51
10	.28990	3.44951	.30891	3.23714	.32814	3.04749	.34758	2.87700	50
11	.29021	3.44576	.30923	3.23381	.32846	3.04450	.34791	2.87430	49
12	.29053	3.44202	.30955	3.23048	.32878	3.04152	.34824	2.87161	48
13	.29084	3.43829	.30987	3.22715	.32911	3.03854	.34856	2.86892	47
14	.29116	3.43456	.31019	3.22384	.32943	3.03556	.34889	2.86624	46
15	.29147	3.43084	.31051	3.22053	.32975	3.03260	.34922	2.86356	45
16	.29179	3.42713	.31083	3.21722	.33007	3.02963	.34954	2.86089	44
17	.29210	3.42343	.31115	3.21392	.33040	3.02667	.34987	2.85822	43
18	.29242	3.41973	.31147	3.21063	.33072	3.02372	.35020	2.85555	42
19	.29274	3.41604	.31178	3.20734	.33104	3.02077	.35052	2.85289	41
20	.29305	3.41236	.31210	3.20406	.33136	3.01783	.35085	2.85023	40
21	.29337	3.40869	.31242	3.20079	.33169	3.01489	.35118	2.84758	39
22	.29368	3.40502	.31274	3.19752	.33201	3.01196	.35150	2.84494	38
23	.29400	3.40136	.31306	3.19426	.33233	3.00903	.35183	2.84229	37
24	.29432	3.39771	.31338	3.19100	.33266	3.00611	.35216	2.83965	36
25	.29463	3.39406	.31370	3.18775	.33298	3.00319	.35248	2.83702	35
26	.29495	3.39042	.31402	3.18451	.33330	3.00028	.35281	2.83439	34
27	.29526	3.38679	.31434	3.18127	.33363	2.99738	.35314	2.83176	33
28	.29558	3.38317	.31466	3.17804	.33395	2.99447	.35346	2.82914	32
29	.29590	3.37955	.31498	3.17481	.33427	2.99158	.35379	2.82653	31
30	.29621	3.37594	.31530	3.17159	.33460	2.98868	.35412	2.82391	30
31	.29653	3.37234	.31562	3.16838	.33492	2.98580	.35445	2.82130	29
32	.29685	3.36875	.31594	3.16517	.33524	2.98292	.35477	2.81870	28
33	.29716	3.36516	.31626	3.16197	.33557	2.98004	.35510	2.81610	27
34	.29748	3.36158	.31658	3.15877	.33589	2.97717	.35543	2.81350	26
35	.29780	3.35800	.31690	3.15558	.33621	2.97430	.35576	2.81091	25
36	.29811	3.35443	.31722	3.15240	.33654	2.97144	.35608	2.80833	24
37	.29843	3.35087	.31754	3.14922	.33686	2.96858	.35641	2.80574	23
38	.29875	3.34732	.31786	3.14605	.33718	2.96573	.35674	2.80316	22
39	.29906	3.34377	.31818	3.14288	.33751	2.96288	.35707	2.80059	21
40	.29938	3.34023	.31850	3.13972	.33783	2.96004	.35740	2.79802	20
41	.29970	3.33670	.31882	3.13656	.33816	2.95721	.35772	2.79545	19
42	.30001	3.33317	.31914	3.13341	.33848	2.95437	.35805	2.79289	18
43	.30033	3.32965	.31946	3.13027	.33881	2.95155	.35838	2.79033	17
44	.30065	3.32614	.31978	3.12713	.33913	2.94872	.35871	2.78778	16
45	.30097	3.32264	.32010	3.12400	.33945	2.94591	.35904	2.78523	15
46	.30128	3.31914	.32042	3.12087	.33978	2.94309	.35937	2.78269	14
47	.30160	3.31565	.32074	3.11775	.34010	2.94028	.35969	2.78014	13
48	.30192	3.31216	.32106	3.11464	.34043	2.93748	.36002	2.77761	12
49	.30224	3.30868	.32139	3.11153	.34075	2.93468	.36035	2.77507	11
50	.30255	3.30521	.32171	3.10842	.34108	2.93189	.36068	2.77254	10
51	.30287	3.30174	.32203	3.10532	.34140	2.92910	.36101	2.77002	9
52	.30319	3.29829	.32235	3.10223	.34173	2.92632	.36134	2.76750	8
53	.30351	3.29483	.32267	3.09914	.34205	2.92354	.36167	2.76498	7
54	.30382	3.29139	.32299	3.09606	.34238	2.92076	.36199	2.76247	6
55	.30414	3.28795	.32331	3.09298	.34270	2.91799	.36232	2.75996	5
56	.30446	3.28452	.32363	3.08991	.34303	2.91523	.36265	2.75746	4
57	.30478	3.28109	.32396	3.08685	.34335	2.91246	.36298	2.75496	3
58	.30509	3.27767	.32428	3.08379	.34368	2.90971	.36331	2.75246	2
59	.30541	3.27426	.32460	3.08073	.34400	2.90696	.36364	2.74997	1
60	.30573	3.27085	.32492	3.07768	.34433	2.90421	.36397	2.74748	0
M.	Cotang.	Tang.	Cotang.	Tang.	Cotang.	Tang.	Cotang.	Tang.	M.
	73°		72°		71°		70°		

	20°		21°		22°		23°		
M.	Tang.	Cotang.	Tang.	Cotang.	Tang.	Cotang.	Tang.	Cotang.	M.
0	.36397	2.74748	.38386	2.60509	.40403	2.47509	.42447	2.35585	60
1	.36430	2.74499	.38420	2.60283	.40436	2.47302	.42482	2.35395	59
2	.36463	2.74251	.38453	2.60057	.40470	2.47095	.42516	2.35205	58
3	.36496	2.74004	.38487	2.59831	.40504	2.46888	.42551	2.35015	57
4	.36529	2.73756	.38520	2.59606	.40538	2.46682	.42585	2.34825	56
5	.36562	2.73509	.38553	2.59381	.40572	2.46476	.42619	2.34636	55
6	.36595	2.73263	.38587	2.59156	.40606	2.46270	.42654	2.34447	54
7	.36628	2.73017	.38620	2.58932	.40640	2.46065	.42688	2.34258	53
8	.36661	2.72771	.38654	2.58708	.40674	2.45860	.42722	2.34069	52
9	.36694	2.72526	.38687	2.58484	.40707	2.45655	.42757	2.33881	51
10	.36727	2.72281	.38721	2.58261	.40741	2.45451	.42791	2.33693	50
11	.36760	2.72036	.38754	2.58038	.40775	2.45246	.42826	2.33505	49
12	.36793	2.71792	.38787	2.57815	.40809	2.45043	.42860	2.33317	48
13	.36826	2.71548	.38821	2.57593	.40843	2.44839	.42894	2.33130	47
14	.36859	2.71305	.38854	2.57371	.40877	2.44636	.42929	2.32943	46
15	.36892	2.71062	.38888	2.57150	.40911	2.44433	.42963	2.32756	45
16	.36925	2.70819	.38921	2.56928	.40945	2.44230	.42998	2.32570	44
17	.36958	2.70577	.38955	2.56707	.40979	2.44027	.43032	2.32383	43
18	.36991	2.70335	.38988	2.56487	.41013	2.43825	.43067	2.32197	42
19	.37024	2.70094	.39022	2.56266	.41047	2.43623	.43101	2.32012	41
20	.37057	2.69853	.39055	2.56046	.41081	2.43422	.43136	2.31826	40
21	.37090	2.69612	.39089	2.55827	.41115	2.43220	.43170	2.31641	39
22	.37123	2.69371	.39122	2.55608	.41149	2.43019	.43205	2.31456	38
23	.37157	2.69131	.39156	2.55389	.41183	2.42819	.43239	2.31271	37
24	.37190	2.68892	.39190	2.55170	.41217	2.42618	.43274	2.31086	36
25	.37223	2.68653	.39223	2.54952	.41251	2.42418	.43308	2.30902	35
26	.37256	2.68414	.39257	2.54734	.41285	2.42218	.43343	2.30718	34
27	.37289	2.68175	.39290	2.54516	.41319	2.42019	.43378	2.30534	33
28	.37322	2.67937	.39324	2.54299	.41353	2.41819	.43412	2.30351	32
29	.37355	2.67700	.39357	2.54082	.41387	2.41620	.43447	2.30167	31
30	.37388	2.67462	.39391	2.53865	.41421	2.41421	.43481	2.29984	30
31	.37422	2.67225	.39425	2.53648	.41455	2.41223	.43516	2.29801	29
32	.37455	2.66989	.39458	2.53432	.41490	2.41025	.43550	2.29619	28
33	.37488	2.66752	.39492	2.53217	.41524	2.40827	.43585	2.29437	27
34	.37521	2.66516	.39526	2.53001	.41558	2.40629	.43620	2.29254	26
35	.37554	2.66281	.39559	2.52786	.41592	2.40432	.43654	2.29073	25
36	.37588	2.66046	.39593	2.52571	.41626	2.40235	.43689	2.28891	24
37	.37621	2.65811	.39626	2.52357	.41660	2.40038	.43724	2.28710	23
38	.37654	2.65576	.39660	2.52142	.41694	2.39841	.43758	2.28528	22
39	.37687	2.65342	.39694	2.51929	.41728	2.39645	.43793	2.28348	21
40	.37720	2.65109	.39727	2.51715	.41763	2.39449	.43828	2.28167	20
41	.37754	2.64875	.39761	2.51502	.41797	2.39253	.43862	2.27987	19
42	.37787	2.64642	.39795	2.51289	.41831	2.39058	.43897	2.27806	18
43	.37820	2.64410	.39829	2.51076	.41865	2.38863	.43932	2.27626	17
44	.37853	2.64177	.39862	2.50864	.41899	2.38668	.43966	2.27447	16
45	.37887	2.63945	.39896	2.50652	.41933	2.38473	.44001	2.27267	15
46	.37920	2.63714	.39930	2.50440	.41968	2.38279	.44036	2.27088	14
47	37953	2.63483	.39963	2.50229	.42002	2.38084	.44071	2.26909	13
48	.37986	2.63252	.39997	2 50018	.42036	2.37891	.44105	2.26730	12
49	.38020	2.63021	.40031	2.49807	.42070	2.37697	.44140	2.26552	11
50	.38053	2.62791	.40065	2.49597	.42105	2.37504	.44175	2.26374	10
51	.38086	2.62561	.40098	2.49386	.42139	2.37311	.44210	2.26196	9
52	.38120	2.62332	.40132	2.49177	.42173	2.37118	.44244	2.26018	8
53	.38153	2.62103	.40166	2.48967	.42207	2.36925	.44279	2.25840	7
54	.38186	2.61874	.40200	2.48758	.42242	2.36733	.44314	2.25663	6
55	.38220	2.61646	.40234	2.48549	.42276	2.36541	.44349	2.25486	5
56	.38253	2.61418	.40267	2.48340	.42310	2.36349	.44384	2.25309	4
57	.38286	2.61190	.40301	2.48132	.42345	2.36158	.44418	2.25132	3
58	.38320	2 60963	.40335	2.47924	.42379	2.35967	.44453	2.24956	2
59	.38353	2.60736	.40369	2.47716	.42413	2.35776	.44488	2.24780	1
60	.38386	2.60509	.40403	2.47509	.42447	2.35585	.44523	2.24604	0
M.	Cotang.	Tang.	Cotang.	Tang.	Cotang.	Tang.	Cotang.	Tang.	M.
	69°		68°		67°		66°		

	24°		25°		26°		27°		
M.	Tang.	Cotang.	Tang.	Cotang.	Tang.	Cotang.	Tang.	Cotang.	M.
0	.44523	2.24604	.46631	2.14451	.48773	2.05030	.50953	1.96261	60
1	.44558	2.24428	.46666	2.14288	.48809	2.04879	.50989	1.96120	59
2	.44593	2.24252	.46702	2.14125	.48845	2.04728	.51026	1.95979	58
3	.44627	2.24077	.46737	2.13963	.48881	2.04577	.51063	1.95838	57
4	.44662	2.23902	.46772	2.13801	.48917	2.04426	.51099	1.95698	56
5	.44697	2.23727	.46808	2.13639	.48953	2.04276	.51136	1.95557	55
6	.44732	2.23553	.46843	2.13477	.48989	2.04125	.51173	1.95417	54
7	.44767	2.23378	.46879	2.13316	.49026	2.03975	.51209	1.95277	53
8	.44802	2.23204	.46914	2.13154	.49062	2.03825	.51246	1.95137	52
9	.44837	2.23030	.46950	2.12993	.49098	2.03675	.51283	1.94997	51
10	.44872	2.22857	.46985	2.12832	.49134	2.03526	.51319	1.94858	50
11	.44907	2.22683	.47021	2.12671	.49170	2.03376	.51356	1.94718	49
12	.44942	2.22510	.47056	2.12511	.49206	2.03227	.51393	1.94579	48
13	.44977	2.22337	.47092	2.12350	.49242	2.03078	.51430	1.94440	47
14	.45012	2.22164	.47128	2.12190	.49278	2.02929	.51467	1.94301	46
15	.45047	2.21992	.47163	2.12030	.49315	2.02780	.51503	1.94162	45
16	.45082	2.21819	.47199	2.11871	.49351	2.02631	.51540	1.94023	44
17	.45117	2.21647	.47234	2.11711	.49387	2.02483	.51577	1.93885	43
18	.45152	2.21475	.47270	2.11552	.49423	2.02335	.51614	1.93746	42
19	.45187	2.21304	.47305	2.11392	.49459	2.02187	.51651	1.93608	41
20	.45222	2.21132	.47341	2.11233	.49495	2.02039	.51688	1.93470	40
21	.45257	2.20961	.47377	2.11075	.49532	2.01891	.51724	1.93332	39
22	.45292	2.20790	.47412	2.10916	.49568	2.01743	.51761	1.93195	38
23	.45327	2.20619	.47448	2.10758	.49604	2.01596	.51798	1.93057	37
24	.45362	2.20449	.47483	2.10600	.49640	2.01449	.51835	1.92920	36
25	.45397	2.20278	.47519	2.10442	.49677	2.01302	.51872	1.92782	35
26	.45432	2.20108	.47555	2.10284	.49713	2.01155	.51909	1.92645	34
27	.45467	2.19938	.47590	2.10126	.49749	2.01008	.51946	1.92508	33
28	.45502	2.19769	.47626	2.09969	.49786	2.00862	.51983	1.92371	32
29	.45538	2.19599	.47662	2.09811	.49822	2.00715	.52020	1.92235	31
30	.45573	2.19430	.47698	2.09654	.49858	2.00569	.52057	1.92098	30
31	.45608	2.19261	.47733	2.09498	.49894	2.00423	.52094	1.91962	29
32	.45643	2.19092	.47769	2.09341	.49931	2.00277	.52131	1.91826	28
33	.45678	2.18923	.47805	2.09184	.49967	2.00131	.52168	1.91690	27
34	.45713	2.18755	.47840	2.09028	.50004	1.99986	.52205	1.91554	26
35	.45748	2.18587	.47876	2.08872	.50040	1.99841	.52242	1.91418	25
36	.45784	2.18419	.47912	2.08716	.50076	1.99695	.52279	1.91282	24
37	.45819	2.18251	.47948	2.08560	.50113	1.99550	.52316	1.91147	23
38	.45854	2.18084	.47984	2.08405	.50149	1.99406	.52353	1.91012	22
39	.45889	2.17916	.48019	2.08250	.50185	1.99261	.52390	1.90876	21
40	.45924	2.17749	.48055	2.08094	.50222	1.99116	.52427	1.90741	20
41	.45960	2.17582	.48091	2.07939	.50258	1.98972	.52464	1.90607	19
42	.45995	2.17416	.48127	2.07785	.50295	1.98828	.52501	1.90472	18
43	.46030	2.17249	.48163	2.07630	.50331	1.98684	.52538	1.90337	17
44	.46065	2.17083	.48198	2.07476	.50368	1.98540	.52575	1.90203	16
45	.46101	2.16917	.48234	2.07321	.50404	1.98396	.52613	1.90069	15
46	.46136	2.16751	.48270	2.07167	.50441	1.98253	.52650	1.89935	14
47	.46171	2.16585	.48306	2.07014	.50477	1.98110	.52687	1.89801	13
48	.46206	2.16420	.48342	2.06860	.50514	1.97966	.52724	1.89667	12
49	.46242	2.16255	.48378	2.06706	.50550	1.97823	.52761	1.89533	11
50	.46277	2.16090	.48414	2.06553	.50587	1.97681	.52798	1.89400	10
51	.46312	2.15925	.48450	2.06400	.50623	1.97538	.52836	1.89266	9
52	.46348	2.15760	.48486	2.06247	.50660	1.97395	.52873	1.89133	8
53	.46383	2.15596	.48521	2.06094	.50696	1.97253	.52910	1.89000	7
54	.46418	2.15432	.48557	2.05942	.50733	1.97111	.52947	1.88867	6
55	.46454	2.15268	.48593	2.05790	.50769	1.96969	.52985	1.88734	5
56	.46489	2.15104	.48629	2.05637	.50806	1.96827	.53022	1.88602	4
57	.46525	2.14940	.48665	2.05485	.50843	1.96685	.53059	1.88469	3
58	.46560	2.14777	.48701	2.05333	.50879	1.96544	.53096	1.88337	2
59	.46595	2.14614	.48737	2.05182	.50916	1.96402	.53134	1.88205	1
60	.46631	2.14451	.48773	2.05030	.50953	1.96261	.53171	1.88073	0
M.	Cotang.	Tang.	Cotang.	Tang.	Cotang.	Tang.	Cotang.	Tang.	M.
	65°		64°		63°		62°		

M	28° Tang.	28° Cotang.	29° Tang.	29° Cotang.	30° Tang.	30° Cotang.	31° Tang.	31° Cotang.	M.
0	.53171	1.88073	.55431	1.80405	.57735	1.73205	.60086	1.66428	60
1	.53208	1.87941	.55469	1.80281	.57774	1.73089	.60126	1.66318	59
2	.53246	1.87809	.55507	1.80158	.57813	1.72973	.60165	1.66209	58
3	.53283	1.87677	.55545	1.80034	.57851	1.72857	.60205	1.66099	57
4	.53320	1.87546	.55583	1.79911	.57890	1.72741	.60245	1.65990	56
5	.53358	1.87415	.55621	1.79788	.57929	1.72625	.60284	1.65881	55
6	.53395	1.87283	.55659	1.79665	.57968	1.72509	.60324	1.65772	54
7	.53432	1.87152	.55697	1.79542	.58007	1.72393	.60364	1.65663	53
8	.53470	1.87021	.55736	1.79419	.58046	1.72278	.60403	1.65554	52
9	.53507	1.86891	.55774	1.79296	.58085	1.72163	.60443	1.65445	51
10	.53545	1.86760	.55812	1.79174	.58124	1.72047	.60483	1.65337	50
11	.53582	1.86630	.55850	1.79051	.58162	1.71932	.60522	1.65228	49
12	.53620	1.86499	.55888	1.78929	.58201	1.71817	.60562	1.65120	48
13	.53657	1.86369	.55926	1.78807	.58240	1.71702	.60602	1.65011	47
14	.53694	1.86239	.55964	1.78685	.58279	1.71588	.60642	1.64903	46
15	.53732	1.86109	.56003	1.78563	.58318	1.71473	.60681	1.64795	45
16	.53769	1.85979	.56041	1.78441	.58357	1.71358	.60721	1.64687	44
17	.53807	1.85850	.56079	1.78319	.58396	1.71244	.60761	1.64579	43
18	.53844	1.85720	.56117	1.78198	.58435	1.71129	.60801	1.64471	42
19	.53882	1.85591	.56156	1.78077	.58474	1.71015	.60841	1.64363	41
20	.53920	1.85462	.56194	1.77955	.58513	1.70901	.60881	1.64256	40
21	.53957	1.85333	.56232	1.77834	.58552	1.70787	.60921	1.64148	39
22	.53995	1.85204	.56270	1.77713	.58591	1.70673	.60960	1.64041	38
23	.54032	1.85075	.56309	1.77592	.58631	1.70560	.61000	1.63934	37
24	.54070	1.84946	.56347	1.77471	.58670	1.70446	.61040	1.63826	36
25	.54107	1.84818	.56385	1.77351	.58709	1.70332	.61080	1.63719	35
26	.54145	1.84689	.56424	1.77230	.58748	1.70219	.61120	1.63612	34
27	.54183	1.84561	.56462	1.77110	.58787	1.70106	.61160	1.63505	33
28	.54220	1.84433	.56501	1.76990	.58826	1.69992	.61200	1.63398	32
29	.54258	1.84305	.56539	1.76869	.58865	1.69879	.61240	1.63292	31
30	.54296	1.84177	.56577	1.76749	.58905	1.69766	.61280	1.63185	30
31	.54333	1.84049	.56616	1.76629	.58944	1.69653	.61320	1.63079	29
32	.54371	1.83922	.56654	1.76510	.58983	1.69541	.61360	1.62972	28
33	.54409	1.83794	.56693	1.76390	.59022	1.69428	.61400	1.62866	27
34	.54446	1.83667	.56731	1.76271	.59061	1.69316	.61440	1.62760	26
35	.54484	1.83540	.56769	1.76151	.59101	1.69203	.61480	1.62654	25
36	.54522	1.83413	.56808	1.76032	.59140	1.69091	.61520	1.62548	24
37	.54560	1.83286	.56846	1.75913	.59179	1.68979	.61561	1.62442	23
38	.54597	1.83159	.56885	1.75794	.59218	1.68866	.61601	1.62336	22
39	.54635	1.83033	.56923	1.75675	.59258	1.68754	.61641	1.62230	21
40	.54673	1.82906	.56962	1.75556	.59297	1.68643	.61681	1.62125	20
41	.54711	1.82780	.57000	1.75437	.59336	1.68531	.61721	1.62019	19
42	.54748	1.82654	.57039	1.75319	.59376	1.68419	.61761	1.61914	18
43	.54786	1.82528	.57078	1.75200	.59415	1.68308	.61801	1.61808	17
44	.54824	1.82402	.57116	1.75082	.59454	1.68196	.61842	1.61703	16
45	.54862	1.82276	.57155	1.74964	.59494	1.68085	.61882	1.61598	15
46	.54900	1.82150	.57193	1.74846	.59533	1.67974	.61922	1.61493	14
47	.54938	1.82025	.57232	1.74728	.59573	1.67863	.61962	1.61388	13
48	.54975	1.81899	.57271	1.74610	.59612	1.67752	.62003	1.61283	12
49	.55013	1.81774	.57309	1.74492	.59651	1.67641	.62043	1.61179	11
50	.55051	1.81649	.57348	1.74375	.59691	1.67530	.62083	1.61074	10
51	.55089	1.81524	.57386	1.74257	.59730	1.67419	.62124	1.60970	9
52	.55127	1.81399	.57425	1.74140	.59770	1.67309	.62164	1.60865	8
53	.55165	1.81274	.57464	1.74022	.59809	1.67198	.62204	1.60761	7
54	.55203	1.81150	.57503	1.73905	.59849	1.67088	.62245	1.60657	6
55	.55241	1.81025	.57541	1.73788	.59888	1.66978	.62285	1.60553	5
56	.55279	1.80901	.57580	1.73671	.59928	1.66867	.62325	1.60449	4
57	.55317	1.80777	.57619	1.73555	.59967	1.66757	.62366	1.60345	3
58	.55355	1.80653	.57657	1.73438	.60007	1.66647	.62406	1.60241	2
59	.55393	1.80529	.57696	1.73321	.60046	1.66538	.62446	1.60137	1
60	.55431	1.80405	.57735	1.73205	.60086	1.66428	.62487	1.60033	0
M.	61° Cotang.	61° Tang.	60° Cotang.	60° Tang.	59° Cotang.	59° Tang.	58° Cotang.	58° Tang.	M.

M	32° Tang.	32° Cotang.	33° Tang.	33° Cotang.	34° Tang.	34° Cotang.	35° Tang.	35° Cotang.	M.
0	.62487	1.60033	.64941	1.53986	.67451	1.48256	.70021	1.42815	60
1	.62527	1.59930	.64982	1.53888	.67493	1.48163	.70064	1.42726	59
2	.62568	1.59826	.65024	1.53791	.67536	1.48070	.70107	1.42638	58
3	.62608	1.59723	.65065	1.53693	.67578	1.47977	.70151	1.42550	57
4	.62649	1.59620	.65106	1.53595	.67620	1.47885	.70194	1.42462	56
5	.62689	1.59517	.65148	1.53497	.67663	1.47792	.70238	1.42374	55
6	.62730	1.59414	.65189	1.53400	.67705	1.47699	.70281	1.42286	54
7	.62770	1.59311	.65231	1.53302	.67748	1.47607	.70325	1.42198	53
8	.62811	1.59208	.65272	1.53205	.67790	1.47514	.70368	1.42110	52
9	.62852	1.59105	.65314	1.53107	.67832	1.47422	.70412	1.42022	51
10	.62892	1.59002	.65355	1.53010	.67875	1.47330	.70455	1.41934	50
11	.62933	1.58900	.65397	1.52913	.67917	1.47238	.70499	1.41847	49
12	.62973	1.58797	.65438	1.52816	.67960	1.47146	.70542	1.41759	48
13	.63014	1.58695	.65480	1.52719	.68002	1.47053	.70586	1.41672	47
14	.63055	1.58593	.65521	1.52622	.68045	1.46962	.70629	1.41584	46
15	.63095	1.58490	.65563	1.52525	.68088	1.46870	.70673	1.41497	45
16	.63136	1.58388	.65604	1.52429	.68130	1.46778	.70717	1.41409	44
17	.63177	1.58286	.65646	1.52332	.68173	1.46686	.70760	1.41322	43
18	.63217	1.58184	.65688	1.52235	.68215	1.46595	.70804	1.41235	42
19	.63258	1.58083	.65729	1.52139	.68258	1.46503	.70848	1.41148	41
20	.63299	1.57981	.65771	1.52043	.68301	1.46411	.70891	1.41061	40
21	.63340	1.57879	.65813	1.51946	.68343	1.46320	.70935	1.40974	39
22	.63380	1.57778	.65854	1.51850	.68386	1.46229	.70979	1.40887	38
23	.63421	1.57676	.65896	1.51754	.68429	1.46137	.71023	1.40800	37
24	.63462	1.57575	.65938	1.51658	.68471	1.46046	.71066	1.40714	36
25	.63503	1.57474	.65980	1.51562	.68514	1.45955	.71110	1.40627	35
26	.63544	1.57372	.66021	1.51466	.68557	1.45864	.71154	1.40540	34
27	.63584	1.57271	.66063	1.51370	.68600	1.45773	.71198	1.40454	33
28	.63625	1.57170	.66105	1.51275	.68642	1.45682	.71242	1.40367	32
29	.63666	1.57069	.66147	1.51179	.68685	1.45592	.71285	1.40281	31
30	.63707	1.56969	.66189	1.51084	.68728	1.45501	.71329	1.40195	30
31	.63748	1.56868	.66230	1.50988	.68771	1.45410	.71373	1.40109	29
32	.63789	1.56767	.66272	1.50893	.68814	1.45320	.71417	1.40022	28
33	.63830	1.56667	.66314	1.50797	.68857	1.45229	.71461	1.39936	27
34	.63871	1.56566	.66356	1.50702	.68900	1.45139	.71505	1.39850	26
35	.63912	1.56466	.66398	1.50607	.68942	1.45049	.71549	1.39764	25
36	.63953	1.56366	.66440	1.50512	.68985	1.44958	.71593	1.39679	24
37	.63994	1.56265	.66482	1.50417	.69028	1.44868	.71637	1.39593	23
38	.64035	1.56165	.66524	1.50322	.69071	1.44778	.71681	1.39507	22
39	.64076	1.56065	.66566	1.50228	.69114	1.44688	71725	1.39421	21
40	.64117	1.55966	.66608	1.50133	.69157	1.44598	71769	1.39336	20
41	.64158	1.55866	.66650	1.50038	.69200	1.44508	71813	1.39250	19
42	.64199	1.55766	.66692	1.49944	.69243	1.44418	.71857	1.39165	18
43	.64240	1.55666	.66734	1.49849	.69286	1.44329	.71901	1.39079	17
44	.64281	1.55567	.66776	1.49755	.69329	1.44239	.71946	1.38994	16
45	.64322	1.55467	.66818	1.49661	.69372	1.44149	.71990	1.38909	15
46	.64363	1.55368	.66860	1.49566	.69416	1.44060	.72034	1.38824	14
47	.64404	1.55269	.66902	1.49472	.69459	1.43970	.72078	1.38738	13
48	.64446	1.55170	.66944	1.49378	.69502	1.43881	.72122	1.38653	12
49	.64487	1.55071	.66986	1.49284	.69545	1.43792	.72167	1.38568	11
50	.64528	1.54972	.67028	1.49190	.69588	1.43703	.72211	1.38484	10
51	.64569	1.54873	.67071	1.49097	.69631	1.43614	.72255	1.38399	9
52	.64610	1.54774	.67113	1.49003	.69675	1.43525	.72299	1.38314	8
53	.64652	1.54675	.67155	1.48909	.69718	1.43436	.72344	1.38229	7
54	.64693	1.54576	.67197	1.48816	.69761	1.43347	.72388	1.38145	6
55	.64734	1.54478	.67239	1.48722	.69804	1.43258	.72432	1.38060	5
56	.64775	1.54379	.67282	1.48629	.69847	1.43169	.72477	1.37976	4
57	.64817	1.54281	.67324	1.48536	.69891	1.43080	.72521	1.37891	3
58	.64858	1.54183	.67366	1.48442	.69934	1.42992	.72565	1.37807	2
59	.64899	1.54085	.67409	1.48349	.69977	1.42903	.72610	1.37722	1
60	.64941	1.53986	.67451	1.48256	.70021	1.42815	.72654	1.37638	0
M.	57° Cotang.	57° Tang.	56° Cotang.	56° Tang.	55° Cotang.	55° Tang.	54° Cotang.	54° Tang.	M.

	36°		37°		38°		39°		
M.	Tang.	Cotang.	Tang.	Cotang.	Tang.	Cotang.	Tang.	Cotang.	M.
0	.72654	1.37638	.75355	1.32704	.78129	1.27994	.80978	1.23490	60
1	.72699	1.37554	.75401	1.32624	.78175	1.27917	.81027	1.23416	59
2	.72743	1.37470	.75447	1.32544	.78222	1.27841	.81075	1.23343	58
3	.72788	1.37386	.75492	1.32464	.78269	1.27764	.81123	1.23270	57
4	.72832	1.37302	.75538	1.32384	.78316	1.27688	.81171	1.23196	56
5	.72877	1.37218	.75584	1.32304	.78363	1.27611	.81220	1.23123	55
6	.72921	1.37134	.75629	1.32224	.78410	1.27535	.81268	1.23050	54
7	.72966	1.37050	.75675	1.32144	.78457	1.27458	.81316	1.22977	53
8	.73010	1.36967	.75721	1.32064	.78504	1.27382	.81364	1.22904	52
9	.73055	1.36883	.75767	1 31984	.78551	1.27306	.81413	1.22831	51
10	.73100	1.36800	.75812	1.31904	.78598	1.27230	.81461	1.22758	50
11	.73144	1.36716	.75858	1.31825	.78645	1.27153	.81510	1.22685	49
12	.73189	1.36633	.75904	1.31745	.78692	1.27077	.81558	1.22612	48
13	.73234	1.36549	.75950	1.31666	.78739	1.27001	.81606	1.22539	47
14	.73278	1.36466	.75996	1.31586	.78786	1.26925	.81655	1.22467	46
15	.73323	1.36383	.76042	1.31507	.78834	1.26849	.81703	1.22394	45
16	.73368	1.36300	.76088	1.31427	.78881	1.26774	.81752	1.22321	44
17	.73413	1.36217	.76134	1.31348	.78928	1.26698	.81800	1.22249	43
18	.73457	1.36134	.76180	1.31269	.78975	1.26622	.81849	1.22176	42
19	.73502	1.36051	.76226	1.31190	.79022	1.26546	.81898	1.22104	41
20	.73547	1.35968	.76272	1.31110	.79070	1.26471	.81946	1.22031	40
21	.73592	1.35885	.76318	1.31031	.79117	1.26395	.81995	1.21959	39
22	.73637	1.35802	.76364	1.30952	.79164	1.26319	.82044	1.21886	38
23	.73681	1.35719	.76410	1.30873	.79212	1.26244	.82092	1.21814	37
24	.73726	1.35637	.76456	1.30795	.79259	1.26169	.82141	1.21742	36
25	.73771	1.35554	.76502	1.30716	.79306	1.26093	.82190	1.21670	35
26	.73816	1.35472	.76548	1.30637	.79354	1.26018	.82238	1.21598	34
27	.73861	1.35389	.76594	1.30558	.79401	1.25943	.82287	1.21526	33
28	.73906	1.35307	.76640	1.30480	.79449	1.25867	.82336	1.21454	32
29	.73951	1.35224	.76686	1.30401	.79496	1.25792	.82385	1.21382	31
30	.73996	1.35142	.76733	1.30323	.79544	1.25717	.82434	1.21310	30
31	.74041	1.35060	.76779	1.30244	.79591	1.25642	.82483	1.21238	29
32	.74086	1.34978	.76825	1.30166	.79639	1.25567	.82531	1.21166	28
33	.74131	1.34896	.76871	1.30087	.79686	1.25492	.82580	1.21094	27
34	.74176	1.34814	.76918	1.30009	.79734	1.25417	.82629	1.21023	26
35	.74221	1.34732	.76964	1.29931	.79781	1.25343	.82678	1.20951	25
36	.74267	1.34650	.77010	1.29853	.79829	1.25268	.82727	1.20879	24
37	.74312	1.34568	.77057	1.29775	.79877	1.25193	.82776	1.20808	23
38	.74357	1.34487	.77103	1.29696	.79924	1.25118	.82825	1.20736	22
39	.74402	1.34405	.77149	1.29618	.79972	1.25044	.82874	1.20665	21
40	.74447	1.34323	.77196	1.29541	.80020	1.24969	.82923	1.20593	20
41	.74492	1.34242	.77242	1.29463	.80067	1.24895	.82972	1.20522	19
42	.74538	1.34160	.77289	1.29385	.80115	1.24820	.83022	1.20451	18
43	.74583	1.34079	.77335	1.29307	.80163	1.24746	.83071	1.20379	17
44	.74628	1.33998	.77382	1.29229	.80211	1.24672	.83120	1.20308	16
45	.74674	1.33916	.77428	1.29152	.80258	1.24597	.83169	1.20237	15
46	.74719	1.33835	.77475	1.29074	.80306	1.24523	.83218	1.20166	14
47	.74764	1.33754	.77521	1.28997	.80354	1.24449	.83268	1.20095	13
48	.74810	1.33673	.77568	1.28919	.80402	1.24375	.83317	1.20024	12
49	.74855	1.33592	.77615	1.28842	.80450	1.24301	.83366	1.19953	11
50	.74900	1.33511	.77661	1.28764	.80498	1.24227	.83415	1.19882	10
51	.74946	1.33430	.77708	1.28687	.80546	1.24153	.83465	1.19811	9
52	.74991	1.33349	.77754	1.28610	.80594	1.24079	.83514	1.19740	8
53	.75037	1.33268	.77801	1.28533	.80642	1.24005	.83564	1.19669	7
54	.75082	1.33187	.77848	1.28456	.80690	1.23931	.83613	1.19599	6
55	.75128	1.33107	.77895	1.28379	.80738	1.23858	.83662	1.19528	5
56	.75173	1.33026	.77941	1.28302	.80786	1.23784	.83712	1.19457	4
57	.75219	1.32946	.77988	1.28225	.80834	1.23710	.83761	1.19387	3
58	.75264	1.32865	.78035	1.28148	.80882	1.23637	.83811	1.19316	2
59	.75310	1.32785	.78082	1.28071	.80930	1.23563	.83860	1.19246	1
60	.75355	1.32704	.78129	1.27994	.80978	1.23490	.83910	1.19175	0
M.	Cotang.	Tang.	Cotang.	Tang.	Cotang.	Tang.	Cotang.	Tang.	M.
	53°		52°		51°		50°		

	40°		41°		42°		43°		
M.	Tang.	Cotang.	Tang.	Cotang.	Tang.	Cotang.	Tang.	Cotang.	M.
0	.83910	1.19175	.86929	1.15037	.90040	1.11061	.93252	1.07237	60
1	.83960	1.19105	.86980	1.14969	.90093	1.10996	.93306	1.07174	59
2	.84009	1.19035	.87031	1.14902	.90146	1.10931	.93360	1.07112	58
3	.84059	1.18964	.87082	1.14834	.90199	1.10867	.93415	1.07049	57
4	.84108	1.18894	.87133	1.14767	.90251	1.10802	.93469	1.06987	56
5	.84158	1.18824	.87184	1.14699	.90304	1.10737	.93524	1.06925	55
6	.84208	1.18754	.87236	1.14632	.90357	1.10672	.93578	1.06862	54
7	.84258	1.18684	.87287	1.14565	.90410	1.10607	.93633	1.06800	53
8	.84307	1.18614	.87338	1.14498	.90463	1.10543	.93688	1.06738	52
9	.84357	1.18544	.87389	1.14430	.90516	1.10478	.93742	1.06676	51
10	.84407	1.18474	.87441	1.14363	.90569	1.10414	.93797	1.06613	50
11	.84457	1.18404	.87492	1.14296	.90621	1.10349	.93852	1.06551	49
12	.84507	1.18334	.87543	1.14229	.90674	1.10285	.93906	1.06489	48
13	.84556	1.18264	.87595	1.14162	.90727	1.10220	.93961	1.06427	47
14	.84606	1.18194	.87646	1.14095	.90781	1.10156	.94016	1.06365	46
15	.84656	1.18125	.87698	1.14028	.90834	1.10091	.94071	1.06303	45
16	.84706	1.18055	.87749	1.13961	.90887	1.10027	.94125	1.06241	44
17	.84756	1.17986	.87801	1.13894	.90940	1.09963	.94180	1.06179	43
18	.84806	1.17916	.87852	1.13828	.90993	1.09899	.94235	1.06117	42
19	.84856	1.17846	.87904	1.13761	.91046	1.09834	.94290	1.06056	41
20	.84906	1.17777	.87955	1.13694	.91099	1.09770	.94345	1.05994	40
21	.84956	1.17708	.88007	1.13627	.91153	1.09706	.94400	1.05932	39
22	.85006	1.17638	.88059	1.13561	.91206	1.09642	.94455	1.05870	38
23	.85057	1.17569	.88110	1.13494	.91259	1.09578	.94510	1.05809	37
24	.85107	1.17500	.88162	1.13428	.91313	1.09514	.94565	1.05747	36
25	.85157	1.17430	.88214	1.13361	.91366	1.09450	.94620	1.05685	35
26	.85207	1.17361	.88265	1.13295	.91419	1.09386	.94676	1.05624	34
27	.85257	1.17292	.88317	1.13228	.91473	1.09322	.94731	1.05562	33
28	.85308	1.17223	.88369	1.13162	.91526	1.09258	.94786	1.05501	32
29	.85358	1.17154	.88421	1.13096	.91580	1.09195	.94841	1.05439	31
30	.85408	1.17085	.88473	1.13029	.91633	1.09131	.94896	1.05378	30
31	.85458	1.17016	.88524	1.12963	.91687	1.09067	.94952	1.05317	29
32	.85509	1.16947	.88576	1.12897	.91740	1.09003	.95007	1.05255	28
33	.85559	1.16878	.88628	1.12831	.91794	1.08940	.95062	1.05194	27
34	.85609	1.16809	.88680	1.12765	.91847	1.08876	.95118	1.05133	26
35	.85660	1.16741	.88732	1.12699	.91901	1.08813	.95173	1.05072	25
36	.85710	1.16672	.88784	1.12633	.91955	1.08749	.95229	1.05010	24
37	.85761	1.16603	.88836	1.12567	.92008	1.08686	.95284	1.04949	23
38	.85811	1.16535	.88888	1.12501	.92062	1.08622	.95340	1.04888	22
39	.85862	1.16466	.88940	1.12435	.92116	1.08559	.95395	1.04827	21
40	.85912	1.16398	.88992	1.12369	.92170	1.08496	.95451	1.04766	20
41	.85963	1.16329	.89045	1.12303	.92224	1.08432	.95506	1.04705	19
42	.86014	1.16261	.89097	1.12238	.92277	1.08369	.95562	1.04644	18
43	.86064	1.16192	.89149	1.12172	.92331	1.08306	.95618	1.04583	17
44	.86115	1.16124	.89201	1.12106	.92385	1.08243	.95673	1.04522	16
45	.86166	1.16056	.89253	1.12041	.92439	1.08179	.95729	1.04461	15
46	.86216	1.15987	.89306	1.11975	.92493	1.08116	.95785	1.04401	14
47	.86267	1.15919	.89358	1.11909	.92547	1.08053	.95841	1.04340	13
48	.86318	1.15851	.89410	1.11844	.92601	1.07990	.95897	1.04279	12
49	.86368	1.15783	.89463	1.11778	.92655	1.07927	.95952	1.04218	11
50	.86419	1.15715	.89515	1.11713	.92709	1.07864	.96008	1.04158	10
51	.86470	1.15647	.89567	1.11648	.92763	1.07801	.96064	1.04097	9
52	.86521	1.15579	.89620	1.11582	.92817	1.07738	.96120	1.04036	8
53	.86572	1.15511	.89672	1.11517	.92872	1.07676	.96176	1.03976	7
54	.86623	1.15443	.89725	1.11452	.92926	1.07613	.96232	1.03915	6
55	.86674	1.15375	.89777	1.11387	.92980	1.07550	.96288	1.03855	5
56	.86725	1.15308	.89830	1.11321	.93034	1.07487	.96344	1.03794	4
57	.86776	1.15240	.89883	1.11256	.93088	1.07425	.96400	1.03734	3
58	.86827	1.15172	.89935	1.11191	.93143	1.07362	.96457	1.03674	2
59	.86878	1.15104	.89988	1.11126	.93197	1.07299	.96513	1.03613	1
60	.86929	1.15037	.90040	1.11061	.93252	1.07237	.96569	1.03553	0
M.	Cotang.	Tang.	Cotang.	Tang.	Cotang.	Tang.	Cotang.	Tang.	M.
	49°		48°		47°		46°		

	44°				44°				44°		
M.	Tang.	Cotang.	M.	M.	Tang.	Cotang.	M.	M.	Tang.	Cotang.	M.
0	.96569	1.03553	60	20	.97700	1.02355	40	40	.98843	1.01170	20
1	.96625	1.03493	59	21	.97756	1.02295	39	41	.98901	1.01112	19
2	.96681	1.03433	58	22	.97813	1.02236	38	42	.98958	1.01053	18
3	.96738	1.03372	57	23	.97870	1.02176	37	43	.99016	1.00994	17
4	.96794	1.03312	56	24	.97927	1.02117	36	44	.99073	1.00935	16
5	.96850	1.03252	55	25	.97984	1.02057	35	45	.99131	1.00876	15
6	.96907	1.03192	54	26	.98041	1.01998	34	46	.99189	1.00818	14
7	.96963	1.03132	53	27	.98098	1.01939	33	47	.99247	1.00759	13
8	.97020	1.03072	52	28	.98155	1.01879	32	48	.99304	1.00701	12
9	.97076	1.03012	51	29	.98213	1.01820	31	49	.99362	1.00642	11
10	.97133	1.02952	50	30	.98270	1.01761	30	50	.99420	1.00583	10
11	.97189	1.02892	49	31	.98327	1.01702	29	51	.99478	1.00525	9
12	.97246	1.02832	48	32	.98384	1.01642	28	52	.99536	1.00467	8
13	.97302	1.02772	47	33	.98441	1.01583	27	53	.99594	1.00408	7
14	.97359	1.02713	46	34	.98499	1.01524	26	54	.99652	1.00350	6
15	.97416	1.02653	45	35	.98556	1.01465	25	55	.99710	1.00291	5
16	.97472	1.02593	44	36	.98613	1.01406	24	56	.99768	1.00233	4
17	.97529	1.02533	43	37	.98671	1.01347	23	57	.99826	1.00175	3
18	.97586	1.02474	42	38	.98728	1.01288	22	58	.99884	1.00116	2
19	.97643	1.02414	41	39	.98786	1.01229	21	59	.99942	1.00058	1
20	.97700	1.02355	40	40	.98843	1.01170	20	60	1.00000	1.00000	0
M.	Cotang.	Tang.	M.	M.	Cotang.	Tang.	M.	M.	Cotang.	Tang.	M.
	45°				45°				45°		

RISE PER MILE OF VARIOUS GRADES.

Grade per Station.	Rise per Mile.	Grade per Station.	Rise per Mile.	Grade per Station.	Rise per Mile.	Grade per Station.	Rise per Mile.
.01	.528	.41	21.648	.81	42.768	1.21	63.888
.02	1.056	.42	22.176	.82	43.296	1.22	64.416
.03	1.584	.43	22.704	.83	43.824	1.23	64.944
.04	2.112	.44	23.232	.84	44.352	1.24	65.472
.05	2.640	.45	23.760	.85	44.880	1.25	66.000
.06	3.168	.46	24.288	.86	45.408	1.26	66.528
.07	3.696	.47	24.816	.87	45.936	1.27	67.056
.08	4.224	.48	25.344	.88	46.464	1.28	67.584
.09	4.752	.49	25.872	.89	46.992	1.29	68.112
.10	5.280	.50	26.400	.90	47.520	1.30	68.640
.11	5.808	.51	26.928	.91	48.048	1.31	69.168
.12	6.336	.52	27.456	.92	48.576	1.32	69.696
.13	6.864	.53	27.984	.93	49.104	1.33	70.224
.14	7.392	.54	28.512	.94	49.632	1.34	70.752
.15	7.920	.55	29.040	.95	50.160	1.35	71.280
.16	8.448	.56	29.568	.96	50.688	1.36	71.808
.17	8.976	.57	30.096	.97	51.216	1.37	72.336
.18	9.504	.58	30.624	.98	51.744	1.38	72.864
.19	10.032	.59	31.152	.99	52.272	1.39	73.392
.20	10.560	.60	31.680	1.00	52.800	1.40	73.920
.21	11.088	.61	32.208	1.01	53.328	1.41	74.448
.22	11.616	.62	32.736	1.02	53.856	1.42	74.976
.23	12.144	.63	33.264	1.03	54.384	1.43	75.504
.24	12.672	.64	33.792	1.04	54.912	1.44	76.032
.25	13.200	.65	34.320	1.05	55.440	1.45	76.560
.26	13.728	.66	34.848	1.06	55.968	1.46	77.088
.27	14.256	.67	35.376	1.07	56.496	1.47	77.616
.28	14.784	.68	35.904	1.08	57.024	1.48	78.144
.29	15.312	.69	36.432	1.09	57.552	1.49	78.672
.30	15.840	.70	36.960	1.10	58.080	1.50	79.200
.31	16.368	.71	37.488	1.11	58.608	1.51	79.728
.32	16.896	.72	38.016	1.12	59.136	1.52	80.256
.33	17.424	.73	38.544	1.13	59.664	1.53	80.784
.34	17.952	.74	39.072	1.14	60.192	1.54	81.312
.35	18.480	.75	39.600	1.15	60.720	1.55	81.840
.36	19.008	.76	40.128	1.16	61.248	1.56	82.368
.37	19.536	.77	40.656	1.17	61.776	1.57	82.896
.38	20.064	.78	41.184	1.18	62.304	1.58	83.424
.39	20.592	.79	41.712	1.19	62.832	1.59	83.952
.40	21.120	.80	42.240	1.20	63.360	1.60	84.480

Grade per Station.	Rise per Mile.	Grade per Station.	Rise per Mile.	Grade per Station.	Rise per Mile.	Grade per Station.	Rise per Mile.
1.61	85.008	1.81	95.568	2.10	110.880	4.10	216.480
1.62	85.536	1.82	96.096	2.20	116.160	4.20	221.760
1.63	86.064	1.83	96.624	2.30	121.440	4.30	227.040
1.64	86.592	1.84	97.152	2.40	126.720	4.40	232.320
1.65	87.120	1.85	97.680	2.50	132.000	4.50	237.600
1.66	87.648	1.86	98.208	2.60	137.280	4.60	242.880
1.67	88.176	1.87	98.736	2.70	142.560	4.70	248.160
1.68	88.704	1.88	99.264	2.80	147.840	4.80	253.440
1.69	89.232	1.89	99.792	2.90	153.120	4.90	258.720
1.70	89.760	1.90	100.320	3.00	158.400	5.00	264.000
1.71	90.288	1.91	100.848	3.10	163.680	5.10	269.280
1.72	90.816	1.92	101.376	3.20	168.960	5.20	274.560
1.73	91.344	1.93	101.904	3.30	174.240	5.30	279.840
1.74	91.872	1.94	102.432	3.40	179.520	5.40	285.120
1.75	92.400	1.95	102.960	3.50	184.800	5.50	290.400
1.76	92.928	1.96	103.488	3.60	190.080	5.60	295.680
1.77	93.456	1.97	104.016	3.70	195.360	5.70	300.960
1.78	93.984	1.98	104.544	3.80	200.640	5.80	306.240
1.79	94.512	1.99	105.072	3.90	205.920	5.90	311.520
1.80	95.040	2.00	105.600	4.00	211.200	6.00	316.800

THE END

www.ingramcontent.com/pod-product-compliance
Lightning Source LLC
LaVergne TN
LVHW010244110826
845151LV00004B/1390

* 9 7 8 1 4 2 5 5 2 3 4 5 9 *